D1061609

HANDBOOK OF LABORATORY WASTE DISPOSAL

IMPORTANT NEW & FORTHCOMING BOOKS IN CHEMICAL SCIENCE

COMPUTATIONAL METHODS FOR CHEMISTS
A. F. CARLEY and P. H. MORGAN, University College, Cardiff
REDUCTIONS IN ORGANIC CHEMISTRY
M. HUDLICKY, Virginia Polytechnic Institute and State University
BIOHALOGENATION: Principles, Basic Roles and Applications
S. L. NEIDLEMAN and J. GEIGERT, Cetus Corporation, Emeryville, California
ORGANOSILICON AND BIOORGANOSILICON CHEMISTRY: Structure, Bonding, Reactivity and Synthetic Application
Editor. H. SAKURAI, Tohoku University, Sendai, Japan
COMPUTER AIDS TO CHEMISTRY
Editors: G. VERNIN and M. CHANON, Université de Droit d'Economie et des Sciences, Marseilles, France
HANDBOOK OF AQUEOUS ELECTROLYTE SOLUTIONS
A. HORVATH, ICI, Mond Division, Runcorn, Cheshire
BIOINORGANIC CHEMISTRY
R. W. HAY, University of Stirling
ORGANIC PHOTOCHEMISTRY
J. D. COYLE, Open University
PHYSICAL PHOTOCHEMISTRY
R. DEVONSHIRE, University of Sheffield
INORGANIC PHOTOCHEMISTRY
A. HARRIMAN, The Royal Institution
PHOTOCHEMICAL AND PHOTOELECTRICAL CONVERSION OF SOLAR ENERGY
A. MILLS, University College, Swansea, and J. DARWENT, University of London
MATHEMATICAL AND COMPUTATIONAL CONCEPTS IN CHEMISTRY
N. TRINAJSTIC, Rugjer Boskovic Institute, Zagreb
CHEMICAL MONITORING OF OCCUPATIONAL TOXICITY
V. FOA, *et al.,* University of Milan
HANDBOOK OF LABORATORY WASTE DISPOSAL
M. J. PITT, Loughborough University of Technology, and E. PITT, University of Aston in Birmingham
COMMUNICATION, STORAGE AND RETRIEVAL OF CHEMICAL INFORMATION
JANET E. ASH, *et al.*
CELLULOSE CHEMISTRY AND ITS APPLICATIONS
T. P. NEVELL, UMIST, and S. HAIG ZERONIAN, University of California

HANDBOOK OF LABORATORY WASTE DISPOSAL

MARTIN J. PITT, M.Phil, Inst.Chem.E.
Particle Technology Group, Department of Chemical Engineering
Loughborough University of Technology

and

EVA PITT, B.Sc., Ph.D.
Cancer Research Campaign Group Laboratories
Department of Pharmacy, University of Aston in Birmingham

ELLIS HORWOOD LIMITED
Publishers · Chichester

Halsted Press: a division of
JOHN WILEY & SONS
New York · Chichester · Brisbane · Toronto

First published in 1985
and Reprinted in 1987 by
ELLIS HORWOOD LIMITED
Market Cross House, Cooper Street,
Chichester, West Sussex, PO19 1EB, England
*The publisher's colophon is reproduced from James Gillison's drawing of the ancient Market
Cross, Chichester.*

Distributors:

Australia and New Zealand:
JACARANDA WILEY LIMITED
GPO Box 859, Brisbane, Queensland 4001, Australia

Canada:
JOHN WILEY & SONS CANADA LIMITED
22 Worcester Road, Rexdale, Ontario, Canada

Europe and Africa:
JOHN WILEY & SONS LIMITED
Baffins Lane, Chichester, West Sussex, England

North and South America and the rest of the world:
Halsted Press: a division of
JOHN WILEY & SONS
605 Third Avenue, New York, NY 10158, USA

© 1985 M. J. Pitt and E. Pitt/Ellis Horwood Limited

British Library Cataloguing in Publication Data
Pitt, Martin J.
Handbook of laboratory waste disposal. —
(Ellis Horwood series in Information Science)
1. Chemical laboratories — Waste disposal
I. Title II. Pitt, Eva
542'.1 QD51

Library of Congress Card No. 85–7615

ISBN 0–85312–634–8 (Ellis Horwood Limited)
ISBN 0–470–20202–5 (Halsted Press)

Printed in Great Britain by Butler & Tanner, Frome, Somerset

Table of Contents

Personal Acknowledgements
Chapter
1   Introduction
2   The Legal Situation
3   Organization
4   Sources of Information
5   Laboratory Drainage
6   Fume Extraction
7   Burning and Incineration
8   Outside Contractors
9   Chemicals
10 Biological Materials
11 Radioactive Substances
12 Special Problems
    — Carcinogens, Teratogens and Cytotoxic Agents
    — Explosives
    — Unidentified Material
    — Clearing out a 'Dead' Store
    — Small Laboratories and Mobile Units
13 Educational Institutions
14 Materials Recovery
15 Emergency Procedures

Appendix A   Addresses in the UK, Eire, USA and Canada
Appendix B   Commercial Products
Appendix C   Chemical Tables
Appendix D   Glossary of Abbreviations

Personal Acknowledgements

It is impossible to list all the people who have contributed information and comments for this handbook. However, we would like to thank the following for their especially valuable assistance.

Prof. M.M. Renfrew, Chemistry Dept, University of Idaho; Dr R.C. Keen, Dept of Construction & Environmental Health, Bristol Polytechnic; R.J. Butler, Pollution Control Division, Waste Disposal Dept, West Midlands County Council; W.H. Norton and W.K. Kingsley, Office of Safety Training, J.T. Baker Chemical Co., Phillipsburg; B.J. Collins, MRC Regional Health Service Infection Research Laboratory, Dudley Road Hospital, Birmingham; P. Behan, Animal House, University of Aston in Birmingham; Dr S.S. Chissick, Safety Officer, Kings College, University of London; N. Musselwhite, Chief Fire Officer, County of Somerset; W.S. Holmes, Technical Dept, Phosphates Division, Albright & Wilson Ltd, Warley, West Midlands; H.J. Matsuguma & R.F. Walker, Energetic Materials Division, U.S. Dept of the Army, Armament Munitions and Chemical Command; Dr D.S. Brown, Radiological Protection Officer, Nuclear Chemistry, Loughborough University of Technology.

CHAPTER 1

Introduction

This book is about all types of laboratory waste, and their practical management. It is intended to provide sensible advice and information culled from a wide range of sources, which may assist those who have to deal with laboratory waste in any way. Such waste may include paper, drainage and fume extraction, as well as the materials considered special to the laboratory. These wastes are important because of their volume and because they are often overlooked. Waste paper can start a major fire as easily as flammable solvents. A simple flood may often do as much damage as an explosion. Faulty ventilation may remove a hazard from the laboratory and present it to people who have no connection with the work.

Regrettably, we have observed many laboratories with virtually no waste disposal organization. This leads to inappropriate disposal routes, dangerous combinations in storage, and often the unnecessary and hazardous accumulation of unwanted material. The theme of this book is that waste can and should be managed. Often, very simple procedures are all that are required, and we have tried to point out practical methods which have been found to work. However, where especially difficult material is involved, then special techniques are necessary.

As it would be impracticable to give all these special methods in a single book, a short bibliography is given for each topic. The selection of items in these lists is a little personal. However, we have chosen publications which seem to be particularly helpful for waste disposal practice. As far as possible, the list is limited to items still in print. The few exceptions are some

7

older publications which may be found in libraries and which have not yet been adequately replaced. We have tried to include publications available on each side of the Atlantic.

The methods, principles and cautionary advice given in this book are derived in part from the sources given in the bibliography, to a lesser extent from other published sources (less widely available or only giving a very small amount of relevant information), but to a major extent from people. We have faced many of the practical problems in our laboratory experience, and visited many other laboratories in order to see actual practices and to talk to managers, scientists and technicians.

This handbook does not set out to be a recipe book to be followed blindly. Laboratories and local situations vary too widely for this, and there are important differences between countries. Instead, it gives information which will assist a technical person to judge the available options in his or her case. Too often, it seems, methods are put forward for the disposal of a particular material (e.g. a chemical) without regard to the amount involved and the situation (for example: Is it an emergency or regular planned occurrence? What facilities and skills can be used? Can the problem be avoided in any way?). For this reason, the book is divided into many subsections dealing with different situations, each of which is titled for easy reference. Many sections refer to other sections, but as we are aware that books of this nature are rarely read cover to cover, a point is occasionally repeated where it is particularly important.

Some data is given in the relevant sections, but there is an additional compilation in the Appendices. These are tables of information which we have found not to be very readily available, or which are not widely known.

As safety is of prime importance in laboratory operations, there are cautionary notes in the text referring to the special dangers of certain procedures. Regrettably, these mainly derive from actual accidents published in the literature or personally known to us.

It is pleasing to note the many initiatives now under way with regard to chemical waste, but this necessarily means that no book can be fully up to date. There is even more difficulty in the rapidly changing pattern of regulations. However, in both these cases an effort has been made to point out the underlying principles, which are liable to remain for some years to come.

CHAPTER 2

The Legal Situation

2.1 INTRODUCTION

This chapter deals with the current situation in the United Kingdom and in the United Statea of America. Of course, laws vary from country to country and will change with time. However, there seems to be a move towards certain common ideas in the laws relating to health and safety and pollution. In particular, countries which are members of the European Economic community are obliged to introduce regulations to satisfy Community Directives, such as the 1978 Directive on Toxic and Dangerous Waste.

There is not at present in the UK or USA any specific codifying enactment relating solely to laboratory waste. In the absence of specific legislation, all the laws applying to the substances and the processes involved may be assumed to affect the laboratory. These are numerous and complex. Most relevant laws were produced to control industrial practices: they may not seem relevant to laboratory situation, but even so may still be strictly enforceable by the appropriate authority.

Hence it is vital that those responsible for laboratories have an accurate knowledge of their legal duties, and it is essential that they consult with the relevant authorities if in doubt. This can prevent a breach of the regulations through ignorance (which is no defence in law!).

It is impractical to list all the laws which might apply to laboratories, and it is for a legal expert to give an exact interpretation for a particular situation. However, there are some basic common factors in both UK and USA legislation which can be readily appreciated. For more detailed help, it is recommended that the enforcement authority (for a particular substance, procedure or place) be consulted in the first instance.

2.2 BREAKING THE LAW

Basically the law can be broken in three ways:

(1) If it can be shown that harm has been done to people or objects (e.g. a sewage works) or the environment, then there may be both criminal and civil liability. (Civil action is where damages are claimed by the injured party.) However, this is often the most difficult type of offence to prove. It is usually necessary to show that damage has been done, that it was directly due to the defendant's actions, and that the defendant was negligent, reckless or malicious in that action, or that the action itself was illegal.

(2) If it can be shown that a specific enactment has been breached — for example if toxic material is deposited on a site not licensed for that material — then a criminal offence may have been committed, even if no identifiable harm results. In some instances the control agencies find that it is very easy to prosecute because the legislation is so specific (e.g. with certain chemical or radioactive substances). In other cases it is necessary to show that the action was illegal because it was not in accord with accepted good practice. If a relevant code of practice exists (e.g. the US National Research Council's 'Prudent Practices for Handling Hazardous Chemicals in Laboratories' or the UK Medical Research Council's 'Guidelines for Work with Potent Carcinogens') and was not followed, then the defence may have to produce technical evidence to show that the procedure actually used was at least as good as the recognized one.

(3) A technical, but nevertheless serious, offence may be committed even if the correct operation was undertaken, but the specified procedure was not strictly adhered to (e.g. labelling, keeping records, notifying authorities). These are often the easiest offences to detect and prosecute. It is unlikely to be a sufficient defence in law to show good intentions. However, if a genuine effort was made and no harm resulted, the authority may decide not to prosecute.

Additionally, it is possible to breach regulations, limits or requirements laid down by officers or authorities (such as those responsible for drainage, fire protection, waste disposal or any form of licensing). In some cases these have the direct force of law, and a breach is of itself an offence. In others the offence may be committed in failing to put matters right when required to do so by the appropriate authority (e.g. remedying a fire hazard). Even if prosecution does not result, such breaches (especially if serious and/or repeated) may result in a licence being revoked (or not renewed) or the refusal to provide a necessary service such as sewerage!

In the USA there exists the possibility that some act may be illegal because of its interference with the constitutional rights of others. In the UK there exist certain rights under Common Law (which is in essence accumulated legal decisions which are used as a basis for 'fair play'.) However, it is not likely that the authorities would need to use these options to control laboratory waste disposal, as there are certain more specific and powerful laws available.

Illegal acts may have penalties other than direct prosecution. They may be grounds for terminating a contract (such as a contract of employment, or with a supplier of goods or services) and often specifically invalidate insurance cover, which could prove costly in the event of an accident or claim for damages.

2.3 GENERAL LEGISLATION

2.3.1 Health and Safety of People

To all intents and purposes, any employer has an absolute duty to his work people, and to a large extent others who may have access to his premises (e.g. a waste collector). Waste disposal is no different in this respect from any other activity of the company or organization.

Under the same legislation, individuals have a duty to take care of their own safety and that of others. Not only can they be held partly to blame for their own mishaps, they can actually commit a crime by failing to fulfill this duty. (UK: Health & Safety at Work Act 1974. USA: Occupational Safety & Health Act 1970).

2.3.2 Protection of the Environment

In recent years, the law has made it clear that the responsibility for any waste (and any injury or pollution it causes) is directly attributable to the company or organization producing that waste. This responsibility cannot be taken away: it remains even when the waste has left the premises and is in someone else's possession

Thus it is not sufficient to pay someone to take the material away without checking that he is likely to dispose of it correctly. If the contractor commits an offence, it is possible that the originator of the waste might be held responsible for not having *made sure* that disposal would be safe and legal. If it can be shown that the contractor might reasonably be supposed to be untrustworthy (e.g. if the arrangement was 'pay the money and ask no questions') then the individual who arranged the contract may be personally liable.

As another aspect to this duty, note that the persons due to transport and/or dispose of the waste must be given adequate information and/or instruction. Failure to provide adequate reasonable information could render both the individual and the institution concerned liable for criminal prosecution and for civil damages. (UK: Control of Pollution Act 1974. USA: Resource Conservation & Recovery Act 1976.)

See Chapter 8 for legal and practical aspects of the use of outside contractors.

2.4 SPECIFIC SUBSTANCES

Apart from the general duties just outlined, particular materials may be regulated by special laws. Pre-eminent among these are the petroleum substances, which are subject to control because of their flammability, toxicity and value in terms of tax revenue. Ethanol is also controlled as a substance liable to duty. Where dutiable substances are to be disposed of as, or recovered from, waste, then the customs' authorities ought to be consulted. They may require certain records to be kept and specific precautions to be taken, and have the right to inspect stores and apparatus.

Materials which have in the past created fire hazards in industry — notably explosives and flammable liquids and gases — are the subject of specific regulations as to their use, which includes disposal. Similarly, some well-known toxic hazards (pharmaceutical drugs and poisons, many metals, asbestos, plus a few industrial chemicals and known carcinogens) have been the subject of special laws. These generally relate to storage, use and exposure of persons to the substance. In some cases it may not be permissible for a young person or woman to deal with the material.

Radioactive substances are closely controlled by law, and their disposal is especially restricted and monitored by the relevant authorities. (See section 11.1.1.) There is some interest in generating similar controls for severe carcinogens and the most harmful pathogens.

An important general point is that waste is included in the total inventory where a substance is licensed (e.g. ethanol, explosives, flammables, petroleum substances, radioactives). Accumulation of waste may cause the total stock to exceed the licensed amount.

See section 6.1.2 for legal controls on airborne hazardous substances. See section 10.3 and Table 10.1 for UK definitions of clinical waste. See section 10.5.1 for some aspects of the laws relating to animals in laboratories. See section 11.1.1 for regulation of radioactive substances. See section 12.1.5 for controls on asbestos.

2.5 THE DISPOSAL PROCESS

Special additional laws relate to the disposal process itself. Basically, all sites or units (such as incinerators) for the disposal of radioactive or chemically dangerous waste need a licence, and usually planning permission. This can apply even to disposal of waste on the producer's own premises. It is necessary to ensure that any company accepting hazardous waste is specifically licensed to accept the material. Note that although a specific waste may be acceptable, and even mentioned by name on the licence, no site is actually obliged to accept it. In particular, landfill sites are becoming more reluctant to accept radioactive wastes.

Other waste needs to be disposed of in such a way that it does not constitute a risk to public health (e.g. putrescible materials), cause a nuisance to others (e.g. paper blowing about) or present any risk of polluting the air (e.g. by smoke) or water (e.g. by rain washing material into a watercourse). Even if a licence is not required, permission may be necessary (under planning and water laws) for a piece of land, water or premises to be used for waste disposal or storage of substantial amounts of waste.

For the purpose of control, lists of categories of substance have been established in different countries. If a given material by its quantity and nature comes within a definition on the list, then it is subject to special controls on disposal and transport. A material is either in or out of the list definition: there are no half measures, though it may ultimately take a court case to decide a particular instance.

The UK list is known as 'Special Waste' and is issued by the Department of the Environment under 1980 regulations under section 17 of the Control of Pollution Act 1974. The USA list is known as 'Hazardous Waste' and was established by the Environmental Protection Agency, and promulgated in the Federal Register in 1980, following the Resource Conservation & Recovery Act 1976. The USA and UK lists are not identical, though they have much in common, including specific reference to laboratory chemicals. It is intended that the USA list will ultimately contain categories for infectious materials. (UK: Control of Pollution Act Section 17 (Special Waste) Regulations 1980. USA: Federal Register May 19, 1980. Resource Conservation & Recovery Act Regulations.)

Appendix C-1 provides more details of UK Special Waste. Appendix C-2 provides a listing of USA Hazardous Waste.

2.6 TRANSPORTATION

If a 'hazardous waste' or 'special waste' is to be transported off the site (except by pipeline) for disposal, then some very specific requirements have to be fulfilled. Chief among these is that certain forms have to be completed in advance, notifying the relevant authorities. These are the authorities responsible for waste disposal and for water systems in the local areas concerned. At the time of writing, there are changes in progress in the UK in the organization of both types of authority.

Information on these notification procedures is available from enforcing authorities. In addition, waste disposal contractors usually provide some assistance with the paperwork. In the USA a transport contractor for hazardous waste must be specially licensed, but this is not currently a requirement in the UK.

During transportation, the waste must be in suitable containers (in the USA approval is given by the Department of Transport) and labels are required of a specific form for many wastes. In addition, the transporter must be given such information as he might reasonably require for safety. The waste disposal authorities have a right to inspect and sample hazardous or special waste in transit.

A vehicle used to transport waste must be roadworthy, suitable for the purpose, and both licensed and insured for the operation. It will probably need to display a sign or signs indicating the main hazard, and of an approved design. A consignment note (as specified by the hazardous waste/special waste regulations) must be carried giving details of the waste.

See section 2.3.2 for the laboratory's duty to the transporter, and Chapter 8 for some aspects of dealing with contractors.

2.7 EMERGENCIES

In a genuine emergency, the strict letter of the law may be relaxed by the authorities, or a different procedure may be permitted. However, it is advisable to contact the appropriate agency as quickly as possible. For example, if a spillage has already entered surface drains, the water authority may issue a consent for it to be diluted with large volumes of water. However, should this be done secretively and subsequently be discovered, the water authority may take punitive legal acton which could conceivably include a claim for a large amount of compensation for damage (to drains, treatment works, wildlife, etc.).

An accidental release of material in excess of that normally permitted (to the air, drains or watercourse) is called an excursion. Small excursions are not uncommon and can usually be tolerated. Large excursions may require prompt emergency action by experts. For this reason, it is important that such situations are notified as quickly as possible with an estimate of the amount of release. As this is normally done by telephone, it is a good idea to take the name of the person who speaks on behalf of the control agency, and who will issue instructions about what (if anything) should be done. Clearly this could be very important if legal action follows.

See Chapter 15 for practical emergency action.

2.8 EXEMPTIONS AND EXCEPTIONS

Generally, any law may be assumed to apply unless there is a specific exemption in the statute itself, or in an enforcing regulation, or by special dispensation of the enforcing agency. A major source of exemption occurs from the legal difficulty of enforcement agencies prosecuting themselves. Thus the laboratories or disposal sites belonging to such agencies may not have to meet certain requirements of the law. In the UK, certain places (notably parts of the National Health Service and Ministry of Defence establishments) are considered 'Crown Property' and are thus immune to prosecution by the Crown (i.e. the State) for certain offences. Note that these exemptions are limited. One agency may be prosecuted by another, and persons on Crown Property are still subject to the law of the land.

A common exception to the law is for extremely small amounts of certain substances, e.g. radioactives. Many laws include a definition of some minimum amount, and it is sometimes possible to avoid complications such as licensing by controlling use and disposal to these limits.

Certain establishments, notably hospitals and some teaching, have specific exemption orders, e.g. for radioactives. However, note that this kind of exemption is restricted to a certain amount for a certain purpose. Larger amounts of different purposes may not be exempted.

For practical reasons, the enforcement agencies may not bother with producers of very small amounts of waste, but this does not imply any true legal exception. The letter of the law may be enforced at any time. At the time of writing, the USA Environmental Protection Agency is considering its policy towards small-quantity generators of waste and their exclusion from control programmes.

It is important to recognize that exemption from control does not mean exemption from liability. Furthermore, an exemption usually only applies to a particular premises. If, for example, waste is transported elsewhere then the full requirements for transport and disposal may become applicable.

Although government-related laboratories are sometimes not bound by law in certain respects, it is the practice in the UK and the USA for government-related organizations to set internal standards and control measures in accordance with the law applied elsewhere. In the USA, the President asked Federal agencies to comply with the Occupational Safety and Health Act, 1970, by issuing Executive Order 12196.

BIBLIOGRAPHY

UK

Barbour Health & Safety Library [Microfiche: Regularly Updated] Barbour Index Ltd, Windsor.
Cooke, A.J.D. (1976) *A Guide to Laboratory Law*, Butterworths.
Dangerous Substances [1983: Loose-Leaf: Regularly Updated] Wolters Samson (UK) Ltd, New Malden.
Haigh, N. (1983) *EEC Environmental Policy & Britain*, Environmental Data Services.
Webster, C.A.R. (1981) *Environmental Health Law*, Sweet & Maxwell.

USA

Code of Federal Regulations [Annual] Office of the Federal Register, Washington D.C.
Firestone, D.B. and Reed, F.C. (1981) *Environmental Law for Non-Lawyers*, Ann Arbor.
Guidebook to Occupational Safety and Health [Regularly Updated] Commerce Clearing House, Chicago.

Hazardous Waste Management: A Survey of State Legislation 1982, National Conference of State Legislatures, Denver.

National Research Council (1983) *Prudent Practices for Disposal of Chemicals from Laboratories*, National Academy Press.

Ream, K.A. (1983) Laboratory Waste, RCRA, and the ACS, *Chem. Tech.* **13,** 572-576.

The Management and Disposal of Hazardous and Chemical Waste [Seminars] J.T. Baker Chemical Co., Phillipsburg.

CHAPTER 3

Organization

3.1 INTRODUCTION

Laboratory waste disposal needs to be managed: it cannot be allowed to 'just happen'. It is in everybody's interest that an efficiently organized system is in operation, otherwise problems and dangers may arise. Rules and regulations are essential, but they are not the only means of management, and can give rise to a seemingly officious and bureaucratic attitude which is likely to be resented and therefore not complied with.

In essence the laboratory manager should consider all waste, not just the obviously hazardous, and ensure that there is an adequate and convenient method for its regular removal. He (or she) should consider all the people who are likely to be involved, including those who are not under his direct control. By means of locations, signs, colour codes and instructions, the possibility of errors should be minimized. In the best situations, the waste disposal methods are so obvious and seem so sensible that the people concerned are not aware that they are being managed.

3.2. FORMALITIES

3.2.1 Licences and Documents

For waste disposal, the laboratory may act as an independent unit, or may be part of another organization. Where formal documents relate to the whole organization or site, the laboratory manager should make sure that they include the special situation of the laboratory, e.g. to ensure that the insurance is not invalidated by a normal waste disposal operation.

Some formal matters relating to waste disposal are as follows.

Firstly the laboratory may be required to register as a producer of hazardous waste. (This is not required in the UK at the time of writing, and small generators are currently exempted in the USA, though this may change in the next few years. Up to date information should be available from the local waste disposal enforcement authority.)

Secondly, waste disposal operations carried out on site may require a licence and possibly planning permission. If these involve burning or venting to atmosphere, specific permission with definite limits may be required under air pollution legislation.

Most laboratories discharge water to the sewers, which will require a trade effluent consent from the local drainage authority (see Chapter 5 and section 9.3.3). Care must be taken if laboratory drainage is part of a site that the laboratory waste is permitted. Sometimes it is advisable to have a separate discharge and separate consent for the laboratory.

It is necessary for all but the smallest employers to issue a policy document on health and safety for their employees. Laboratory activities must be included (in a special section if necessary) with specific reference to waste disposal.

Licences are required for operations with radioactive substances, ethanol, petroleum spirit, explosives, some poisons and animals. The licence holder is subject to inspection, and waste disposal is an important part of the requirements. (See section 2.4.)

Certain pathogens and highly toxic substances are generally prohibited except by special permission for specified laboratories.

Unless the organization has its own facilities, arrangements will need to be made as a formal contract with a waste disposal contractor or a local authority for removal of general refuse ('trade waste'). Movement of hazardous wastes to a disposal site elsewhere will require a contract with the disposer and the transporter, and formal notification to authorities in advance. (See sections 2.6 and 9.3.2.)

3.2.2 Record-keeping

If the laboratory has a licence for the use and disposal of particular materials, it is usually required by the enforcing agency to keep adequate records. In addition, records relating to waste disposal can be a useful management tool. They may help to point out areas of bad economic practice or where there might be problems of safety. It is often salutary to compare the records of purchase of selected materials and the records of disposal. If they do not match even approximately, then where is the waste actually going?

A knowledge of the volumes and frequencies of waste arisings is vital for an informed decision about future waste disposal. For example, this information is necessary in planning for an incinerator (see Chapter 7) or a materials recovery programme (see Chapter 14 and section 3.12).

The existence of a definite and trustworthy record can be invaluable in disputes with contractors or control agencies. It may be very important in an emergency such as a fire or breach of containment of a hazardous material. (See Chapter 15.)

A typical system will have cards or books at suitable locations (e.g. beside a sink for radioactive disposal — see sections 5.5.1 and 11.8, or a collection point for biohazard waste — see sections 10.2, 10.3, 10.5, 10.6). The notes in these locations will be copied into a central record at intervals (daily, weekly or monthly, depending on the type of waste and the degree of hazard). For example, it may be necessary to add up radioactive disposal in various rooms to check that the institution does not exceed its permitted amount of discharge.

The more important this central register is, the more care must be taken. A duplicate record is a sensible precaution. If possible a fire-resistant cabinet should be used to store records of materials which are radioactive (see section 11.1.3), subject to duty, 'special waste' or 'hazardous waste' as defined by legislation (see sections 2.5 and 2.6, and Appendices C-1 and C-2), precious metals, or any waste which is of major legal or economic significance.

Control agencies may specify certain forms or methods of keeping records for particular waste.

3.3 PEOPLE

3.3.1 Responsibilities

Everybody deals with waste, even if it is only a cup from a coffee machine. It is therefore advisable to inform all new staff of the general practice on the site, and particular details of their place of work. It is helpful if the general pattern of segregation can be made into a list or lists, which can be given to a new employee. It is unwise to rely on others passing on the information in the course of work.

In principle, everyone should have adequate training in waste disposal or collection as far as it affects his or her job. Even highly qualified scientists may need instruction to ensure that they can operate a machine or procedure correctly. Non-technical staff who collect waste from laboratories should be strongly urged not to touch anything they are unsure of, or which they know to be wrong (e.g. glass in a paper bin) but to report the matter instead.

Certain specified people should be the only ones allowed to perform some waste disposal procedures (e.g. burning explosives, purifying mercury, operating a macerator, chemically neutralizing carcinogens). Certain jobs should only be carried out by certain groups, or there should be special

1

2

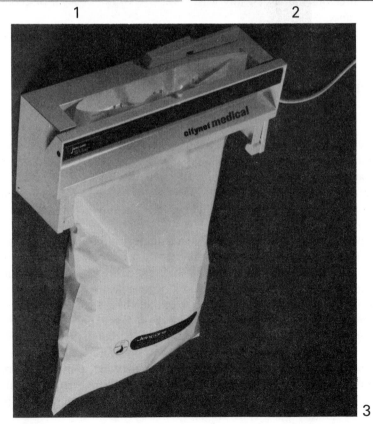

3

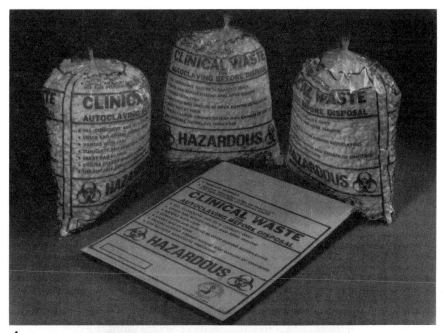

4

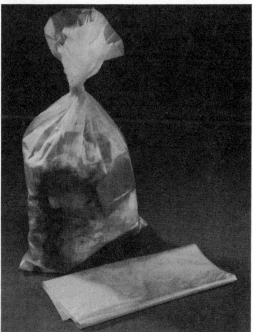

5

① A simple unguarded sack holder, rather full. The bag could easily be torn by a passing trolley. There is nothing to restrain the contents if the bag fails in any way. ② A commercial bag for biological waste, which comes already labelled. The bag is temporarily closed with adhesive tape for transport to the autoclave, but must be opened during sterilization. ③ A simple wall-mounted device for heat-sealing plastic bags of waste for security in transport. Photo courtesy of Jencons Scientific Ltd, Leighton Buzzard, Beds. ④ Clinical waste bags. Photo courtesy of Jencons Scientific Ltd, Leighton Buzzard, Beds. ⑤ A heavy gauge reinforced polypropylene sack which is autoclavable and tear-resistant.

arrangements. For example, laboratory personnel may be expected to carry out certain cleaning and housekeeping operations which would be done by other staff elsewhere. If non-technical staff do enter a laboratory, for example to clean benches or sinks, then it must be fully understood that it is the responsibility of the laboratory staff to ensure that all hazardous or delicate items have been removed from the area to be cleaned.

Difficulties can occur when non-technical people act outside their own area of competence. In our experience, greater difficulties occur when highly technical people act irresponsibly because they feel waste disposal or clearing up are beneath their dignity. It must be recognized that these problems occur, and should be taken as a challenge to good management.

In a large institution, it may be worth while appointing some technical person (usually on a part-time basis) to have special responsibility for waste disposal, since this can otherwise be too much of a burden for a site safety officer in addition to his other duties. It is generally recommended that for each laboratory there is a specific person who looks after the day-to-day waste disposal, and sees that site policy is adhered to. This most emphatically does *not* mean someone to take on the chore of other people's waste disposal and cleaning up. Each laboratory worker must be responsible for the safe disposal of the materials that he or she works with. However, it is a great help to have a specific person who will keep an eye on the disposal facilities and report any problems promptly.

The person assigned to waste disposal in a laboratory or group of rooms may be a senior scientist or may be the youngest technician present. However, whatever the grade or status, he or she must have sufficient practical experience and laboratory commonsense to take a realistic attitude. Management must make it clear that the person concerned is implementing company policy, with which his or her more senior colleagues must comply.

3.3.2 Hygiene

Laboratory personnel must be provided with facilities for washing hands (and possibly face), toilets (in number and style as required by the legislation relevant to the institution, or by local regulations), and some arrangement for refreshment and meal breaks. In addition, management must make allowance for the fact that some employees will wish to smoke, and some will on occasion suffer from colds, coughs and hay fever.

3.3.2.1 Handwash Facilities

There should always be a wash-basin in the laboratory area so that people can wash their hands before leaving. If at all possible, this should be reserved for the purpose, and not used as a general working sink. If it is impracticable or unsafe to have a sink within a laboratory, then there should be one in the changing room or nearby, and the layout should ensure that people pass the basin on their way out, to the toilet or to the canteen, coffee room, etc.

Paper towels are the most widely used hygienic method of hand drying. They have the advantage that they can be used to wipe oil or other dirt from the hands, and can be used for cleaning or small spillages. (See section 15.2.2.3.) It is worth looking at several types and consulting the staff, as some are more acceptable than others. In fact, some brands are positively unpleasant or annoying to use.

Where paper towels are used, there must of course be a waste bin provided of a suitable type (see section 3.4) and arrangements made for it to be emptied. It is usually advisable for this (relatively innocuous) waste to be treated separately from laboratory waste as such. In particular, contaminated towels, packages, glass, metal, etc. should not go into a bin reserved for handwash towels. (see section 3.8)

An ordinary cloth towel or roller towel has no place in a laboratory since hygiene cannot be assured. A mechanical roller towel (which provides a clean fresh portion for each user) is acceptable in some laboratories, so long as there are efficient arrangements for replacing the towel

A hot air blower is favoured in some institutions. It is hygienic and greatly reduces the amount of waste. However, it should *not* be used in laboratories working with unsealed radioactive sources or severely toxic chemicals since there is a risk that harmful substances could be blown into the laboratory air.

If a mechanical roller towel or a hot air blower is used, there should still be paper towels avilable in case the normal hand-drying method fails.

If the water pressure is too high (or there is a poor combination of sink and tap design) then a wash-basin may splash. This can be eliminated by regulating the pressure or by installing an in-line flow restrictor. These are very inexpensive and easy to fit. An excessively hot water supply should not be tolerated. The supply system must be correctly adjusted to give a comfortable temperature. If hotter water is required for washing apparatus etc., it can be generated locally with a suitable heater.

See Chapter 5 for further details on sinks and drains. See section 11.8.1 for comments on basins in radioactive laboratories.

3.3.2.2 Toilets

In some circumstances, owing to location or specialized working conditions, it is obvious that laboratory staff will require separate toilets. In many more instances the toilets are communal, but are predominantly used by one group of employees. Management should pay some attention to the possibility of contamination if the institution has very dirty areas or very sensitive ones (ultra-clean rooms or special care units in hospitals, for example).

If a toilet is used for the disposal of laboratory waste, then the use of that toilet should be restricted as far as possible, and there should be an agreed system for cleaning that toilet (whether by cleaning staff, laboratory staff or both).

See sections 5.2 (drain functions), 5.3.2 (sluices), 5.6 (effluent quality), 9.3.3 (chemicals to drain), 10.2.3.12 (urine), 10.5.2 (animal excreta), 11.6.5 (low activity radioactives), and particularly 11.8.2 (sluices and latrines for radioactive waste disposal).

See previous section (3.3.2.1) for comments on wash-basins and hand drying. A convenient combination is a hot air blower and a paper towel dispenser. The hot air blower greatly reduces the quantity of waste to be collected from the toilet. The paper towels are provided for those who prefer them, for face drying and in case the blower fails.

The waste bins in a toilet are expected to contain only paper towels and domestic-type refuse. They should never be used for the disposal of waste from the laboratory.

3.3.2.3 Paper Tissues

In most laboratories it is inadvisable for staff to use their own cloth handerkerchiefs. It is therefore necessary to provide paper tissues, and insist that they be used instead. If the laboratory does not use such tissues in its work, then wall-mounted containers can be used to dispense tissues, and greatly reduce the chance of boxes being misappropriated. (Where tissues are used in the work, some laboratories use soft toilet rolls, which are more economic than boxes of tissues.)

3.3.2.4 Refreshment Facilities

It is now universally accepted that food and drink should not be consumed or stored in any laboratory, no matter how innocuous the work or how careful the staff. The principal reason is to protect the staff, but management should remember that food and drink may cause contamination or even damage if spilled or carelessly handled. For this reason, staff should be able to leave the laboratory to visit a canteen, rest-room or even an outside cafe as appropriate. A minimum arrangement might be a vending machine in the corridor outside the laboratory. In this case, there should of course be a bin and notices to the effect that drinks may not be taken into the laboratory.

Where a staff room is provided for lunch and refreshment breaks, there should be a sink and drainer. The most satisfactory method of providing drinks appears to be a vending machine subsidized by the institution. If the drinks are very cheap then staff are unlikely to make alternative (less hygienic) arrangements. The need for a small coin reduces the chance that someone will empty the machine out of idle mischief.

If staff provide their own tea and coffee, then it is most hygienic for disposable cups to be provided. (The most acceptable ones provide a disposable container inside a plastic holder something like a tea-cup, so that they can be held comfortably even when the drink is hot.) Otherwise, there must be proper facilities for washing and drying cups and utensils to good kitchen standards.

If cooking and food storage facilities are provided (e.g. cupboards, a refrigerator) then there must be a definite arrangement for periodic cleaning (and defrosting of refrigerators). Waste bins from refreshment areas will usually be emptied along with office bins. It is therefore important that they are not used for laboratory waste as such.

3.3.2.5 Smoking

It is to be hoped that smoking is now effectively banned from all scientific labotatories. However, provision must be made in other areas such as rest-rooms (unless smoking is prohibited altogether in the building, or staff choose to have these rooms no-smoking zones for the comfort of the majority). Separate ashtrays should be provided in adequate numbers and suitable locations, or there is a danger of fire from cigarette ends discarded into paper bins. It is best if the ashtrays are emptied in the morning, so that any embers will have been extinguished overnight and will therefore not ignite other refuse.

SMOKING — A CAUTIONARY TALE

A cigarette was discarded into a bin of paper towels, in a washroom adjacent to a pharmaceutical research laboratory. The resulting fire ignited fumes from a leaking distillation apparatus, which had penetrated to the washroom. The fire flashed back and exploded the solvent storage vessel causing severe damage to the laboratory.

Health & Safety Information Bulletin No. 60 (Dec. 1980), pp. 12–13.

3.3.3 Safety and Security

The laboratory has a duty to its employees and others to ensure their safety in all aspects of its activities, which include waste disposal. (See section 2.3.1.) It also has a duty to itself to take precautions against negligent or malicious action by people, which is the security consideration.

It is found in practice that people tend to take the easy option, hence waste disposal should be so organized that the correct and safe method is the easy one. Any procedure where people have to go out of their way or take extraordinary steps is likely to fail. However, if the correct receptacle is to hand, and the disposal procedure is facilitated by the use of helpful devices, then safety can become routine.

The most common unsafe practice in waste disposal seems to be the obstruction of gangways and exits. If the waste is combustible (e.g. paper) or flammable (i.e. solvents) then the danger in the event of a fire is obvious. This is in addition to any injuries or spillages from people knocking into or tripping over waste receptacles.

Another common unsafe practice is the accumulation of too much waste in the work area. Even a pile of empty cardboard boxes is a fire hazard, but keeping large amounts of harmful substances (chemicals, infected materials, radioactives) exposes staff to an unnecessary risk.

Laboratory staff must have due concern for any other people who handle the waste (refuse collectors, cleaners, contractors, incinerator operators). This means, for example, not overfilling receptacles, keeping glass out of paper bins, using the correct colour code and ensuring adequate labelling.

The security aspects of waste are rarely considered. A simple error is to place drums or a pile of solid by a fence or building where they can aid trespassers to climb over or up to a window. Many trespassers are not only thieves but also vandals, and it is relatively common practice to set fire to refuse. Therefore it is obvious that combustible waste should not be left next to a door or wooden structure. If the site is not fully secure, then drums or cans of flammable solvents should not be left outside. Where they are kept in a stockade (rather than a building) they should be far enough away from the perimeter so that they cannot be punctured from outside.

Where dangerous drugs or poisons are used, it is generally understood that they must be kept secure from thieves or would-be suicides. However, the waste disposal arrangements for such materials must be equally secure. This applies, for example, to out-dated pharmaceuticals awaiting collection for destruction elsewhere.

Hypodermic syringes and needles are coveted by a certain class of person for the misuse of drugs. They have often been taken from hospital and laboratory waste. In other cases, they have been deposited on a tip with public access, to be found by children, with consequent adverse publicity for the institution.

As a general rule, hypodermics used for non-hazardous materials should be made unusable by bending the needle and/or breaking off the tip. Special devices are available to aid this practice. However, where more dangerous substances (pathogens, radioactives, very toxic drugs and chemicals) are used, then staff should not take even the slight risk of accidental injection and should discard them whole into appropriate containers. (See section 10.6) These containers should then be removed promptly for disposal, and kept secure in transit. For example, they should not be left on a trolley in a hospital corridor.

Care must also be taken for recoverable waste with some value, such as platinum or mercury. Many laboratory institutions (e.g. hospitals, universities) are relatively open to outsiders. However, there is no need to expose too many employees to temptation!

3.4 BAGS AND BINS

Solid waste is commonly collected and transported in plastic bags. A convenient procedure is for a bin to be lined with a bag which is removed altogether and replaced, thereby reducing handling and keeping the bin clean. Waste paper bins in offices may not be lined, but it is preferable for a bin in a laboratory to be lined even if it is emptied out to a larger sack, and even if its contents are supposed to be innocuous.

The bags must be strong enough for the purpose. In general, the greater the hazard, the stronger the bag required.

High density polyethylene is stronger than low density polythene, and thus can achieve the same strength with a lower thickness. Polypropylene is slightly stronger still. Note that the resistance to tearing is not the same as simple strength. Laminated plastic bags are available which have superior resistance to puncturing. Multi-layer paper sacks are sometimes as good as or better than plastic for relatively sharp items such as disposables. No plastic or paper bag should be used directly for sharp objects such as needles or scalpel blades.

It is recommended that for general laboratory purposes a minimum thickness of 25 microns (high density polythene) is adequate, but 100 microns should be used for high risk material. These figures should be double for low density polythene. Polypropylene should be 20 and 80 microns respectively.

A good supply of small plastic bags is useful for collecting waste in the form of powder or small objects. If the waste is not to be autoclaved (see section 10.2.2.3) then the small bags are closed with tape or simple ties. This provides additional security if the collection bin is knocked over, or a sack splits in transport.

All laboratory staff should be shown the use of a plastic bag for picking up a contaminated object. The method is as follows. A bag of suitable size (but at least 150 mm square) is turned inside-out, then placed on one hand. The bagged hand then grasps the object, and the bag is everted so as to contain the object. This prevents direct contact, and avoids contaminating gloves. However, if the contamination is injurious, then both hands should be gloved and standard hygiene observed.

A wall-mounted heat sealer can be used to seal up plastic bags of waste. This is particularly valuable for wastes which are offensive rather than harmful. It can also be used for extra security in transport (see above).

CAUTION: if sealed packets are sent for autoclaving, it is essential that they be slit open to allow steam to penetrate. See section 10.2.2.3 and as follows.

Where infectious material is involved, the technique of 'double bagging' is routinely used. That is, the items are collected in an autoclavable and labelled bag, which is placed inside another bag of similar size to provide double security. (Note that it is essential to permit steam to penetrate the contents: see section 10.2.) The inner bag is best of the type which opens during steaming and closes again during cooling. The outer bag should be of normal type. It is opened just prior to autoclaving and carefully sealed just after.

Colour coding is particularly easy with plastic bags. It is strongly recommended that a colour code be agreed throughout the organization (not just in an individual laboratory). The Health and Safety Commission of the UK recommends yellow bags for clinical (e.g. hospital) waste, but contrary colour codes have been reported in use in hospitals (Carr 1981; GLC 1983).

Table 3.1 is a suggested colour code including the HSC recommendations but extending them for other waste. If a different scheme is in well-controlled operation, there is no need to change. On the other hand if there is no colour code, or it is poorly followed, then it would be sensible if new arrangements kept to a general pattern.

In particular, it is strongly recommended that black sacks are only used for fairly innocuous material, because of their widespread usage for domestic refuse. Further, the colours yellow and red are used in several conventions for hazardous waste, so they should not be used merely to indicate a different collection point or some other arbitrary difference. Tie-on tags or stick-on labels can be used to categorize as well as, or instead of, a sack colour code.

It is very bad practice to use old bags (e.g. from supplies) especially if they are labelled for a specific contents. The savings are negligible, and there should be no need if the laboratory has available a good supply of the right sort of bags.

Plastic bags can be used to line conventional bins, but there are now widely available units specially made to hold bags. The minimum is the so-called 'sack holder' in which a collar is used to freely suspend a plastic bag. The collar may have a lid or may snap closed (usually for small bags). These sack holders should not be used for anything but innocuous waste, and should not be used in areas with a serious fire risk or frequent nearby traffic, because of the chance that the bag may be damaged and release its contents. A slightly better version has a strong mesh basket to protect against damage from (say) passing trolleys and to contain the waste if the bag should fail. These may be adequate in many workshops and some engineering laboratories.

The best sack holders for laboratory waste have the bag enclosed within a metal body which both contains and protects the waste. These usually have a lid which is self-closing and operated by a foot pedal, to give the maximum hygiene in use.

Some small bag holders have the top kept flat closed by a spring mechanism, which must be pulled open to add waste. They are fairly effective in reducing smells and access by flies for catering and some biological waste. However, there is some risk of contamination in opening and closing them.

As a general rule, all waste receptacles should be normally covered. A possible exception is bins for paper towels in a hand-wash area. However, if open bins are used for this purpose, care must be taken that other materials (such as glass, contaminated paper or gloves) are not put in. Some

Table 3.1. Colour code for plastic sacks

The following colours are suggested for segregation of waste collected in plastic sacks. They conform with the UK Health and Safety Commission recommendations for clinical waste, and with the most general practice. Smaller bags of other colours may be used within the laboratory, but it is recommended that these be collected inside a sack of standard colour. Care should be taken that existing colour codes (e.g. for linen handling) do not conflict.

Colour of bag	Type of Waste
Black	Paper, and domestic-type waste from offices, rest-rooms, etc. Free of chemicals, sharp objects, infectious material.
White or clear or neutral	Non-toxic, non-infectious industrial-type waste, e.g. from workshops. May contain sharp items (e.g. metal pieces) if the bag is strong enough and/or the item is packed. Broken glass should be free of chemicals or infection. If used for foodstuffs, to be labelled 'Catering waste, burn without delay'.
Yellow	Infected or potentially infected material destined for incineration. Preferably with a biohazard symbol and wording.
Yellow plus black band	Waste preferably to be incinerated, but which may be disposed to land if suitably controlled. May have biohazard labelling.
Yellow plus radiation symbol	Waste contaminated with radioactivity. The symbol is essential. The utmost care should be taken that it is not confused with non-radioactive biohazard.
Light blue (or light blue inscription on transparent)	Material which is for autoclaving before disposal. If the bag itself is to be treated it must be of suitable material.
Red	Hazardous waste not mentioned above. Includes asbestos, chemically contaminated paper. All items in a single bag must be compatible, and bags must not be combined. (Used in UK hospitals for foul or infected linen which is to be laundered.)

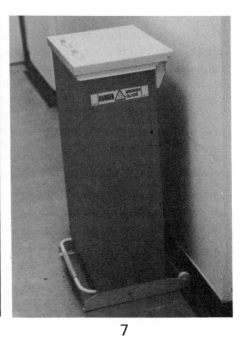

6   7

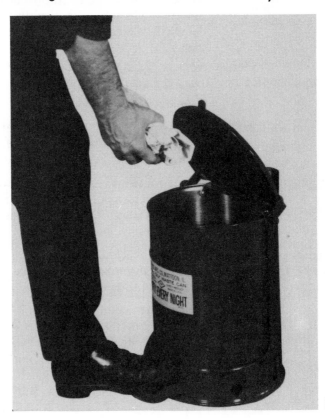

8

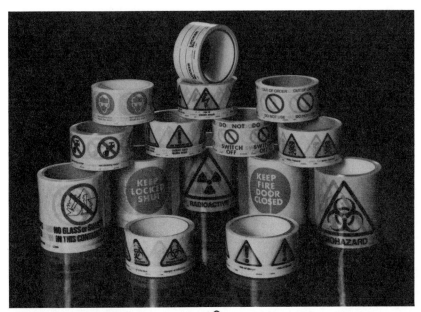

9

10

⑥ A swing-top bin for hazardous waste, showing typical contamination of the swing-lid. ⑦ A good design of bin, clearly labelled. It has a foot-operated self-closing lid and contains a multi-layer paper sack which has been tested and found suitable for rejected glassware. ⑧ A metal pedal bin for the collection of combustible waste, with self-closing lid, labelled 'Empty every night'. Photo courtesy of Justrite Manufacturing Co, Des Plaines, Ill., USA ⑨ A wide range of self-adhesive labels are available to mark waste receptacles and containers. ⑩ A simple home-made label can ensure that washing-up facilities are correctly used, and reduce the risk that dirty labware will contaminate clean items.

laboratories have a set of open bins together. It is virtually inevitable that some waste will go in the wrong bin, owing to carelessness or the temptation to aim in from a distance.

A lift-off lid is not convenient for frequently used bins, though it may be acceptable for a container used only once or twice a day (e.g. to collect bench containers such as 'sharps' boxes). The very best arrangement is a foot-operated lid, which leaves both hands free and gives the maximum hygiene. A swing-top may be better than no top, but gives some risk of contamination in use.

An unusual design of open bin has a special top which will, in the event of a fire, direct combustion products back down and thus (it is claimed) smother the fire. These may be useful in areas where little segregation is required and there is a need for maximum convenience in disposal of paper, plastic or other combustible waste. However, note that a steel pedal bin with a steel lid will also tend to smother a fire.

For certain purposes, a rigid disposable container (usually a cardboard box) may take the place of a bin. These are commercially available in waterproof reinforced cardboard, ready labelled for aerosol cans and for glass waste. Ordinary cardboard boxes have been used for these purposes, but pose the risk of physical failure and the problem of misleading labelling.

As the bags may be colour coded and labelled, so may the bins be. There is absolutely no technical reason why every waste bin cannot be clearly labelled as to its proper contents. Many factories have the position of a bin marked on the floor and both the bin and a plaque on the wall are clearly labelled. This practice could well be adopted by laboratories.

The location and size of waste bins need to be considered for maximum convenience. People should not have to walk too far for disposal, and there should be an adequate number to take a day's waste without being more than two-thirds full. It is of course helpful for cleaning staff if the bins from which they collect are near the door (or in some cases bags are tied up and put out by the laboratory staff).

3.5 BOTTLES AND CANS

Liquid waste is collected in glass or plastic bottles and metal or plastic cans or drums. The majority of laboratories make use of old reagent containers, but there is increasing use of specially-bought containers, including so-called 'safety cans' for flammable liquids.

The most common error is to have too large a container in the work area: too many laboratories keep 5-gallon drums (or larger) of waste solvent, which it may take them more than a year to accumulate. Generally speaking, the collection container for waste liquids should be no larger than the largest reagent container. A larger container may be kept in a suitable secure store, to which the laboratory vessels are brought.

Except where very small amounts of very special waste material (e.g. highly expensive material for recovery) are involved, liquid waste containers should be emptied on a weekly basis. Where larger amounts of flammable or toxic liquids are involved, then daily removal may be advisable. Note that the presence of larger quantities of flammable liquids (even as waste) may often be illegal, which might invalidate insurance in the event of a fire.

Chlorinated or brominated solvents are normally collected separately from general hydrocarbons and other flammable solvents. However, note that some chlorinated solvents are in fact flammables. Moreover, all halogenated solvents are quite reactive and corrosive. Unless specially made for this purpose, metal cans tend to corrode away quite quickly.

CAUTION: aluminium can react violently with some halogenated solvents. Chloroform can react explosively with acetone, and other reactions between halogenated and other solvents are possible. See Table 14.1 for other solvent hazards.

Polythene has a limited life with halogenated solvents — it will tend to crack after prolonged exposure. However, special high-density heavy-gauge containers specially made for corrosive solvents collection have an acceptable life.

Acids and caustics are best collected in stout-walled polyethylene or polypropylene bottles. Wide-mouthed vessels with a self-closing (but pressure-relieving) lid are also commercially available for this purpose.

If glass bottles are used for the collection of liquid waste they should stand in a suitable outer container — a plastic bottle carrier, a plastic bucket or a deep tray. Plastic film on the outside should not be relied upon to retain the contents in the event of breakage. For narrow-necked bottles (e.g. winchester pattern) a funnel will be required — preferably a plastic one with external ribs on the stem.

For obvious reasons, metal cans are preferred for flammable liquids. If use is made of old oil cans etc. then the can should be completely painted red, and a 'flammable' symbol affixed, so that there is no misunderstanding. However, red-painted tinplate cans are easily available in a range of sizes for a very modest price, so there is really no need for such economies. Standard cans can be improved by the addition of a flash arrestor, which is commercially available at modest cost and can be transferred to other cans.

The safest is of course the purpose-made 'safety can' for flammable liquids, which is apparently expensive. However, quite a small one will be sufficient for a typical laboratory, and its contents can be safely transferred to the drum in the store on a regular basis, encouring good housekeeping. Note that there is a difference between cans intended for waste collection and those for dispensing. The collection cans have a wide mouth to enable them to be filled without use of a funnel.

Small amounts of water and traces of chemicals will inevitably cause corrosion, so ordinary cans should be discarded altogether at least every year. Purpose-made cans should be inspected at least every year, but may be expected to last about 5 years, and stainless steel ones almost indefinitely.

If old reagent containers are used for waste collection, then the original label should be completely removed — even where the waste is similar. (Writing 'waste' on the label of a toluene bottle may mean that waste toluene is inside or it may not.) Self-adhesive labels with the necessary warnings are commercially available, but there is no objection to a careful home-made label which gives sufficient information. In the case of a drum, it may not be possible to remove the labelling, in which case it should be completely painted over, and new information put on. If this is done, then it will be clear that all the signs are intentional and not just left over from previous use.

If a laboratory does not have its own source of drums or large plastic containers, these may be obtained from waste disposal contractors or certain companies which specialize in drum reconditioning. Note that transport off site in the USA will generally require containers conforming to Department of Transport regulations.

While collection of hazardous waste into very large containers will generally reduce the unit cost for disposal by a contractor, care should be taken that this does not mean an unwarranted hazard due to the keeping of large amounts of waste on site.

3.6 DUSTBINS AND SKIPS

A small laboratory may get rid of its more innocuous refuse in dustbins or sacks along with local collections from houses, shops and small business premises. Extra-large bins are used in some areas for blocks of flats and business premises. These are preferable because they are often cheaper per volume, are mechanically handled and thus pose less risk of the collectors and are large enough for bigger items to be put in. (See section 3.8.)

For larger amounts, or if the laboratory waste is unacceptable as normal refuse, a skip service is used. A skip is essentially a large steel container for waste. A lorry brings an empty one and removes the full one as often as required. There is a hire charge for the skip and a removal charge (usually per skip load irrespective of whether it is full or not). The skip lorry driver is legally entitled to refuse to remove a skip which is overloaded by weight or volume, and the company may make a charge for a wasted visit.

Open skips are only really suitable for material such as builders' rubble. For the vast majority of laboratories a closed skip should be used. This prevents material being blown away and greatly reduces the risk of fires, health hazards and public nuisance. If the site is not totally secure, it is possible to padlock some skips to prevent arson or the addition of other rubbish outside working hours.

3.7 BENCHWORK

It is not enough merely to have some collection containers: the work practice should be arranged to facilitate easy and safe disposal of materials.

Large pieces of apparatus should incorporate drain taps, ports, etc. for safe emptying and cleaning. Apparatus for work with especially hazardous materials may incorporate the means for neutralizing at the end of the experiment.

For the majority of laboratories, it is generally useful to have small receivers for specific wastes in the working area. For example, a jar of disinfectant may be handy for discarding swabs in biological work. A plastic bag supported by a beaker may be used to collect disposable tips from an automatic pipette. A student experiment may be provided with an individual bottle for the waste from that work, as well as the original reagents. Someone using a spectrophotometer intensively may put a flask or beaker handy to collect the discarded fluid from the cells, instead of taking each cell to a main solvent can.

Where repetitive routine work is involved, there is much to be said for appropriate waste receivers being located permanently by the workplace. For less routine work, individuals should be encouraged to make up temporary containers for their short-term needs. Screw-capped glass jars, lidded polypropylene beakers and a supply of self-adhesive labels are very useful for this purpose.

3.8 SEGREGATION

One of the major rules of laboratory safety is to keep incompatible materials apart. However, a more refined system of segregation is required even for materials which do not actually react together. Broken glass should never be put into a waste paper bin, because the collection staff will not expect it, and can therefore be injured (though the glass may ultimately end up on the same refuse tip as the paper). Contractors usually charge more for disposal of chlorinated solvents than for hydrocarbons etc. and may take samples to check drums of waste removed. If a small amount of chlorinated solvent is put into a large drum of hydrocarbons, the whole drum will be charged at the higher rate, which is an unnecessary expense. Waste destined for recovery should be kept well apart from similar material for disposal — for example a random mixture of waste solvents added to single solvent waste can make distillation uneconomic. In the case of gas cylinders, there have been many wasted journeys because an empty cylinder has been put with the full ones, and later taken away for use by an unsuspecting worker.

A common bad practice is to put all the waste receivers together — usually under a sink. This may put paper, flammable liquids, corrosives, reactives, infective and radioactives in close proximity, which has been described as 'an accident waiting to happen'.

In setting up a laboratory, one should ask what wastes can arise, including those from breakages and spills, and provide receptacles accordingly. A broken glass bin may not be used for months, but by its existence reduces the chance of glass going into a paper bin. It will often be possible to gather various wastes together for common treatment (e.g. routine biological

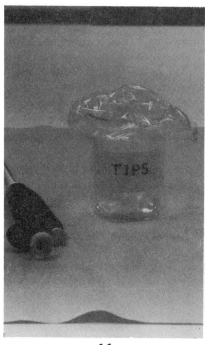

11

12

13

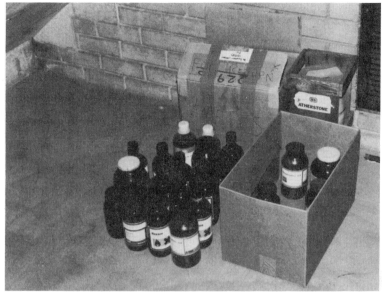

14

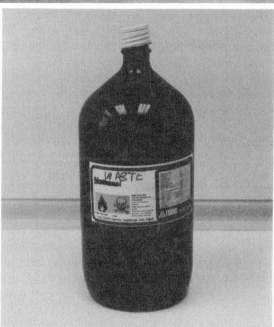

15

⑪ A collection vessel for disposable pipette tips, made from a plastic bag supported by a beaker. Much better than discarding onto the bench.   ⑫ A commercial disposable box for the collection of broken glass and similar waste. The box is waterproofed, ready labelled and is supplied packed flat. Photo courtesy of Lawtons Ltd, Liverpool.   ⑬ For many routine procedures it is worth making a specific waste receptacle for on-the-spot use. This aids both safety and efficiency.   ⑭ Empty chemical containers and packaging, put ready for collection by the supplier on his next delivery. Most chemical suppliers are pleased to accept glass jars and bottles, wooden or plastic crates and similar packaging, if returned in clean condition.   ⑮ An unsuitable waste collection vessel. The cap does not fit, the old label has not been removed, and there is no actual information as to the contents.

tissue for autoclaving/incineration; flammable solvents for removal to the main store; paper towels from the handbasin bin; and waste paper from the desk bin). However, attention should be given to incompatible materials occurring within these classes. For example, paper towels or tissues contaminated with chemicals, radioactives or infectious micro-organisms should not be put into routine paper waste; broken thermometers should be treated separately from other broken glass; some solvents should be treated separately. Special procedures are often needed for a single chemical or a single micro-organism used in the laboratory.

In some circumstances, wastes will be collected separately initially, but may later go into the same place — for example different bottles of chemicals into the same drum for landfill, different types of biological waste all to be incinerated, glass, paper and workshop waste into a common skip. It can be appreciated that the needs of collection are sometimes different from the needs of final disposal, and the two should not be confused.

Segregation can be of economic benefit, as some wastes are more expensive. By keeping to a minimum the bulk of especially difficult waste (up until a decision to dilute is taken) there are usually savings in storage requirements and general protection of employees. It is particularly vital to keep hazardous, difficult, or unusual waste out of the ordinary refuse handled by cleaning staff and domestic refuse collectors. Large pieces of metal or concrete can cause injuries or damage machinery if hidden in a pile of paper. Fluorescent light tubes and cathode ray equipment (including TV sets) can also cause injury when they break — possibly quite serious. Thus it is necessary to remember hazards other than the obvious laboratory ones in order to maintain good relations with the refuse collection authority or a trade waste collector.

See section 3.4 (Table 3.1) for colour coding of bags, and comments on bin labelling. See section 10.3 (Table 10.1) for categories of clinical waste.

3.9 SPECIAL AIDS

Throughout this book, reference is made to commercial devices which can aid waste collection and disposal. It seems that many laboratory managers are not aware of some of the items on the market to improve the safety and effectiveness of waste operations, so a short summary follows.

3.9.1 Labels

Clear and unambiguous labelling is a vital safety aid. Signs and self-adhesive labels are commercially available ex-stock with most of the waste classifications one might want, including some surprisingly specialist ones. Pre-labelled bags and boxes are widely available as an enormous help in waste segregation control. Furthermore, it is not very expensive to have your own labels or signs printed, and the major suppliers are willing to receive suggestions for additions to their ranges.

3.9.2 Handling Equipment

Adequate provision of trolleys, drum carriers, bottle carriers and trucks is vital to prevent accidents in transport of waste. A particularly useful item is the 'bin truck', a low platform on wheels, with sides of about 50 cm high. This will take a large number of waste sacks upright, and contains spillages. If the volume justifies it, then chute systems can be a safe way of transporting certain kinds of waste, e.g. from an upper floor

3.9.3 Mechanical Converters

Industrial units for mechanically treating waste are generally too large and expensive for a laboratory alone, and there are safety problems about shared units. However, a grinder fitted into a sink drain (sometimes known as a 'kitchen waste disposal unit') can be invaluable for laboratories handling food and certain other biological material. A larger unit is made for macerating disposable bedpans (and flushing the resulting particles to sewer) in hospitals, but has been adapted for several other purposes by laboratories. See sections 5.3.5, 10.4 and 10.5.4.

Other crushing and shredding devices are usually too large, but it is possible to buy a hand-operated device for crushing glass bottles and broken glassware into relatively fine pieces, which gives a smaller volume of waste glass and less chance of injury (particularly with respect to other methods of breaking down glass).

A few laboratories find it worthwhile having a small baling unit. This compresses voluminous waste such as paper or agricultural produce into smaller sealed packages which are easier to transport and more hygienic.

CAUTION: waste which is to be incinerated should not be baled or compressed, as this can cause serious difficulties. See section 7.4.5.2

3.9.4 Recovery Equipment

Although home-made apparatus can be perfectly satisfactory, for regular recovery of some valuable material (notably solvents and silver and mercury metals) then consideration should be given to some of the commercial devices (e.g. automatic stills) which are manufactured especially for this purpose. See Chapter 14.

3.9.5 Spillage Control

There are now several kits or pre-packed absorbents sold for dealing with spillages of hazardous chemicals. Like the fire extinguisher, their capacity is limited, but they have the great advantage of being available when needed urgently in an immediately usable form. (See sections 15.2.2 and 15.2.3.)

Suction pumps and 'wet' vacuum cleaners can be useful for larger amounts of material, having regard to any corrosive or flammable hazard.

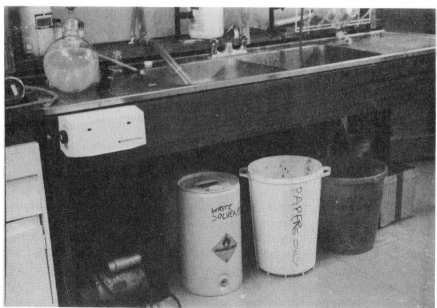

16

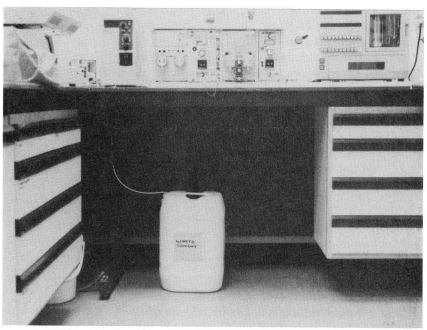

17

⑯ A typical waste collection point under a sink. The solvent can and glass bin are much too large for the size of laboratory. The two open bins together greatly increase the risk of material going into the wrong bin.   ⑰ A convenient collection of solvent from an automatic machine. However, the tubing should be better secured, and the very large size of the container presents an unnecessary hazard, especially if it were to leak or there was a fire.

Detectors can also be bought to detect leaks on the floor or loss of fluid flow (e.g. coolant to a condenser) and sound an alarm or shut down apparatus. A flammable gas detector may be useful to prevent fires in the event of leaks, via an alarm or shut-off mechanisms. With suffocating gases or vapours, an oxygen monitor may indicate an unobserved leak before entry into (say) a small room where the air has been partially displaced.

3.10 BROKEN GLASS, SHARP METAL AND AEROSOL CANS

These classes of waste can be especially hazardous if they are included in with other categories. Sharp objects can cause injuries; metal can damage machines for processing waste; and aerosol cans are very dangerous in incinerators. If significant amounts of these materials are likely to arise, then a regular disposal procedure must be developed.

Firstly, the collection process must be obvious and safe. Single-thickness plastic bags are rarely good enough, though some workshops use them for lathe turnings etc. and accept the occasional need to sweep up after a bag failure. Reinforced plastic or paper sacks are available which usually have adequate puncture resistance. Some manufacturers offer specially constructed cardboard boxes, ready-labelled for broken glass and for aerosols, which are a particularly good idea where arisings are frequent (e.g. in a hospital or in a laboratory testing aerosol products).

Where arisings are only occasional, e.g. from accidental breakages, then a plain metal or plastic bin with a lid (of not too large a size) can be located in each laboratory suite. This bin (or its liner if this is of the bucket type) can then be taken (say) weekly and emptied into the central bin by a responsible person, e.g. the technician in charge of the laboratory. It is extremely important that the central bin is of stout metal construction and is clearly labelled 'BROKEN GLASS ONLY'. Arrangements must be made with the waste disposal service for it to be emptied before it is more than two-thirds full, as a full bin can be very dangerous for the operative to handle.

It must be the responsibility of each laboratory to decontaminate broken glass (from toxic, corrosive, radioactive or infective agents) so that there is no significant hazard apart from the physical one. If this is impractical (e.g. for arisings of radioactive sharps and disposable glassware) then the materials must be collected in a secure container (e.g. a lidded metal bucket) sealed up and disposed as hazardous waste.

For a discussion on the use of 'sharps' containers for the collection of disposable glassware, syringes, scalpel blades, etc. see section 10.6. Where such items arise but without any biological contamination, it may still be useful to have the commercial containers for collection, and to dispose of them by incineration. (See also section 7.4.5.2.)

Sharp metal can be dealt with in a similar fashion to broken glass, unless there is reason to collect more valuable metals for salvage. However, note that turnings and dust from light metals (aluminium, magnesium, titanium,

18

19

20

21

⑱ A commercial collection vessel for waste solvents. The wide mouth permits filling without a funnel. The mouth is fitted with a flame arrestor and self-closing cover. Photo courtesy of Justrite Manufacturing Co, Des Plaines, Ill., USA. ⑲ Waste solvent storage in a safety can kept in a ventilated fire-resistant cupboard. ⑳ A metal safety can for the collection of flammable radioactive waste. The beaker is kept to aid pouring. ㉑ An open skip like this is mainly suitable for builder's rubble and the like. It is not suitable for paper or light plastic items which could blow away, and should never be used for chemical or clinical waste. If the site is not totally secure, then outsiders may dump other waste in the skip, and vandals may start fires.

zinc, light alloys) can be quite a severe fire hazard, and even moist iron has been known to generate sufficient heat to ignite paper.

Where no chemical, radioactive or infectious hazard exists, these wastes are all generally acceptable for disposal to municipal or commercial landfill sites. An occasional aerosol-can (e.g. one per dustbin or bag) may be tolerated by an incinerator dealing with domestic refuse, but a box or bag full of aerosols (even supposedly empty cans) should not be sent to any incinerator without prior warning. A domestic refuse incinerator can usually tolerate the amounts of broken glass generated by a laboratory, but smaller incinerators (e.g. a hospital unit) may be damaged by glass melting inside. Sharp objects in the ash can also be a hazard to operators of small incinerators. (See also Chapter 7.)

3.11 RETURNS TO MANUFACTURERS

3.11.1 General Comments

Manufacturers often request the return of packages, but few laboratories take the trouble to comply fully. Packages for chemicals (and some other goods) are often charged for, if not returned. This may apply (for example) to pallets, crates and protective containers for bottles of liquid. Thus many laboratories may be incurring expense in disposal of items for which they are being charged!

Even where a charge is not made, return of cardboard boxes and empty bottles can assist the manufacturer, while reducing the volume of waste for the laboratory.

A very common waste of money is the unnecessary retention of gas cylinders. The rental of a cylinder is often more costly then the price of the gas it contains. Management should therefore ensure that empty cylinders are returned promptly. Furthermore, cylinders which are part-used should not be kept unnecessarily unless the contents are particularly expensive. With many gases it will be cheaper to return a half-full cylinder and buy in another one later, instead of keeping one unused for a few months. In a large institution it is not difficult to overlook a cylinder, unless there is a proper record system and an individual with special responsibility to keep an eye on the stock. (See section 9.4 for further comments on gas cylinders.)

It is economically vital to ensure that any goods which are delivered in a faulty condition (or not as specified) are returned promptly. Failure to notify the supplier within a few days will make it more difficult to claim that the error was on his part.

Where the laboratory made an error in ordering, then the supplier may still take the goods back if this is requested immediately. It is reasonable for the supplier to make a charge for the delivery and/or a 'restocking charge'. Chemicals and certain other products will not be accepted for return if the supplier's seal is broken or if they have been kept for a significant time.

There are a few cases in which a supplier will accept goods back for disposal. This mainly applies to chemicals containing valuable elements, but can include instruments or parts containing recoverable material such as gold filaments. Occasionally a manufacturer will assist a customer in waste disposal on the basis of goodwill.

3.11.2 Boxes, Crates and Pallets

When dealing with a regular supplier, these can normally be collected when a further delivery is made. The laboratory should note any damage on the delivery note when signing for goods.

With a one-off delivery, the goods should be removed on the spot (if reasonably possible) so that the delivery driver can take away the carrying item. If this is not done it is unlikely to be worthwhile returning the item.

Boxes, crates and pallets which have been affected by chemicals (owing to a breakage etc.) should not be returned, but should be disposed of as hazardous waste.

3.11.3 Chemical Containers

The supplier will usually indicate which containers may be returned. Metal containers and large plastic containers are usually returnable and often carry a charge. Plastic jars for solids and ampoules for liquids are generally single-trip containers.

All containers to be returned should have the original label. The outside should be clean and free of chemicals. Liquid containers need not be cleaned, and may contain traces of the liquid providing this is the one originally supplied. If a bottle has been used for any other liquid, it must be well cleaned out, though it may be left wet with water inside. It is absolutely vital that any added material (e.g. sodium wire) is removed from a bottle before return. All caps should be secure.

A similar practice can be adopted with jars of solid, though with deliquescent substances it may be preferable to rinse out the jar.

CAUTION: substances kept wet for safety should be well rinsed from the jar and the cap. See Appendix C-6, class (1).

Substances kept under other liquids for safety (e.g. sodium) should not have their containers returned.

3.12 ECONOMICS

Waste disposal costs money, both in direct costs and in the time spent by staff. Since this expenditure is seen as non-productive, there may be some pressure to minimize it. However, it is more important to see that the money is wisely spent. Generally, any device or procedure which makes waste disposal simpler and more reliable will benefit the safety and smooth working of the laboratory. For this reason, commercial items such as pre-labelled sacks and boxes are recommended for careful consideration, although they will add to the consumables cost.

It is difficult to be enthusiastic about waste disposal, so it often receives scant attention in planning. The net result can be that time is wasted in inefficient disposal activities, and there may be considerable problems and even unforeseen expense. It is therefore important that the question of waste disposal be satisfactorily answered when planning any research project, new building, new instrument or operating procedure. Provision for waste disposal should be made in budgets, along with provision for chemicals and other supplies.

It is usually possible to cut down on waste production. The simple expedient of buying smaller quantities or using a less wasteful procedure can often be adopted. (See section 13.3.3 for low-waste chemical procedures.) Solvents and other materials may be salvaged. (See Chapter 14 for various methods of recycling.) Modern management techniques can and should be applied to laboratory procedures to look at the efficient use of labour and of resources.

However, reducing waste is not the same as reducing waste disposal costs. For example, if a senior scientist spends several hours recovering a relatively cheap solvent, then the laboratory has made a loss (though the scientist often feels a sense of satisfaction). Another false economy is the salvaging of items made for a single use. The time lost in cleaning is a major penalty, but there may be risks to staff handling toxic, infective and often sharp objects. Furthermore, many disposable items do not decontaminate well, and residues can spoil subsequent experiments. The laboratory should either buy re-usable items and set up proper facilities (washing machines etc.) or budget and plan for a regular re-stocking of disposables. Best of all, the laboratory manager should critically evaluate both proposals, taking into account staff time and safety aspects.

Note that a change to disposables will normally mean an increase in the volume of waste, and a change in its character. The arrangement of bins and the procedures for emptying them and removing waste may need to be altered. Similarly, a change from solvent disposal to solvent recycling will require a change in the strategy of collection and storage.

Section 7.4.4 illustrates how unrealistic costings can be produced for an incinerator, though equally false forecasts could be derived for other items or procedures. It is better to determine the true need for waste disposal and then to find the cost of providing it. The cheapest method may not be the one of choice, if another can show significant advantages in reliability, safety or legality! In particular, it is necessary to include provision for emergency situations, such as fires and spillages, as failure under these circumstances could prove extremely expensive. (See Chapter 15.)

Section 3.11.1 illustrates a few cases where lack of waste management can lead to unnecessary expense. Chapter 8 includes some important financial matters in dealing with outside contractors. Costs may also be controlled in dealing with drainage (Chapter 5), ventilation (Chapter 6) and disinfection (section 10.2.3 and Appendix B-5).

3.13 PROBLEMS

The following is a checklist of some of the items which can go wrong, and where some kind of provision is required.

(1) *People.* Would the sudden absence of any individual cause difficulties? For example, the laboratory manager, the incinerator operator, the person who authorizes certain documents. Can another person be found if necessary?
(2) *Contractors.* How dependent is the laboratory on efficient prompt removal by one agency? How long could waste be accumulated in the event of a strike? Do you know of an alternative disposal service if your contractor fails altogether?
(3) *Spillages.* What are the most serious spillages which could occur? Are there plans and facilities for dealing with minor and major spillages? Does your regular contractor offer an out-of-hours emergency service?
(4) *Drains.* Have you established a good relationship with the drainage authority? Does your present consent meet your needs, or would it be useful to have some specification relaxed to cover occasional problems (this may be negotiable)? Do you know who to telephone at any hour if there has been an accidental release of hazardous material to the drains?
(5) *Law.* Are your present waste disposal and storage procedures fully within the law? Would a tightening of legal control cause any insurmountable difficulties?
(6) *Knowledge.* Are you aware of all waste disposal?

BIBLIOGRAPHY AND REFERENCES

British Standards Institute (1980) BS 5906 *Code of Practice for Storage and On-site Treatment of Solid Waste from Buildings.*
Carr, P. (1981) The Hospital Laundry Bag, *Nursing Times*, **77**, No. 23, 983–4.
Greater London Council (1983) *Final Report of the Working Party on the Disposal of Clinical Waste in the London Area.*
UK Dept of the Environment (1983) *Waste Management Paper No. 25: Clinical Waste*, HMSO.
UK Health & Safety Commission (1982) *The Safe Disposal of Clinical Waste*, HMSO.

CHAPTER 4

Sources of Information

4.1 GENERAL COMMENTS

This book gives reference at the end of chapters to publications which contain information of relevance to the waste disposal operation. Details of some agencies and commercial organizations are given in the Appendices, along with some tabulated data. The following notes are merely intended to draw attention to other sources of information which might be overlooked.

4.2 LABORATORY KNOWLEDGE

The first and most valuable source is the technical knowledge of laboratory staff. The fact that some material has just become waste in no way alters its properties. It is just as flammable, toxic, radioactive, explosive, infective, corrosive, volatile or inert as before. More important, at the moment it becomes waste, someone (the user) ought to know what these properties are. If he or she does not, then the previous laboratory usage was unsafe.

More detailed information, which cannot be given off-the-cuff, should be available in the notes, books or data sheets gathered for the original use.

For example, if a micro-organism is involved, is it noted for its temperature sensitivity or its heat resistance? Will the laboratory culture have produced a more chemically tolerant strain? What is its life cycle and needs for proliferation?

If a pharmaceutical is involved, what is its favoured mode of entry to the body? How is it activated or de-activated in the body — for example, what does it do in stomach acid?

For any chemical, what sort of reactions does it undergo readily? Can it be hydrolysed, oxidized, reduced or neutralized to a more innocuous form?

All these sorts of question should be readily answered by scientists working with hazardous materials, and can give useful clues to a safe means of disposal.

4.3 MANUFACTURERS AND SUPPLIERS

It is surprising how many people are prepared to guess about some technical matter (such as a detail of incinerator servicing, or the properties of allergens) when a telephone call to the supplier may give accurate information. Labels, catalogues, charts, data sheets and instruction books are frequently ignored.

Conversely, the quality of some suppliers' information is less than perfect, and some companies are more willing than others in providing assistance. Regrettably, major suppliers of hazardous materials have been quite unable to suggest a disposal method for their products. It is increasingly rare for the supplier to undertake disposal on behalf of a customer.

In general, suppliers of goods have a legal duty to supply information relevant to the safe use of their products, but this does not mean they are obliged to state the ingredients of commercial formulations. There is clearly some question as to whether waste disposal is part of the normal use. However, paying customers are advised to use their supplier's technical service as an early source of information. Possibly it may be easier to get suggestions for waste disposal if this is done prior to purchase, and perhaps as a condition of ordering.

Another expedient is to get the name of another user of the goods and enquire as to his disposal method.

4.4 GOVERNMENT AGENCIES

Many people are reluctant to inform the local or national enforcement agencies of a waste disposal problem, because they fear prosecution or at least an insistence that a very expensive procedure be adopted.

In fact, it is our experience that the agencies are positively helpful, though it is always possible to fall foul of a zealous individual. The avowed aim of the agencies is to prevent pollution and danger, rather than to gain prosecutions. If the laboratory treats the agencies as well-intentioned, they are likely to regard the laboratory in the same light.

Many problems that seem insuperable have in fact occurred before, and the agency will know of them (or can find out). In addition, they are usually well aware of commercial facilities or devices which may be used. At the least, they can warn the laboratory manager of possible breaches of the law he or she might not have appreciated.

There is always a chance that (owing to local politics) an agency may insist on the strict letter of the law, no matter how impractical this may be. However, there is a significantly greater chance that in cases of real difficulty the agency may come to an accommodation with the laboratory over a particular course of action

4.5 INSTITUTIONS

If the laboratory or a staff member belongs to a trade association or professional institute, then it is possible to make enquiries. Sometimes direct help is available, more usually the institution will refer to another source, which may be very helpful.

It is often worth contacting a local body that may have special knowledge or facilities, such as a university or research institute. In fact, many people who work at universities never think to explore the facilities (physical and mental) which may be open to them. It is possible that a new problem in one department has been routinely solved in another. A telephone call to the safety officer is a probable starting place.

REFERENCES

[section 4.3]

Frankel, M. (1981) *A Word of Warning: The Quality of Chemical Suppliers' Health and Safety Information*, Social Audit.
Frankel, M. (1982) *Chemical Risk: A Worker's Guide to Chemical Hazards and Data Sheets*, Pluto Press.

CHAPTER 5

Laboratory Drainage

5.1 INTRODUCTION

This is not intended to be a complete treatise on building practice, but a note of some of the important matters which a user of laboratory drains should know.

However, it should be appreciated that few architects are familiar with the special requirements of laboratories, and there is a possibility that inadequate systems may be provided, though they meet with building regulations and may be excellent for other purposes. This particularly applies where laboratories occupy a minor part of a major building programme. In this case (unless the architects have considerable experience of laboratory design) it is recommended that a specialist consultant or laboratory fittings company be brought in at an early stage. The special requirements of the laboratory drainage may have a considerable impact on the overall services design, and may actually cause the position of the laboratory to be relocated.

For example, rooms where there is a risk of flooding should not be sited above sensitive areas such as main computer rooms. Drains from a laboratory handling pathogens should not pass through areas where a leak would be disastrous, such as a food preparation area, or an intensive care ward. Laboratories handling substantial quantities of radioactive liquids should generally be sited on the ground floor or basement.

Where new buildings are involved, it is of course vital for the end users to communicate their special needs. These should include reasonable possibilities, such as the future use of radioactives, or plans to bring in special machines at a later date (for example, macerators feeding a solid slurry to the foul sewer) which may need particular drainage arrangements.

It is common nowadays to build relatively flexible designs allowing some degree of relocation, but replacing drains can be an expensive operation which it is as well to avoid.

Where alterations are made to existing drains, or there is a significant change in use (which may only mean a laboratory adopting new techniques) then careful consideration should be given to the effect on different users. For example, it is quite possible for a small amount of effluent from a laboratory to seriously interfere with the treatment process operated by industrial premises. Some effluents react together dangerously, and some changes of material can cause leaks through corrosion which may be unexpected. (See section 9.3.3 for disposal of chemicals to drain, and Table 9.1 for possible hazardous reactions in drains.)

5.2 FUNCTIONS

Drains are needed to carry away four distinct categories of material, which should be separately considered in the planning of a comprehensive system.

5.2.1 Clean Water

The majority of material carried in most laboratory drains is in fact clean water, or water with only trivial contamination. This includes cooling water from stills and condensers, and much rinse water from cleaning apparatus or handwashing. This category mainly defines the volume requirements of the system. It can be useful in flushing away other categories, but may need to be excluded from expensive or special drainage arrangements such as those for radioactive effluent.

5.2.2 Foul Water

This is biologically contaminated water (or slurry) which is potentially obnoxious, infective, or both. It is normally taken direct to the foul sewer — that is, the sewer which handles the discharge from toilets — by as direct a route as possible. Modern domestic drains tend to combine the foul sewer with drains for other purposes (sinks, baths, rainwater) but industrial premises usually have two or even more segregated drains (for example: foul sewer, rainwater, process water). Older building practice was to refer to this drainage as 'soil pipes'. The modern tendency is to speak of 'sanitary piping'.

5.2.3 Effluent

By this is meant dirt and chemicals deliberately discharged to the drains as a means of disposal. It includes unwanted chemicals, water used for washing apparatus, and the outflows from many instruments which are commonly taken direct to drain. This is generally the category which causes the most difficulty, particularly if the amounts, nature or combinations of materials involved are unsuitable.

5.2.4 Flood Water

Almost every laboratory with a water supply runs the risk of flooding, yet a relatively small number have adequate arrangements. Prevention is good policy, but it is always possible for glass to crack, hoses to come off taps, tanks to overflow. Quite amazingly, many laboratories have open safety showers but no floor drain at all. Clearly it is wise to have floor drains in areas using lots of water. A grid can often be fitted very conveniently under a sink in combination with its drain. Pilot laboratories often have gridded channels and a slight slope on the floor. To give proper protection, floor drains must be well sealed around the edges to ensure that spills go down the inside, not the outside of the pipe.

Where floor drains are not fitted, or may not be fitted (e.g. in a radiochemical laboratory with a fully sealed floor) then it is essential to make drainage arrangements for the safety shower. This will normally be a shower tray, with its own waste pipe running above the floor sealing level. A similar tray can also be used below any other possible source of floor water which does not need to be retained. However, it is best to eliminate such sources as far as possible from sealed floor areas.

5.3 COMPONENTS

Some of the important components of a drainage system are described below, with some features of their functions and design considerations.

5.3.1 Pipes

Except for certain industrial waters (which may be carried in a concrete duct with removable covers) drainage is carried in pipes. There have been major changes in materials and jointing methods in recent years, notably the increasing use of plastics. See section 5.4.

Drainage engineers normally design pipe layouts with a sufficient slope for water to run at 0.75 m s^{-1} even when part full. This reduces the chance of sediment accumulating, thus the drains are said to be 'self-cleaning'. The gradient required varies with pipe diameter. For example, it is 1 in 20 for a 50 mm pipe and 1 in 40 for a 100 mm pipe (running a quarter full). For other technical reasons, it is preferable that the velocity be kept below 3 m s^{-1}. Where the liquid volume is regularly sufficient to substantially fill the pipe, or where very little sediment is expected, then a shallower gradient is possible, for example 1 in 60 for the 50 mm pipe, 1 in 100 for the 100 mm one.

Drains within laboratories are often constructed on the assumption that there will be very little solid in the waste. If it is likely that silt and slurries will be flushed away, then the greater slope will be necessary. In fact, it is best to make a disposal point with as short a connection as practical to a well-used main drain for disposal of solids. This applies, for example, to a sink fitted with a grinder disposal unit, or the connection from a sluice or 'slop-hopper'.

Plastic drainage systems are very often constructed on the assumption that water will normally be less than 70 °C, and only briefly approach 100 °C on rare occasions. If any apparatus is to be used which gives extremely hot water for extended periods, then local modifications may be necessary. These not only relate to the resistance of the material, but also to its supports, which may need to be more frequent and allow greater expansion. The most effective solution is often to run a separate line direct to the nearest main drain.

The usual arrangement is for sinks and other receptors to feed to a sloped pipe which connects more or less directly to a vertical large diameter pipe. The top of this pipe continues to the roof of the building where it is open to the air, to prevent syphoning. There should not normally be more than 6 m of graded pipe leading from the vertical. If absolutely necessary, a separate vertical vent (known as a 'stack') may be provided, or the pipe can be made oversize, which may mean increasing the diameter as additional sink connections are made.

5.3.2 Sinks, Basins, Sluices, Drip Cups and Troughs

Sinks for laboratories are usually of square design and made of stainless steel or vitreous china. Polypropylene or PVC are sometimes suitable. Enamel is perfectly acceptable until it becomes chipped, when it is likely to rust and harbour contamination. China sinks are virtually impervious to all chemicals and are easy to disinfect. However, when they become crazed the fine cracks can harbour radioactive contamination, so they are not suitable for this purpose.

The usual problem with china sinks occurs around the edge where they are sealed to the bench or drainer. Leaks and accumulations of dirt may result. Stainless steel sinks usually avoid this by having the sink and drainer pressed in one piece. Note that for microbiology laboratories, a smooth surface is preferred to the corrugations normally supplied on the drainer, as this can be more effectively disinfected. Satin finish stainless steel is adequate: there is no advantage in mirror finish. Note that it is attacked by dilute hydrochloric acid.

Basins are purely for hand washing and should be kept for this purpose. Sinks should be large enough for the intended use. A double sink with a double drainer is very useful for the washing-up area in any busy chemistry or biology laboratory. (See also section 11.8.1.) A sluice is a large low sink, usually connected directly to the foul sewer, which is used for washing away filth and floor washings. (See sections 10.2.3.11, 10.2.3.12, 11.8.2.)

Drip cups are essentially funnels leading to the sink drainage pipe, placed under water taps. They prevent floods if a tap is left (or a pipe comes loose) and provide a return for water used for cooling (e.g. condensers) or for

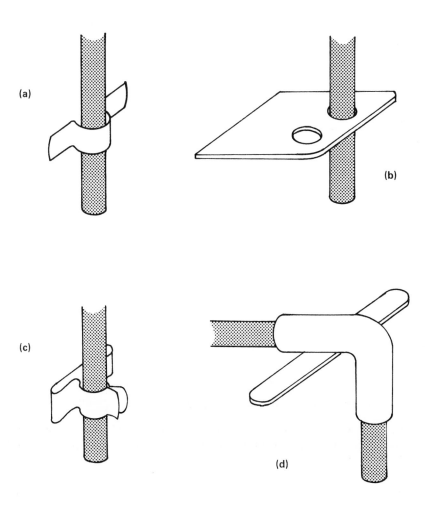

Fig. 5.1. Some methods of securing tubing to sinks, troughs or drain cups. (a) A brass or plastic saddle will enable tubing to be located easily when required. (b) A plate with several holes can be placed at a strategic point to accommodate more than one drainage tube. (c) A spring clip provides greater security, providing it is of a size to match the tubing. (d) An elbow fixed in place provides an easily used and surprisingly secure location for tubing into a sink or trough or drain cup.

water-powered items (vacuum pumps, mixers). It is convenient to fix some spring clips by the cup to hold flexible tubing in place. An even more secure arrangement is an elbow of metal or plastic pipe screwed in place to which the hose can be clipped. (A hose passed through a slightly larger elbow is also fairly secure.) See Fig. 5.1.

Drip cups are excellent for piped waste such as that from some machines. They are commonly used in fume cupboards to pour away liquid waste. A funnel and a length of flexible hose is helpful to avoid splashing.

Troughs are essentially common drip cups which are often used as sinks.

They have many disadvantages. They tend to splash; hoses placed in a trough often move and jet upwards; they have been known to carry flammable vapours (which are heavier than air) some distance to a source of ignition; they expose noxious or harmful vapours to the air. Probably one of the worst features is that, as common property, no-one looks after them. A trough can be converted to a series of drip cups by simply placing a pipe in it, as shown in Fig. 5.2. This is also an effective way to deal with leaking troughs.

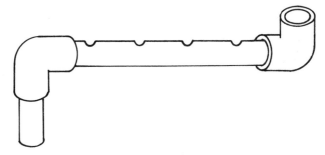

Fig. 5.2. Conversion of a trough to drip cups. The above device can be made from standard PVC pipe and fittings, and laid in an existing trough. The small holes allow for flexible tubing from condensers etc. The elbow fitting gives a larger entry for waste disposal or under a tap. T-pieces can be used if a larger entry is required in the middle of a long trough.

5.3.3 S-bends, Bottle Traps and Catchpots (Fig. 5.3)

Most handbasins and many laboratory sinks are fitted with s-bends, also known as running traps. Their main function is to remain filled with water and thus provide a seal against vapours in the drain. Where a sink or basin is little used, then the bend may actually dry out, allowing vapours (which may be offensive, toxic or flammable) to escape into the room. This can be prevented by occasionally running the tap on a little-used sink. The water in an s-bend can also provide a breeding place for infection, so it should be treated with disinfectant on a routine basis (for example, last thing at night, or at least weekly) in laboratories where this may be important.

S-bends are often fitted with a drain plug, which is mainly for the convenience of the plumber, but will sometimes allow small objects to be retrieved. However, a more convenient device for this purpose is the bottle trap, which is used in place of the s-bend. The base of this can be unscrewed to get out any small objects which have accidentally gone down the drain. Bottle traps in many laboratories accumulate mercury from broken thermometers and manometer spillages. The removable part is usually plastic, but systems are available with a glass bowl. This has the considerable advantage that it can be inspected without dismantling, and is strongly recommended despite its extra cost.

Catchpots are larger receivers (usually floor mounted) which receive liquid from sink waste pipes and overflow into the main branch pipe of the

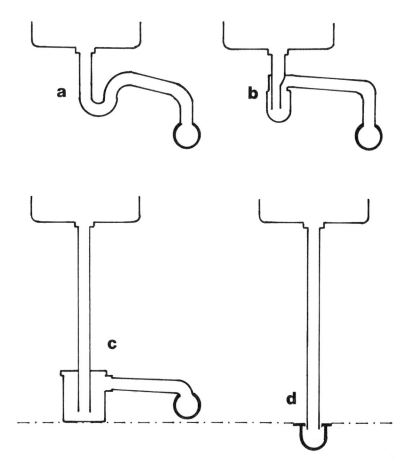

Fig. 5.3. Some methods of connecting a sink to drains: (a) via an s-bend to a horizontal branch drain; (b) via a bottle trap to a horizontal branch drain; (c) via a floor-mounted catchpot to a horizontal drain; (d) directly to a gully floor drain.

(a) The s-bend shown is of the wide type, with a steeply sloping discharge, to ensure that any solids are carried away. this should be used where a sink grinder or other source of solids is fitted.
(b) The bottle trap is convenient for retrieving small items lost down the drain, but can become blocked by debris if not regularly emptied of solid.
(c) The catchpot will retain solids and small items but needs less frequent attention than the bottle trap.
(d) Direct discharge does not provide a water seal against vapours from the drain, and is therefore contrary to building regulations in most laboratories. However, it is useful in some pilot plants and engineering laboratories where solids are collected in an interceptor, and where the gullies do not carry harmful materials.

drains. They provide a liquid seal and a collection point for solid objects. In some older laboratories they take the form of open-top sinks. These are a frequent cause of flooding: typically because a paper towel has blown in and blocked the overflow. If given a loose cover to prevent this accident, and if regularly cleaned (which is an unpopular job) they may be tolerated in (say) chemical or metallurgical laboratories, but are unacceptable in biological laboratories.

Modern catchpots are sometimes all-plastic, but as with bottle traps the most convenient design has a glass chamber. One catchpot may serve one sink or several.

It should be noted that bottle traps and catchpots are reasonably effective in collecting solid. If it is truly intended to dispose of solids (i.e. a slurry) down the drains, then an s-bend is preferred. For this reason it may be necessary to designate certain sinks for certain wastes, and see that they are equipped accordingly.

Traps and catchpots should be inspected at least cursorily several times a year, and cleaned out as often as necessary. Regular disinfection may be necessary for biological laboratories — on a daily basis for the most critical standards.

5.3.4 Interceptors, Dilution Tanks, Delay Tanks and Neutralizer Tanks

Before waste leaves the vicinity of the laboratory, whether to another part of the site or to the public sewers, it will often pass through an underground chamber. This has different designs for different purposes, and which should not be confused.

An interceptor is so arranged that particles much heavier or lighter than water are largely excluded by flotation and sedimentation. A major use is to prevent petrol and oil spills from entering the sewers. However, this is very much a last line of defence. A vehicle engineering laboratory will have interceptors (which will be regularly cleared by a contractor) as will a pilot plant using immiscible solvents, but most laboratories will be expected to have a more reliable primary means of preventing quantities of immiscible liquids entering the drains.

A dilution tank is, as its name suggests, simply a large enough vessel so that effluents are diluted to an acceptable level by relatively clean water. This may mean combining laboratory effluent with other sources of waste water, or just means that occasional concentrated wastes are diluted by the more usual run of water from handbasins, condensers, etc. It can be used where there are known to be predictable regular discharges or where there is a possibility that staff may not be careful enough in their disposal practice. As the latter applies to nearly all laboratories, it is a useful precaution.

Generally speaking, additional volumes of fairly clean water are welcome in a dilution tank. This is not the case for a delay tank. These are used mainly for radiochemical waste, but sometimes for biological waste. Entry to a delay tank should be carefully limited to only the drainage which would most benefit from it. The point of a delay tank is that (on average) waste spends a considerable time (hours to days) before being discharged to the public sewer. This commonly allows short-lived radioactive isotopes to decay so that the discharge is acceptable. Other applications include time for a disinfectant (usually dosed in automatically) to act, or time for biological degradation to occur. In the latter case, there may need to be control over the temperature and aerobic conditions. This can substantially control some pathogens and reduce the biological oxygen demand, which may be a requirement of the sewage authority.

A delay tank can be useful to give extra time in an emergency, and should be considered where large volumes of difficult materials are handled (e.g. in some biological and pharmaceutical larger-scale work). This normally requires some kind of bypass pipe which can be used when the fault is detected, leaving the difficult waste isolated in the delay tank.

Neutralization tanks were installed originally by relatively enlightened universities and research establishments well before modern pollution control standards. A simple and common example was a bed of marble chippings through which the institution's effluent passed. This was on the assumption that a considerable quantity of chemicals would be routinely discarded untreated to the drains, and that this would at least mitigate the acidity.

It is the authors' opinion that a neutralizer for a whole institution represents an admission of failure, and an encouragement to bad practice. If there is some process or procedure which regularly produces acidic waste water, this waste should be treated at the point of origin, before it enters the general drainage system. It would be reasonable to expect a pH probe and automatic dosing pump to produce effluent within the limits laid down by the local sewage autority.

A marble chip neutralizer will of course do little for solvents, soluble poisons and alkalis. Industrial experience of these units is that the chips can very readily lose efficiency owing to surface sulphate, slime formation and general fouling. They therefore require considerable attention.

Even if there are many laboratories producing similar wastes, there are sound reasons for having local treatment rather than a final common one. Firstly, if one local unit fails or is put out of action by mistreatment, then the whole institution does not have to stop. Secondly, it means that there are not long lengths of piping carrying harmful material, with the hazards of corrosion, leakage and danger to maintenance workers.

Some institutions with neutralizers have had to discontinue their use, because the corroded piping can no longer be trusted to carry acid waste.

Where a laboratory is part of a major institution, then there may be an effluent treatment plant. As part of a properly designed and operated modern plant, neutralization tanks are of course valuable, but even this level of facility should not excuse the laboratory from controlling its effluent at source.

5.3.5 Macerators, Grinders and Sink Waste Disposal Units

There are two main sorts of electrically powered machine for shredding solid and discharging it to the drains. It must be appreciated that the laboratory's usage may differ considerably from the designer's intentions, and due allowance must be made.

22

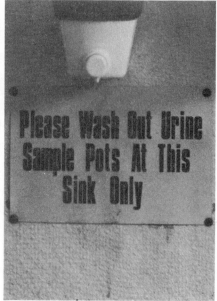

23

24

25

26

㉒ The polypropylene waste pipes from three sinks are fed into this floor-mounted catchpot. The lid is removable for cleaning. The glass pot allows objects and trapped solid to be seen. A sensible amount of space has been left around the catchpot to aid inspection and cleaning.   ㉓ Sinks for special uses should be clearly labelled, as an aid to good practice and hygiene.   ㉔ This water-still has been mounted above a sink, to reduce the risk of flooding if a pipe were to block or break.   ㉕ A roof-mounted fan for a fume cupboard extract duct. This ensures that all the duct inside the building is under suction, so leaks will only be into the system. A vertical discharge like this (with a drain for rainwater) is generally better than a horizontal or hooded exit pipe, though this one is a little short for the situation.   ㉖ A gas-fired pyrolysis incinerator, used for the destruction of pathological and other waste at a medical research centre. The chamber on the right is a loading port with air curtain. Photo courtesy of Robert Jenkins Systems Ltd, Rotherham, S. Yorks.

The large free-standing units are macerators, made to shred disposable bed-pans (and certain other items) with their contents and flush them to the sewer. They are of course used in hospitals. They usually operate under vacuum which prevents the release of infective aerosols, and are resistant to common disinfectants. They will normally tolerate certain plastics and paper or cardboard, but should not be loaded with textiles (e.g. cloths, bandages, cotton wool), metallic objects or plastic film (e.g. plastic bags, cling-film). They will generally dispose of animal tissue (including complete carcasses up to a certain size) though modifications may be required for frozen tissue. They will take most plant tissue, but can have difficulty with some woody or fibrous items. The manufacturer's experience should be sought for different applications, but a 2 kW unit with 200 mm cutters is suitable for many laboratories.

Smaller units are usually fitted directly into a sink and discharge via the waste pipe. These are grinders, and were made for kitchen use. They generally tolerate somewhat tougher material than the large macerators, but provide less protection against the release of aerosols. They should therefore not normally be used for infected tissue. Flexible plastic and cloth may cause the machinery to fail, but they will in fact process broken glass, though this may not be a suitable material for the drains.

In general, macerators require more skill to use, but give less chance of physical injury because of the lid interlocks. Neither macerators nor sink grinders are intended for chemicals. However, they will usually tolerate ordinary levels of contamination, providing the unit is well flushed (as they are normally designed to be).

It is essential that the piping from these units is of the size specified by the manufacturer, and that this is taken direct (via an s-bend) to a well-used main drain. If too large or too long a pipe is used, then blockages may occur. The s-bend should be of the 'long sweep' type, and the entry into the main line should be taken at an acute angle.

If it is necessary to install the unit on a horizontal branch which is not regularly flushed (e.g. by a washing machine or WC) then some arrangement must be made for a source of water up-stream from the macerator or grinder. A simple solution is a toilet cistern feeding directly to the drain line. The cistern is operated after each use of the grinder or macerator to ensure that the drain is kept clean. If desired, this can also be a way of adding disinfectant for overnight. (See section 5.5)

5.4 MATERIALS OF CONSTRUCTION

5.4.1 General Comments

The following notes mainly apply to common materials used for waste pipes. In the past, great store was set by corrosion resistance. Lead was a favourite material, and there was criticism of plastic piping because of the damage it could suffer from organic solvents.

However, today one would not expect a properly managed laboratory to pour strong corrosives or neat solvents down the ordinary drains. Even in the past, well-managed laboratories have operated for many years with only galvanised iron pipes, simply by controlling their effluent. The use of very special materials can be limited to those laboratories or disposal points which truly need them e.g. a high-level radiochemical laboratory, or a pilot scale process where effluent is piped away from treatment.

Problems today are mainly those of special types of corrosion due to prolonged action at a low level of chemicals. For example, a concrete drain can be severely damaged by a neutral solution of sodium sulphate if prolonged for some months. New materials and methods of construction have given opportunities for cost saving and many improvements, but they have of course brought new opportunities for errors in construction.

5.4.2 Concrete

Concrete compositions are coming into increasing use for large drains and sewers. Some grades are more chemical resistant than others, but it is also possible to paint or line ducts and pipes, particularly to improve acid resistance. Unless specially mixed, concrete is susceptible to corrosion by moderate amounts of sulphates, and to a lesser extent by chlorides. The limits laid down by a drainage authority are often determined in some degree by whether or not local piping is concrete.

A cement mortar was the traditional method of joining clay sewer pipes, though other methods are now more common.

5.4.3 Glazed Clay

This is the traditional material for sewer pipes, which is the most likely one to be used for buried drains leading to a public sewer. A 'chemically resistant' grade (e.g. according to British Standard 1143) should normally be specified for laboratory service. With sensible management, it is very unlikely that corrosion will be a problem. The most usual method of joining these pipes now involves rubber and plastic seals. These make for quicker construction and add a little flexibility, but could conceivably leak after prolonged chemical abuse.

5.4.4 Cast-iron

This is an economic and perhaps surprisingly robust material for larger drains within a building, particularly if a more resistant and/or mastic coated pipe is used. The joints do on occasions give trouble. The pipe should therefore be reasonably accessible (e.g. in a service duct or in a false ceiling) and it would not be recommended for untreated pathogenic waste.

5.4.5 Lead

This is the traditional material for troughs and drains in chemical laboratories. If the more resistant 'chemical lead' was used rather than domestic piping, then an old system may still have many years of life. However, it would not be the choice today (except in a few highly specialized circumstances) and there are few tradesmen who can adequately make repairs and modifications. It should be remembered that the scrap value of a lead system may go some way towards its replacement with a plastic system.

CAUTION: if azides or picrates have been discharged to a lead drain over a number of years (even at very tiny concentrations, e.g. as bacterial inhibitors in aqueous solutions) then there is a very real danger that sufficient lead azide or lead picrate may have accumulated to render part of the pipework liable to explode. Actual incidents have occurred, so advice should be taken on methods of treating the pipework before removal.

CAUTION: failure of a lead pipe may be due to corrosion by liquid mercury. This could give off very toxic fumes when heated during the course of a repair.

5.4.6 Copper and Brass

Copper is widely used for water and other supplies, and has been used for waste pipes. Brass or chromed brass fittings are often supplied with sinks. Copper is strong and can be easily joined by a number of methods including brass couplings. It is, however, susceptible to corrosion by a wide range of common chemicals and biological fluids, even when dilute. It is thus unsuitable for a general drain system

However, it may be the material of choice for carrying relatively clean water (and a few solutions) from a fixed machine, especially if the water is too hot for the main plastic piping system.

CAUTION: copper can form explosive compounds with azides or picrates as readily as lead, and also reacts with mercury metal. See section 5.4.5.

Note that stainless steel is sometimes joined by brazing, that is by fusing the two parts together with brass. The joint is therefore as susceptible to corrosion as brass.

5.4.7 Stainless Steel

This is now the material of choice for most laboratory sinks, but has only limited application for piping. In the past, its use was restricted because of the difficulty of jointing and the cost compared with other metals. Good compression joints are now available, and other metals have in general increased in price more sharply, so it may find increasing use. In radiochemical laboratories, pilot plants and food laboratories it should be considered for piping because of its corrosion resistance and ability to be decontaminated and sterilized by steam if necessary.

There are many grades (i.e. compositions) of stainless steel available. Piping selected for a particular process waste should be of a grade which is most resistant to those conditions. For sinks, a grade containing molybdenum is generally most widely satisfactory for laboratories — for example, grade 316 (18% Cr, 12% Ni, 2% Mo). Grade 347 (18% Cr, 9% Ni, 0.5% Nb) has been very successful for piping.

The main defect of stainless steel is its corrosion by dilute hydrochloric acid (which includes other acid solutions containing chlorides). To a lesser extent it is susceptible to sulphuric acid, and it can be weakened by other acids, but prolonged exposure to HCl is the most likely cause of failure.

5.4.8 Glass

In the UK, borosilicate glass is rarely used for laboratory drains, yet it is relatively common in parts of the USA. If correctly mounted it is not much more likely to break than rigid PVC. It has the considerable advantage of being easy to inspect, and the smooth surface both aids decontamination and inhibits the formation of slimes. Although the initial installation cost is higher than for a plastic system, a glass pipeline will probably last longer. It is a little less easy for the amateur to dismantle or modify, though this may not always be a disadvantage.

Glass has a very wide chemical resistance and can be steam sterilized if necessary. It can absorb certain radiochemical species, notably ^{32}P (which is however a short-lived nuclide and thus rarely a real problem in the waste line). If the surface becomes etched (by very strong alkali, hydrofluoric acid or a few other materials) this is probably an indication of extremely bad waste disposal practice.

5.4.9 Polyvinyl Chloride (PVC)

This plastic has been very successful for waste pipes. It can also be readily fabricated into troughs etc. by almost any workshop. It is impervious to most aqueous chemicals at reasonable dilutions (and often at unreasonable ones). It can be joined by heat welding, solvent welding or a variety of mechanical joints. This gives the potential for systems which have the highest integrity where required, but are adaptable where necessary. The smooth surface is relatively easy to keep clean, and does not encourage microbial growth. It is one of the easier materials to decontaminate from radioactive spillages.

A typical PVC drainage system will tolerate hot water and short periods of very hot water, but if exposed to prolonged near-boiling water will deform and may fail. This is one of the main possible restrictions. PVC pipework may also fail where someone uses it for disposal of certain organic solvents, which is of course appallingly bad practice.

PVC pipework is favoured by building engineers because it is very easy to fabricate and may be expected to last a very long time. However, in recent years it has become somewhat expensive compared with polythene and polypropylene.

PVC is not combustible, and therefore does not add to the fire load of the laboratory. However, in a fire the pipes will fail and release their contents. In addition, excessive heat will decompose PVC into very toxic fumes.

5.4.10 Polythene (Vulcathene™)

A black-filled high-density polythene has been widely used (under the trademark Vulcathene) for drip cups, troughs, catchpots, bottle traps and pipework. Commercial items are normally made with screwed joints which can be made watertight by only hand force, but lengths of pipe are typically joined by heat fusion. This joining was not always done correctly, and in many laboratories provides a restriction to the flow. Many systems exist incorporating polythene parts with other materials.

The chemical resistance of this plastic is very good. It is not as rigid as PVC, but more resistant to mechanical impact. Its major drawback is that it will not tolerate high temperatures. If significant volumes of waste in excess of 60 °C are likely, then special care must be taken in providing extra support and allowing for expansion, which is considerable. For wastes running much above 80 °C, it would not be the material of choice. It cannot be steam sterilized.

Against this one limitation must be set the considerable economic advantage compared with most other systems. Many laboratories find that limiting the exposure to heat is very little trouble. (Of course, in a fire it will melt and burn, which may be unacceptable in some laboratories.)

5.4.11 Polypropylene

This plastic is the first choice for drain pipes in many new laboratories. It is strong, with a wide chemical resistance, and can even tolerate boiling water for extended periods, if properly installed.

All routine components are commercially available, and there are several jointing methods. However, the only method which the true amateur can successfully use is the 'push-fit' connectors developed for domestic plumbing. In general, these are not suitable for laboratory waste pipes, and should not be specified for waste of any hazard at all.

A properly fitted system should have the pipework joined by fusion welded sockets and tee branches. (Sometimes a screw joint is reinforced with fusion welding.) There should be screw couplings at key points (e.g. catchpots, ends of benches) to enable pipework to be partially dismantled when necessary for maintenance. Although an amateur can undo and re-make a screwed joint, the initial installation should be left to a skilled person. Fusion welding requires some skill and the proper equipment. Although these are widely available, modifications to pipework or the manufacture of troughs and sinks may not be possible within every workshop or laboratory institution.

The principal disadvantage of polypropylene is that it is combustible, and therefore adds to the fire load of the laboratory. Fire resistant grades may be available.

5.4.12 Other Plastics

Many other materials have been used for individual systems, particularly where there is a large volume of special material to be passed. Nylon flexible tubing was used for some small-bore applications, though this has now been supplanted by polypropylene which has superior acid and alkali resistance. Polyvinylidene fluoride (PVDF) is similar to PVC but with superior chemical resistance, better heat stability and higher price.

Phenylene oxide (Noryl^{TM}) is used for plumbing drinks and food products (including pump parts) and has been used for waste from the processes. It has also been used in medical facilities, as it has excellent resistance to most biological fluids and can be steam sterilized.

Other plastics, notably acrylic, are used for sinks and baths. However, these are intended for domestic use only, and should not be used in laboratories unless expert advice has been taken as to the suitability of the particular material for a particular purpose.

5.5 LAYOUT AND MANAGEMENT

5.5.1 Arrangement (Fig. 5.4)

Many problems can be avoided by sensible arrangement of drainage. Firstly, it is important to decide what drainage is actually required for operation of the laboratory, and see that the large volumes or more difficult materials are catered for.

The most common building arrangement is a set of small (36 mm or so) pipes leading from sinks etc. to a larger horizontal pipe which connects into the vertical drain. The system may of course be more, or less, branched than this.

A simple rule is therefore this: items producing relatively clean water should be placed near the ends of the branches, but operations producing very dirty waste should be located near the vertical drain. For example, water from condensers (e.g. on a water still) or from a hand basin should be connected to the far end of a branch, since it will flush out the branch drain and dilute the waste put down sinks further along the branch. Conversely, the effluent from a sink grinder unit should have the shortest and steepest

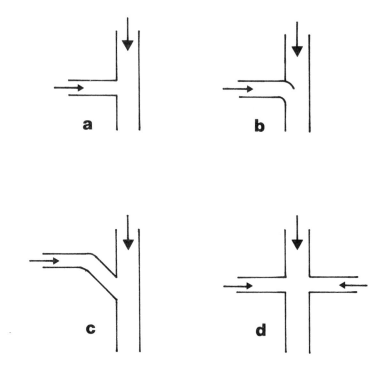

Fig. 5.4. Some methods of connecting waste pipes to a horizontal drain: (a) simple T-piece; (b) swept or pitcher T-piece; (c) angled connection; (d) cross-piece or double branch.

The simple T-piece (a) is adequate for supply pipes, but a swept T-piece (b) is less likely to cause trouble in a waste system. The angled connection (c) is recommended for pipes carrying hazardous waste, or where high solids concentrations may occur. The simple cross-piece (d) should be avoided; two swept or angled connections should be used instead.

possible run towards the main vertical drain pipe, to ensure that the solid slurry is carried into the sewer.

Where other requirements mean that a source of dirty or very hot effluent must be located far away from the vertical drain pipe, then some consideration should be given to a separate drain being laid. As the flow is usually predictable in such circumstances, it is likely that a relatively small diameter of pipe will be required, which will be less difficult and expensive to install. To ensure good flow and final cleansing of this kind of pipe, it may be worth while plumbing in an extra tap to add a flow of clean water to the waste. However, this should not be done in such a way that contamination could leak back into the water supply, e.g. if pressure were lost.

Floor drains should be (as far as possible) located close to items which are most likely to flood (tanks, stills, washing machines, items left with a water supply overnight). This may be by location of the drain in a new laboratory, or by location of the apparatus in an established place. A convenient place for a floor drain is often under a sink where the sink waste pipe goes below floor level. The drain can be a grid around the sink waste, which can often be installed at very little extra cost. Safety showers should always have drains both for actual emergency use and in case of failure.

In the absence of suitable floor drains, it may be possible to locate certain items above a sink or drainer to catch most of the water. Light-weight rigid plastic may be used to assist in directing the flow from, for example, a cracked water still. Some items of equipment can be stood in a tray of suitable size to contain likely flooding. This is essential if the apparatus contains large amounts of material which must be kept out of the drains at all costs.

It is always good practice to specifically decide that certain drainage points will be used for certain wastes. It is preferable if a sign is put up by such points, and for licensed radioactive disposal it is mandatory. The location of disposal points can be chosen with regard to convenience and drain fittings. It may sometimes be necessary to change some detail — for example the provision of a catchpot, or conversely its removal. This both ensures correct operation of the drain and minimizes the cost of providing special fittings.

Quite sophisticated management of drains is possible. For example, a valve controlled by a time switch may be installed to periodically flush a drain. Disinfectant may be dosed in automatically. The pH and/or conductivity may be monitored and indicated, or even corrected. Monitoring and recording of radioactivity is not uncommon. Of course, if these systems are used it is essential that the appropriate waste is always disposed via the correct drain.

Drains carrying hazardous material (pathogens, radioactives, chemicals) have two conflicting requirements: firstly, they must be secure from human contact, and secondly they must be readily accessible for maintenance. To achieve this compromise, it may be necessary to treat laboratory drains separately from other drains in the building. For example, drains from medical laboratories should be kept clear of kitchen drains or those serving infection controlled rooms. (See section 10.3.)

Connections and layout should be made to reduce the hazard of cross-contamination due to blockages or leaks. In the event of a blockage, waste will fill up the pipes and 'back up' to an open point such as a sink. If there is a partial blockage (or if the rate of discharge is sometimes too great for the pipework) then 'backing up' may be an occasional occurrence, possibly unnoticed. For example, a high rate of discharge from an upper room may cause waste to rise up into a sink in a lower room, thereby contaminating it. This might happen when no-one was around to observe it.

The hazard can be reduced by placing 'dirty' drain connections (e.g. sluices and sinks for radioactive waste), on a different horizontal branch to that used for 'clean' connections (e.g. hand-wash basins, coffee-room sinks). Drains which are well-flushed out, and which have a sensible programme of disinfection, are less likely to harbour contamination if 'backing up' should occur.

Devices such as washing machines, sterilizers, ice-machines or refrigerators with a drain point for plumbing into the drains must be connected via a stand-pipe or other method which leaves an air gap. This means that pressure in the drain will not force contamination back into a supposedly clean chamber.

Right-angle (T-piece) connections should not be used for drains carrying significant amounts of solids. An angled or 'swept' connection should be used. (See section 5.3.5.)

Within the laboratory, much of the drains can be carried above floor level behind benches. This keeps the piping out of the way, but allows access when necessary by the removal of panels or bench units. This fact should be remembered as the laboratory is developed. It is not uncommon for access to be completely cut off by some heavy machine, which may even be secured to the fabric of the building!

Sink waste pipes and any other bottle traps or catchpots should always be accessible by no more than opening a door. Transparent units can be quickly inspected quite often, but others should be opened and emptied of accumulated debris at least once a year. Under-bench piping can often be inspected with a torch from one end of a bench. A regular look to see that there are no obvious drips is a worthwhile precaution. In many institutions the horizontal drains either run at floor level or are accessible from the floor below (usually above a false ceiling). These should also be viewed each year, to look for leaks or evidence of corrosion.

For the benefit of maintenance staff, drain pipes carrying harmful material (i.e. biologically infected, radioactive or unneutralized chemicals — say to a treatment process) should be clearly identified by colour, labels or periodic marks. Where laboratory waste pipes run in ducts or are buried in trenches with other pipes and conduits, the laboratory waste pipe should be lowest, so that small leaks are less likely contaminate other services.

If a leak from a particular drain would be disastrous or highly inconvenient, then a trough may be run underneath it. A 'blind' trough — that is, one with sealed ends — will prevent occasional drips from soaking into the fabric of the building. A trough with its own drainage will cope with more serious leaks. This practice is recommended where drains run in less desirable routes because of changes in use — for example if sensitive and expensive equipment is installed in a room underneath existing drains, or where an office or rest-room has drains from upstairs rooms mounted below the ceiling.

5.5.2 Maintenance

A good arrangement of drains will reduce the need for complex maintenance procedures, but it is important that there is some plan in each room for regular attention to the drains. Where problems are unlikely, this may mean no more than an annual inspection for leaks. In biological laboratories, there may need to be a disinfection procedure carried out each day. See section 10.2.3.

Good organization of waste disposal practice will mean that likely problems can be anticipated, and provision made. For example, if substantial amounts of solid or of fatty material (e.g. from disposal units in food laboratories) are discharged, then two provisions should be made. Firstly the drain location or construction should be such as to allow easy application of mechanical devices (i.e. 'rodding through') to remove blockages. Secondly, there should be some regular cleaning procedure to prevent blockages. Hot water and washing soda (sodium carbonate) will clear most biological build-ups. Acid or proprietary cleansers may be needed for scales and some deposits, but these will require carefully supervised application. Periodic emptying of traps may be required to remove inert solids such as sand (e.g. from hand washing in a civil engineering laboratory).

Note that blockages do not only occur in the discharge pipes. They can also happen in vent pipes, preventing them performing their function. This is a not uncommon reason for the strange behaviour of older drainage systems.

One of the simplest but most effective safety procedures for drains is to flush them out each night. This is done in two parts. Firstly, water is run down each sink or other unit so that all the traps are filled with clean water. If appropriate, a disinfectant can be added when this has been done. Secondly, sufficient water is put down the furthest end of each horizontal branch to give a good full-bore flow. (See section 5.3.5 for one way of achieving this.)

This procedure greatly reduces the chance of corrosion, deposit formation and the incubation of micro-organisms. Furthermore, it minimizes the consequences of any unobserved leaks. It is therefore strongly recommended for all drains carrying hazardous materials.

5.6 LIMITS ON EFFLUENT QUALITY

The waste from a laboratory drains will usually end up in the public sewers. Often it will be mixed with other wastes from the same institution. Sometimes it will pass through an effluent treatment plant before leaving the site. In the latter case, the effluent plant will have to meet limits on the quality of the water discharged, under pollution control legislation and as part of the arrangement with the local drainage authority.

Table 9.2 gives the quality limits which might be laid down by a drainage authority in an industrial area of the UK. Note that these are not absolute. They depend on the judgement of the authority as to the local situation (e.g. other effluents, and the type of sewage treatment). A small-volume producer may get agreement for higher levels of some chemicals, but will probably be charged extra for the privilege. conversely, some authorities set low or even zero limits for some materials which are acceptable elsewhere.

It is not always wise to assume that laboratory waste will be sufficiently diluted by other waste, or that it will average out over the day. An industrial plant with a laboratory may have its process effluent close to the limit, so that there is essentially no dilution available. Moreover, checks are usually made on the basis of single samples taken at random times. If a sample is taken just after a laboratory has poured away a bottle of chemical, then the effluent may be measured to be so grossly contaminated that there may be a penalty charge or even prosecution. (In other words, it is assumed that the single sample is typical of the general effluent, and the institution is a gross polluter.)

The actual limits set will normally be based on the chemicals used in the main institution or process. Some chemicals used by the laboratory may therefore not be mentioned. This does not mean that they can be discarded freely to the drains — quite the contrary in fact. It is expected that the laboratory will notify the drainage authority of any materials it intends to put down the drains in any significant quantity, especially if it is likely to be harmful to the environment or the drains.

Where a limit is set, then it is perfectly legitimate to use it. Knowing the daily volume of the drainage, the laboratory manager can calculate what amount of chemical can be discharged, and can use this to arrange controlled disposal of materials. However, the method of charging usually includes a price element proportional to the Chemical Oxygen Demand (COD) or Biological Oxygen Demand (BOD). This means that the same volume of waste will be charged for by the drainage authority at a higher rate if it has a higher COD or BOD.

See section 9.3.3 for information on chemical disposal to drains, 10.2.3.11 for biological liquids, 11.6.5 and 11.8 for radioactive substances.

BIBLIOGRAPHY

American National Standards Institute (1978): ANSI/ASTM D2949, *Specification for 3 inch Thin Wall Poly(Vinyl Chloride) (PVC) Plastic Drain, Waste and Vent Pipe and Fittings.*

American National Standards Institute (1979): ANSI/ASTM D2749–68, *Definitions of Terms Relating to Plastic Pipe Fittings.*

American National Standards Institute (1979): ANSI/ASTM D3311, *Specification for Drain, Waste and Vent (DWV) Plastic Fittings.*

American National Standards Institute (1979): ANSI/UL 430, *Safety Standard for Waste Disposers.*

American National Standards Institute (1980): ANSI A112.21.M, *Floor Drains.*

American National Standards Institute (1980): ANSI 1008, *Plumbing Requirements for Household Food Waste Disposer Units.*

American National Standards Institute (1981): ANSI A13.1, *Scheme for the Identification of Piping Systems.*

American National Standards Institute (1983): ANSI/AHAM FWD-1, *Performance Evaluation Procedure for Household Food Waste Disposers.*

British Standards Institution (1959): BS3202, *Recommendations on Laboratory Furniture and Fittings*.

British Standards Institution (1969): BS4514, *Unplasticized PVC Soil and Ventilating Pipes, Fittings and Accessories*.

British Standards Institution (1971): BS3456: 2.30, *Food Waste Disposal Units*.

British Standards Institution (1975): BS1710, *Identification of Pipelines*.

British Standards Institution (1976): BS5255, *Plastics Waste Pipe and Fittings*.

British Standards Institution (1978): BS5572, *Code of Practice for Sanitary Pipework above Ground*.

British Standards Institution (1979): BS3943, *Specification for Plastics Waste Traps*.

British Standards Institution (1980): BS2598, *Glass Plant, Pipeline and Fittings*.

CHAPTER 6

Fume Extraction

6.1 INTRODUCTION

6.1.1 General Comments

The discharge of material to the atmosphere (inside or outside the laboratory) requires careful consideration. Sensibly used, fume extraction can remove a major health risk for laboratory staff. It can also be a safe and acceptable method of disposal of certain volatile substances.

Conversely, it is possible to expend a great deal of money on fume extraction with little, if any, improvement to the laboratory. An inappropriate, misused or misunderstood system may actually create a new hazard to health.

Fumes extracted from a laboratory constitute its greatest volume of waste. This chapter therefore discusses ventilation systems, firstly from the point of view of the protection of people, and secondly their utility as a disposal route for unwanted materials.

6.1.2 Exposure Limits for Airborne Substances

A number of agencies have produced suggested limits for human exposure to airborne substances. The best known is the 'Threshold Limit Value' or TLV. This is in fact a registered trademark of the American Conference of Governmental Industrial Hygienists. In general, this limit has been derived from expert opinion to give an estimate of the airborne concentration which can be safely tolerated by the average person throughout a normal working

Governmental Industrial Hygienists. In general, this limit has been derived from expert opinion to give an estimate of the airborne concentration which can be safely tolerated by the average person throughout a normal working week and during a working lifetime. For some materials, this is a 'ceiling' value, that is, it should never be exceeded, not even briefly. For others, it is permissible to allow some fluctuation providing the Time Weighted Average (TWA) is below the limit. There are special restrictions on the periods and degrees of exposure where a TWA is used. In addition, the same organization defines 'Short Term Exposure Limit(s)' or STEL(s), which are higher limits permitted for exposures of no more than 15 minutes no more than 4 times with at least an hour between.

Both the TLV and the STEL are for guidance: they have no legal force in the USA or UK. In the USA, there is a list (derived from TLVs) of 'Permissible Exposure Limit(s)' or PEL(s), which may be regarded as legal limits. These were first published in the Code of Federal Regulations 29 CFR 1910.1000 (1972) 'OSHA regulations on toxic materials', but there have been amendments since. The PEL list is never as complete or as up-to-date as the TLV list. (Note: old publications may refer to the MAC, or 'Maximum Allowable Concentration' which is no longer used.)

There is no legal equivalent of the STEL, but it may be taken into account when assessing work practice. A further limit is the 'Immediately Dangerous to Life or Health' concentration, or IDLH. This was intended to aid in respirator selection and usage — that is, persons using high-quality air-supplied apparatus should not be exposed to greater levels except when escaping. However, it may be useful for assessing emergency procedures.

In the UK (as from April 1984) TLVs have been replaced by Occupational Exposure Limit(s)' or OEL(s). These are similar (often identical) to TLVs, but come in two versions: a long-term limit (8 hour TWA), and a short-term exposure limit (10 minute TWA). Note the limit of 10 minutes as compared with 15 minutes for the US STEL. Where there is no short-term limit, the long-term limit is effectively a ceiling value. These are recommended limits with no absolute legal requirement. However, they are used as criteria for good practice under the Health & Safety at Work Act (1974) and may be the basis for legal action.

In addition, the Health and Safety Executive publishes 'control limits' for certain substances which are the subject of particular legislation. These limits have direct legal force, and should not be exceeded. See Table 6.1.

All the above limits relate to materials which are harmful but not significantly infective or radioactive. The use of airborne concentration limits is not practical for the control of infection, but it is important to have effective extract systems, so other standards are used. (See sections 6.2.6–8, 6.3.1, 6.4, 6.5.3.1, and Chapter 10.)

For radioactive substances, the main route of entry to the body is via the respiratory tract, so control of air in the breathing zone is vital. References to limits on airborne concentrations are given in the bibliography to Chapter 11.

Table 6.1. UK control limits for human exposure to airborne materials

Substance	Long-term TWA (8 hours) mg m^{-3}	Short-term TWA (10 minutes) mg m^{-3}	Notes
Acrylonitrile	4	—	
Asbestos — blue	0.2	—	1
— brown	0.2	—	1
— white	0.5	—	1
Carbon disulphide	30	—	
Coal dust			2
Ethylene oxide	10	—	
Isocyanates (all)	0.02	0.07	
Lead and its compounds (except tetraethyl lead)	0.15	—	
Styrene	420	1050	
Tetraethyl lead	0.10	—	
Trichloroethylene	535	802	
Vinyl chloride	8	22	3

Notes
1. Numbers of fibres per cubic centimetre of air, *not* mg m^{-3}.
2. Control limits only apply to mines.
3. Limits are more complex in practice.

6.1.3 Practical Control of Human Exposure

The limits mentioned in section 6.1.2 principally apply to the factory situation, where there is repeated daily exposure to a few substances only. Where a laboratory is engaged in routine work with daily exposure to certain materials, then similar principles apply. Examples are many quality control laboratories and many medical laboratories where solvent fumes are routinely present. If this is the case, a person knowledgeable in occupational health should be consulted for advice on monitoring the workplace air.

In the majority of laboratories, however, the problem is one of frequent short exposures to a great variety of substances, often as complex mixtures. It is usually impossible to make any real measure of the health risk, because there are so many unknowns. This does not excuse the laboratory from taking action: on the contrary, it means that even greater care must be taken to prevent the staff (and others) from being exposed to the substances which are used.

CAUTION: it is important to identify any individual who has regular exposure to a particular substance, even if the work of the laboratory is mixed. Examples are a technician responsible for mercury recovery, or a

Ph.D. student spending 3 years working with specific materials. These people and their work environment can be studied by the same methods as are used for industrial exposure assessment.

With extremely hazardous substances, then the only practical way to protect people is by the use of isolation techniques such as glove boxes or remote handling. In the more usual situation, some kind of controlled ventilation is used.

The importance of good technique cannot be over-emphasized. A piece of 'ventilated furniture' (that is, a fume cupboard or safety cabinet) made to the highest standards cannot protect the sloppy or foolish worker — such as the one who puts his head into the work-space over a source of fumes, or who spills material outside the protective enclosure.

Unfortunately, many conscientious workers are not adequately protected because they have not been taught good techniques, or because the ventilated furniture is inappropriate to their work. For this reason, section 6.2 describes the different kinds of fume extract available, and their general characteristics.

Management should look towards means of fume suppression if at all possible. Where resources are limited, then fume extracts should be used for the processes which most need them — for example, if a fume cupboard is used as a cupboard for unwanted apparatus then it is not available for experimental purposes.

The work practice must allow for both ordinary conditions and accidental releases. For example, if a quantity of harmful liquid is in use in a fume cupboard it is usually advisable to have a tray to retain spillages within the protective enclosure. Work with micro-organisms should follow normal practice of keeping containers covered and avoiding mechanical action likely to create aerosols, even where the work is carried out in a biological safety cabinet. The function of the cabinet is to provide protection against unobserved failures of the work technique or accidental release, not to permit a more open technique (unless crucial to the experiment).

For work with particularly insidious materials, such a some carcinogens, a laboratory worker may prefer to have a further level of protection. That is, the work is carried out in closed apparatus, in a fume cupboard or safety cabinet, and the individual wears a suitable respirator. Respirators may sometimes be used as a precaution against accidental release, but they should not be used as a substitute for maintaining the quality of the general laboratory air.

CAUTION: respirators should only be used by staff who have received proper instruction in their use. They require great care in selection and maintenance.

There are major differences in protection against chemicals and micro-organisms. In general, some fume removal is preferable to none in dealing with chemicals. However, for infective agents, a partial removal gives virtually no useful protection. Conversely, diseases are generally recognized and understood, whereas ill-health due to chemical exposure is less likely to be correctly diagnosed. This particularly applies to mixed exposure of indeterminate amounts of laboratory chemicals.

6.1.4 Practical Problems

A new laboratory should be equipped with a sufficient number of safety cabinets and/or fume cupboards for the type and amount of work intended. These should be of a construction and design which meets current national standards, which fact should be certified by the supplier or builder. New units in older establishments should also meet current standards, and some thought given to the possibility of upgrading old units. (See section 6.4.5.)

As with drains (see section 5.1) there is a good chance that a general firm of architects will not be familiar with the particular needs of the laboratory. This can lead to the wrong kind of unit being supplied (for example, a fume cupboard where a microbiological cabinet is required, or vice versa). At the very least, the location may not be convenient for daily work.

It is therefore recommended that a genuine expert be consulted before equipping new laboratories or making significant alterations to the extracts from old ones. In addition, the person responsible for the laboratory should learn sufficient of the technology involved to be able to discuss the needs of the laboratory with the installers. The notes in this chapter are intended to help this understanding, as well as to promote the effective use of existing extract systems.

It is most important to realize that units will not be used in isolation. If an extract unit is to remove air at its specified rate, then that volume will have to come into the room from somewhere. Thus if a small room is fitted with several large volume extraction devices, the likelihood is that they will make the room uncomfortably draughty when working together and, what is worse, none of them will achieve the promised efficiency. (See sections 6.2.2 and 6.4.1.)

Furthermore, fumes extracted must be carried away, usually to the roof. This can require a long, expensive and sometimes noisy duct. Considerable savings are sometimes possible by careful location. For example, a common duct may be provided in the building structure, and the laboratories grouped near it. (See section 6.3.)

The efficiency of a particular unit can be greatly reduced by the situation. Nearby walls or bulky apparatus will prevent the correct airflow being set up. This airflow can also be upset by the proximity of doors and open windows, by cross-currents from other units or even the movements of people nearby. Even the presence of experimental apparatus within the protective enclosure will change the airflow, usually giving less protection.

See section 6.4.6 for methods of dealing with sub-standard fume extracts.

6.2 TYPES OF EXTRACTION

As there is sometimes confusion between different types of fume handling equipment, and a misunderstanding of their functions and properties, a short description of the main methods of dealing with fumes is appropriate.

6.2.1 General Ventilation

For comfort in an ordinary office or classroom, it is necessary that the stale air be replenished with fresh air. Typically, a movement of air corresponding to one or two times the room volume per hour will be achieved by ordinary movement of people, opening doors, etc. This measure of ventilation is 'air changes per hour'.

Additional movement of air is achieved by fans, and the air may also be warmed, cooled, moistened or dried by additional equipment (air conditioning). An industrial mechanical ventilation system will usually achieve a minimum of 4 air changes per hour, ranging up to 16 or more for special situations such as operating theatres.

In a laboratory, the natural ventilation is commonly supplemented by switching on fume cupboards, but most units nowadays have mechanical ventilation as well, and possibly air conditioning.

It is generally reckoned that biology and physics laboratories require general ventilation of the order of 5 air changes per hour, whereas chemical laboratories require about 10 air changes. In a chemical preparation room, or a dispensing stores, about 12 to 15 air changes are required. However, these traditional figures do not take into account changes in work practices over the last 10 to 20 years. It is quite feasible that an electronics laboratory would be using much more dangerous chemicals than a chemical laboratory. In addition, there has been a considerable move away from work on the open bench in chemical laboratories, so that harmful materials are being contained in special cabinets or fume cupboards or glove bags.

It is important to determine the main purposes of general ventilation in each laboratory. These are typically as follows: (a) to replace stale air, i.e. carbon dioxide, (b) to remove heat — from people and from apparatus such as ovens; (c) to remove water vapour, and prevent condensation; (d) to dilute offensive odours, e.g. from biological material; (e) to dilute harmful vapours, e.g. poisonous chemicals, which may not be offensive, to below the permitted air concentration; (f) to reduce the level of dust and infective organisms — in fact general ventilation is not very effective for this purpose.

Obviously, different laboratories will have very widely differing requirements, and these may vary with the work. For example, a schoolroom laboratory may be perfectly adequate for lectures and demonstrations, but may require the fume cupboards to be switched on (even though they are not specifically in use) to increase ventilation when bench work is in progress.

The two most common problems with general ventilation are the removal of warm air, giving a cold, draughty room, and the lack of sufficient air inlets to match the intended air outlets. For example, a couple of high velocity fume cupboards may require the equivalent of a door left open. It has been known for staff not only to keep doors closed, but to block up ventilator grilles to prevent draughts. This could be crucial if the ventilation is necessary to provide protection from harmful but undetectable vapours.

6.2.2 Air Conditioning

Full air conditioning aims to give a supply of air to a room of a comfortable temperature and humidity, regardless of the outside conditions. It often combines a degree of energy conservation by cooling hot places and using some of the heat to warm other areas.

Air conditioning may be specifically set up for a laboratory facility (e.g. an animal research station) incorporating certain safeguards. However, in many cases it is supplied as a general service to a building, in which the laboratory is just considered as one more room. This can cause problems and even some dangers. It is recommended that the following general points and the specific nature of the laboratory be discussed with an engineer specializing in air conditioning.

Firstly, air conditioning usually involves some recycling of air (up to 75%) which could mean laboratory fumes being passed into another room. (Dust filters will not prevent this.) Secondly, the humidifiers are areas where it is possible that infectious organisms could breed, and then be disseminated through the system. (This happened with the famous Legionnaires' Disease). Thirdly, varying use of fume cupboards etc. could seriously unbalance the system.

As a rule of thumb it is recommended that laboratories may take conditioned air in, but should not contribute to the system. A separate exhaust method will of course be essential in the event of release of harmful material.

6.2.3 Local Extract Ducts — Captor Systems

An efficient means of removing harmful airborne substances can be to suck them into a small duct as they are produced. This can greatly reduce the amount of air to be removed and often removes the vast majority of the substance almost instantly.

Local ducts can be very small, and only need to operate when a particular piece of work is carried out. They have been greatly used in industry for grinding and similar dust-producing activities. Some laboratories are now tending to adopt this method for routine systems of work.

It is important to realize that the local duct is called a captor system because it relies on a forced current of air to drag vapours and particles into it. It is thus essential that the source is as close as possible to the duct. Typically, the velocity of air is reduced to 10% by a distance of only one duct diameter from the throat. The monetary placing of a hand or piece of equipment between the duct and the source may cause serious escape of material. The efficiency of extraction can also be greatly affected by stray air currents due to thermals or the movement of the operator or draughts. These effects can be reduced by sensible placing and usage of the extract system. They can also be mitigated by placing shields around the duct work area. These shields should not be misunderstood. They do not convert a local extract duct into a sort of fume cupboard, but only screen from draughts.

A common example is a duct placed just above the flame of an atomic absorption spectrophotometer. Here a major function is the removal of large amounts of heat: the materials of construction and the airflow must be specified with this in mind. All manufacturers of these instruments can give advice on these ducts, and they are available as a standard option from ventilation engineers.

A horizontal slit extract is sometimes used at the rear of baths to remove harmful vapours, but still allow open access. These work best when the bath is undisturbed, and may not entirely remove the need for respiratory protection. A low-level extract on a bench is often quite successful in keeping the operator's zone substantially (but not completely) clear of vapours, fumes and dust from the work.

Additional ports are often provided on fume ducting for the attachment of flexible tubing or the provision of canopies above a bench. These are often useful as a means of directing general ventilation, i.e. the flow of air is from the door (or inlet grille) towards the offensive or hazardous material. If a flexible hose is placed close enough, then it can be efficient enough to permit work on the open bench (with suitable protection against accidents). For a canopy, it is usually necessary to use screens (e.g. plastic explosion screens) bench-standing or hung from the canopy, to ensure that airflow passes over the apparatus before going into the duct.

6.2.4 Local Extract Ducts — Receptor Systems

A receptor system is defined by ventilation engineers as one in which the contaminant occurs within the system (or arrives there somehow without help). The fan is then used to transport material elsewhere and prevent escape in the wrong direction. Laboratory fume hoods and safety cabinets are all receptor systems if used as intended, and will be dealt with separately. However, a paint spray booth is also a receptor system, and a captor system such as a canopy can be converted to a receptor by arranging screens around the work.

The key feature of any proper receptor system is that the operator can never place himself in the airflow between the work and the extract duct. For example, a booth with an open bath should have the duct behind the bath and not above it to prevent someone leaning over into the fumes. A further general point is that the more open the system, the more liable it is to be disturbed by draughts or movement across the face.

6.2.5 Fume Cupboards (Fume Hoods)

These are the traditional means of protecting the laboratory worker from harmful airborne material. They work by drawing air in through the working face and discharging it through a duct to a safe place far away. They are most effective against gases and vapours which can be diluted down to a negligible concentration. They are only effective to a limited extent for dusts and microbes.

As they work by keeping a considerable flow of air moving across the workplace, this can disturb fine powders, filter papers and flames. This sometimes results in the fan being switched off in order to carry out specific tasks.

Older fume cupboards did not always have sufficient face velocity to completely contain harmful vapours, and gave a lower velocity when the sash (window) was raised. Cupboards built to more modern specifications are commonly referred to as 'high velocity' (i.e. conforming to the highest velocity required by the most recent standards), 'balanced design' (i.e. with an arrangement to keep airflow relatively constant over different sash heights) and 'aerodynamic' (i.e. designed with an effective flow arrangement inside). See section 6.4.1 for a discussion of face velocities.

6.2.6 Microbiological Safety Cabinets Class I

These cabinets are designed to protect the worker and his colleagues by drawing air in through the open front (like a fume cupboard) over the work. Unlike a fume cupboard, there is usually no movable window (though it is a good idea to have a removable piece to close off the work area when not in use and for fumigation). The key feature of this design is that the air extracted from the cabinet by the fan is drawn through a high efficiency particulate air filter (often called a HEPA filter) which substantially removes dust and microbes. In the past, this 'cleaned' air was exhausted to the room. However, the modern tendency is to duct it outside in case the filter should not be totally effective. A short length of duct is generally used, and it is not necessary to site the outlet as remotely as for fume cupboards. (Some people disagree.)

The above description applies to many cabinets which have been sold under various names, but may not actually comply with the present standards. Note that the British Standard 5726 gives a minimum face velocity requirement of 0.5 m s^{-1} in use, whereas the Howie Report recommended that filters should be changed when the flow fell to 0.75 m s^{-1}.

6.2.7 Microbiological Safety Cabinets Class II

As with class I cabinets, these are open-fronted units which discharge air to an extract duct via a High Efficiency Particulate Air filter. The difference is that some of this air is recirculated to provide a clean environment in the cabinet. Most of the incoming air bypasses the work and goes direct to the filters, reducing contamination of the work from the laboratory air. They are sometimes called 'vertical laminar flow cabinets', but this is poor practice as they could be confused with 'horizontal laminar flow cabinets' which offer absolutely no extraction and no protection to the operator. (See section 6.2.10.)

Class II cabinets offer no more protection to the operator than class I cabinets, but a lot more protection to the work. However, they are more complex and expensive. Fig. 6.1. gives the essential features, but designs vary considerably. Generally, more modern ones have two sets of filters.

The original class II cabinets are now referred to as class IIA or class II, type 1. They recirculated about 75% of the air and discharged 25% to the duct. Although perfectly satisfactory for biological hazards, they could allow a concentration of chemicals (and some spillage from the front), so an alternative type has been developed, called a class IIB or class II, type 2, cabinet. This recirculates about 25% of the air, and exhausts about 75%. The class IIB cabinet can be used for work involving hazardous chemicals as well as micro-organisms.

6.2.8 Microbiological Safety Cabinets Class III

These are glove boxes with a safe air supply so that living organisms can be handled and there is minimum risk of contamination either way between the work and the operator. Air is drawn into the unit via a HEPA filter, and air also leaves the unit via another route filter. If it is necessary to add or remove material during the course of the work, this can be done via airlocks, but is most commonly via a 'dunk tank' of disinfectant solution. As it is used for the most dangerous pathogens, it is essential that the exhaust air be vented outside the building, even if two filters are used in series.

6.2.9 Self-contained Portable Fume Cupboards

Several manufacturers now offer units which do not require ducting. These are essentially fume cupboards which exhaust air back to the room, having first passed through particle and adsorbent filters. They have a range of filter cartridges available, which should be selected as appropriate for the intended work. They are effective in controlling odours and toxic vapours. Their effectiveness against dust depends mainly on the airflow being adequate during the working operation (when the user may disturb the air with his or her body and arms). Because of the different nature of biological hazards, they should not be considered adequate protection against microbes.

It should be noted that toxic material is simply retained in the filters. Thus if a solvent is evaporated down, it will accumulate in the filter. For maximum

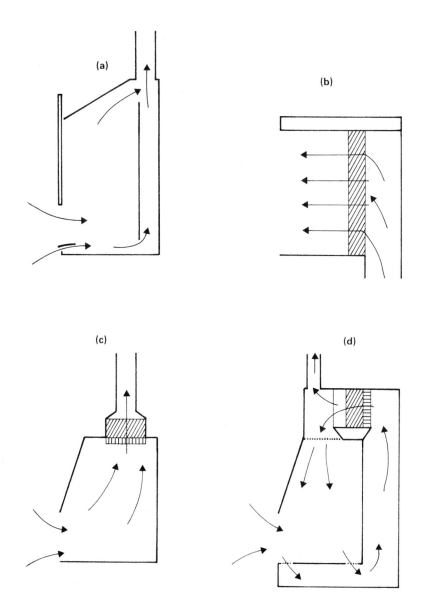

Fig. 6.1. Airflow patterns in ventilated furniture. (a) Fume cupboard (fume hood).
(b) Clean air station (horizontal laminar flow cabinet). (c) Microbiological safety
cabinet class I. (d) Microbiological safety cabinet class II.

filter life, these kind of operations should be avoided (as should the use of
steam baths). The filters should be disposed of as hazardous material.

6.2.10 Clean Air Stations (Horizontal Laminar Flow Cabinets)

These are the opposite to fume cupboards, as is shown in Fig. 6.1. The principle of operation is that air is drawn in (usually under the bench) through a coarse filter then blown horizontally through a HEPA filter across the bench, i.e. at the operator. Unlike fume cupboards and safety cabinets, there is no extraction at all. The purpose is solely to provide a clean (particle-free) working area. No protection is given to the operator, and these units must never be used for venting or evaporation, nor for harmful micro-organisms.

Some clean air stations also use a vertical laminar flow, but these must not be confused with class II cabinets, because they do not provide any protection to the operator.

6.3 EXTRACTION DUCTS

6.3.1 Location of outflow

It may be argued that the exhaust air from a microbiological safety cabinet has been filtered, or that a fume cupboard will give such a dilution that the exhaust will be at a safe breathable concentration. However, what happens if there is an accident — perhaps not realized — which damages the filter? What happens if there is a major release in a laboratory which the fume cupboard is clearing from the atmosphere? What happens when the safety cabinet has been fumigated with formaldehyde? In these cases, a normally 'safe' outflow will become very hazardous. It is therefore necessary to arrange that all outflows are sited where they have least chance of coming in contact with people.

The normal place for an exit pipe is the roof of a building. The pipe should project at least 3 metres above the roof and any items on it. This is because natural airflows tend to follow the shape of the building, so that a discharge at a lower level could travel down and be drawn into a window etc. Horizontal discharges are particularly liable to return into the building, as are discharges from vents with rain covers known as 'chinese hats'. Additional height should be allowed for these discharges, and smoke tests should be carried out (see section 6.5.2).

Some protection from rain can be achieved by the provision of an outer concentric cylinder shield, but there is normally a requirement to make some provision for drainage.

It is preferable to have the fans on the roof, because this means that all the ductwork in the building is under suction (negative pressure) and thus will not release harmful vapours even if it leaks slightly. Where a building has several roof levels, then the fan may be on a lower one, but the discharge should always be made from a pipe leading to a height above the highest roof.

Where it is considered impracticable to take a duct to the roof, then discharge is made through a wall or window. This should be limited to less hazardous extracts, should be as far away as possible from air inlets (including opening windows), should not emit onto accessible catwalks or flat roofs or public places, and should be sited and constructed with regard to the prevailing wind and local weather effects.

A common error appears to be locating discharges together, with the result that material may be emitted from a working fume cupboard and blown back into a non-working one. This can be partially prevented by the fitting of flap vanes. A similar effect can occur when several extract points use a common duct. When they also use a common fan, it is vital that damper vanes are accessible and understood, so that unused cupboards can be isolated, and working ones have maximum efficiency.

The monograph by Hughes gives specific examples of errors in the construction and location of fume outflows. It should be consulted by anyone installing a major fume system. In addition Lees and Smith, *The Design, Construction and Refurbishment of Laboratories*, gives a lot of up-to-date advice on most aspects of fume extraction. For existing systems, the question is: do they work under the prevailing weather conditions and work practices? This should be checked (see section 6.5.2) and a common sense decision taken as the need for restrictions on their use (see section 6.4.6). It is sometimes possible to upgrade the fan by changing the blade angle and to add additional height to a duct or fit a special end which tends to carry the plume higher. However, the latter arrangement is less effective in strong winds.

6.3.2 Treatment of Outflow

Ideas are occasionally advanced for the treatment of exhaust air from safety cabinets and fume cupboards, including chemical treatment and heat treatment. It is generally found that chemical treatment requires such complex equipment and such attention in use that it is far easier to deal with the hazardous material at source. The energy costs for heat treating volumes of air (in order to adequately sterilize it) turn out to be enormous and therefore impracticable. UV light is ineffective.

The two treatment methods that are used are filtration (as a back-up to cabinet HEPA filters or where 'zero discharge' is intended from fume cupboards) and water scrubbing.

The main problem with large duct filters is that they may become clogged if unattended, and will require special care in maintenance because of the accumulated hazardous materials.

Water scrubber systems are best fitted to an individual fume cupboard. They are particularly used where perchloric acid work is carried out, to prevent deposition of very high fire-risk perchlorates in the duct. However, note that a water scrubber will condense and absorb many other hazardous materials, so that evaporation or venting in this type of cupboard is essentially a discharge to water rather than air. The resulting high humidity in the duct can also cause corrosion.

Practical water scrubber systems vary greatly in design and effectiveness. The most efficient are probably those placed closest to the point of release in which the air is drawn through a very fine spray (produced by water under high pressure passing through a special nozzle, called an atomizer or nebulizer), preferably followed by passage through a wetted mesh. This can be a permanent fixture, in which case the water is passed to drain, or a portable recirculating unit, in which case the hazardous substance accumulates in a water tank.

More commonly, a spray is used, but the drop size and contact time are such that it does not absorb much vapour. However, the water running down the walls of the duct or fume cupboard protects them against corrosion, and prevents accumulation. In a very long vertical duct, a so-called 'wetted wall' system may give adequate absorption: a chemical engineer should be consulted for the necessary calculations. Note that there is little advantage in using chemical solutions in most cases. They do not have much greater rate of absorption than clean water, and may cause their own problems of corrosion or blocking of spray jets. However, in a recirculating system it is worth considering a solid or dissolved neutralizing agent to reduce the concentration of the harmful agent or inactivate micro-organisms. In all cases, the effluent from scrubbers must be correctly dealt with as hazardous waste.

Scrubber systems are of course only effective when properly maintained, and when actually switched on, which is sometimes overlooked.

6.3.3 Duct Materials

6.3.3.1 Asbestos-cement

This was the most widely used material, though it has now largely been replaced by PVC. It is of course fire-proof and resistant to many fumes, but can suffer from acid vapour. Material tends to accumulate on the inner surface. Because it is very rigid it is liable to crack near joints if not fully and properly supported.

6.3.3.2 Polyvinyl Chloride (PVC)

This is the most usual material nowadays. Resistant to a wide range of chemicals, but may sag if exposed to too much heat. If strained or struck a blow, it can in fact crack, but is generally resilient and easy to repair. The smooth surface does not encourage microbial growth and normally does not accumulate dust except by static charge in some cases. It does not burn, but gives off toxic black smoke if involved in a fire.

6.3.3.3 Glass Reinforced Plastic (GRP)

This is occasionally used instead of PVC. It is strong and resilient, but its chief drawback is that it is combustible, so that a fire in a duct could spread elsewhere. It is reasonably chemical resistant, though some solvents may eventually cause cracking.

6.3.3.4 Galvanized Steel

This is used for general ventilation, boiler chimneys, etc.; it is not a very good choice for fume cupboards. Acid vapours in particular tend to cause failure of joints and seams. When used for extracting hot gases (from an oven or atomic absorption unit) it has been known for soldered seams to melt.

6.3.3.5 Stainless Steel

Because of its high cost, this is mainly used for short ducts where high temperature resistance is required (e.g. taking gases from a flame atomic absorption spectrophotometer). Resistant to a wide range of chemicals, it can be attacked by wet fumes of hydrochloric, nitric, perchloric and hydrofluoric acid. The smooth surface does not encourage the build-up of dust. Note that although heat-resistant and fire-proof, it will still get hot. It should not be placed near material which could be damaged by heat if used for this purpose, and unprotected ducts should be clear of places where people might reach.

6.3.3.6 Acrylic (Perspex^{TM}, Plexiglass^{TM})

This clear plastic is sometimes used for short pieces of duct where visibility is important (e.g. for routine checking of dust efficiency). A special fire-resisting grade should be specified. Solvent vapours may cause it to fog or crack. (Clear PVC is available, but is less transparent.)

6.3.3.7 Flexible Hose

Flexible hose usually consists of a steel helix supporting a cloth or plastic. It is normally only used for a final connection between a fixed duct and the extract point. It is relatively cheap and allows for close proximity of the fume generation point (e.g. vehicle exhaust, welding) to the throat where work is carried out over a large area. It also gives literal flexibility to laboratories with constantly changing bench projects, some of which will sometimes need extract.

The main disadvantage is that dust and smoke tend to collect on the interior, which could be hazardous. The hose is also likely to have a limited life, especially if much used. The materials should be chosen as appropriate to the temperature and chemical nature of the fumes. Where hot gases are not involved, then flexible PVC is widely applicable. Cloth should always be fire-resistant and preferable reinforced with fibreglass.

Care is needed that flexible take-off points for a common duct are always closed when not required, or the whole system can suffer a serious loss of efficiency.

Flexible stainless steel is available, which can make for easy installation of special local ducts for hot gases.

6.3.4 Dampers

Dampers are movable flaps which control or stop the flow in an extract duct. There are several kinds for several purposes.

6.3.4.1 Flow Set Dampers

These are partial closures fixed by the installing engineer in order to have the correct flow rates and patterns in a system. They may need to be changed if the system is modified, but otherwise should not be altered.

6.3.4.2 Manual Dampers

These are movable vanes so that flow can be cut off from one part of the system. They are commonly used where one fan and a common duct draw from several extract points, not all of which are used at once. It is important that the handles for these should be accessible, and that it should be obvious whether they are open or closed. Some bright orange paint will often make the handle more visible against the grey duct. Motorized vanes are used in some institutions. It is preferable if there is some kind of simple indication of an adequate flow of air, in case there is something wrong with these dampers.

6.3.4.3 Automatic Dampers

Some sophisticated systems have movable motorized dampers which open or close to maintain a set flow (or pressure) despite variations in wind conditions or work practice. Any automatic system can fail, but this is quite rare.

6.3.4.4 Fire Dampers

These are flaps with springs which are held open by means of a fusible link. In the event of a fire, the link melts and the damper closes. They are normally installed in very long horizontal ducts, or where a duct passes through a wall which is intended to isolate one area from another in the event of a fire. They may also be fitted for especially hazardous work.

There is usually a manual method of opening the damper when closed (which may be required in order to clear fumes from an area after a fire). This can also be used to check that the damper has not closed when the fume extract system suddenly fails. Fire dampers sometimes close owing to corrosion of the fusible link, and sometimes owing to dust fires in the duct which the laboratory staff are unaware of.

6.3.4.5 Flap Vanes (Wind Vanes)

These are vanes which open one way, so that the flow of air holds them open. Their main purpose is to prevent the wind blowing back into a system when the fan is off. They cannot be counted upon to seal against fan pressure, when several fans use a common duct. They can, of course, jam open or closed or part-way. They may well accumulate dust and debris.

6.4 MAINTENANCE AND SERVICING OF FUME SYSTEMS

6.4.1 General Requirements — Airflow

No fume handling system can be trusted to cope with hazardous materials unless it has been tested and is regularly maintained. Regrettably, there are many institutions using fume cupboards, extract systems, and biological cabinets which have never been tested for efficiency, and which receive only haphazard maintenance (usually repairs after breakdown).

All fume handling systems should have the airflow measured by a hand-held device. It is best if the measurement is done to a standard technique by an experienced and knowledgeable person. Manufacturers of fume cupboards, and many independent consultants may offer a service to check on fume cupboards etc. for air movement. In a larger institution it is almost certainly worth while buying the necessary instrument and having someone attend a course to learn the correct technique. Laboratories on different sites but belonging to the same group can make use of the same tester, preferably on a regular safety audit.

Each fume cupboard, biological cabinet or extract canopy should be labelled when it is tested, with the date, the tester's initials and preferably the mean measured air velocity.

It is usually found that older units do not meet modern standards for linear air velocity. For a local captor duct a velocity of 10 m s^{-1} (2000 ft/min) may be needed to capture dust released by a vigorous mechanical process such as grinding. However, for fume cupboards and biological cabinets there is no advantage in an airflow above 1 m s^{-1} (200 ft/min) and there may even be disadvantages due to the severe disturbance in the work area if much higher draughts are used. Thus the most modern units which operate at this velocity usually have some automatic bypass system when a fume cupboard front screen is pulled down. For virtually all work not requiring total containment, then a minimum velocity of 0.75 m s^{-1} (150 ft/min) is adequate. A velocity of 0.5 m s^{-1} (100 ft/min) is now the recognized standard for laboratory fume cupboards and biological cabinets. However, this is the absolute minimum under the most adverse conditions (heavily loaded filter, unfavourable wind, closed doors, other extractors in use). For work with hazardous pathogens (Category A, B1, B2) or very hazardous chemicals, then it is best to have a minimum reading of 0.75 m s^{-1} (150 ft/min) under ordinary conditions.

It is considered that for limited work with chemicals of less severe hazard (e.g. in schools) a face velocity of 0.3 m s^{-1} (60 ft/min) is sufficient. This cannot be considered adequate protection for any kind of biological work.

For the most critical work it is also important that the air velocity does not vary greatly (i.e. by more than 20%) over the open area. To check this, the unit is operated with the front fully open, with doors and windows in the room closed and with all other extract systems on. The open area is divided into 9 imaginary rectangles (i.e. 3 horizontal and 3 vertical divisions) and a reading taken in the centre of each. For fume cupboards and class I biological cabinets there should not be much vertical variation. For class II cabinets the inward airflow is usually much greater at the bottom than the top. It is suggested that the minimum velocity should be 0.5 m s^{-1} (100 ft/min) and the average should be at least 0.7 m s^{-1} (140 ft/min). As well as the horizontal inflow, the vertical curtain in a class II cabinet can be tested by an imaginary horizontal grid. Official standards for face velocities do vary, as is shown in Table 6.2.

For biological work, the most important function of the airflow is not actually the velocity, but the 'protection factor', i.e. how much it prevents particles coming out of the work area and into the breathing zone of the worker. This is more difficult to measure, and several variants are in use or under development. British Standard 5726 gives a biological method (which has been criticized) and the US National Sanitation Foundation Standard 49 gives a related test. Many people prefer non-biological methods such as that of Clark and Goff, and these tend to be used by manufacturers in commissioning cabinets.

6.4.2 General Requirements — Physical

All electrical equipment (fans, lights, power points, control gear, flow indicators, alarm systems) should be checked by a qualified electrician at least once a year. Local regulations may require more frequent inspections. A record should be kept of the inspections and it is preferable if a note is attached with the date of inspection and the initials of the electrician.

The ductwork should be visually inspected at least once a year for any evidence of cracks, leaks or other failures. Access panels should be removed to inspect service pipes and wire conduits for evidence of leaks or corrosion. A particular hazard is corrosion of gas pipes by chemical vapours. Mercury vapours may attack electrical and structural metals. Dust and fluff should be removed from accessible areas, and windows (including light covers) cleaned — taking due acount of any materials in use. Fan blades should be viewed and any accumulated material removed at least every 2 years — more often if there is a serious build-up. A common reason for loss of efficiency is the build-up of material on fan blades until there is only a sticky ball of fluff being driven by the motor.

The duct exit should be viewed each year. If there is a protective grille, this must be cleaned. At least every 10 years the duct should be opened in several places for inspection and possible cleaning, to remove the hazard of a dust fire or explosion in the duct.

Table 6.2. Face velocity standards for safety cabinets, fume cupboards
(fume hoods) and similar protective enclosures with fume extraction

Velocity $m \ s^{-1}$	Tolerance	Comments
0.25	Minimum	US Public Health Service (1976) minimum for class 1 cabinets. Most authorities agree this is insufficient, but these cabinets may be in use.
0.3	±30 %	UK Dept of Education & Science (1982) minimum for fume cupboards in schools. Assumes normal school use, where fumes are released in small quantities and infrequently from a limited number of chemicals. Most authorities agree this is the minimum to give useful protection.
0.5	±15%	British Occupational Hygiene Society (1975) minimum under working conditions.
0.5	Minimum	Ionizing Radiations (Unsealed Radioactive Substances) Regulations 1968. All measurements of velocity must be above this minimum.
0.5	Minimum	British Standard BS 5726 (1979) for class I microbiological safety cabinets. Minimum under all conditions of use. When velocity drops to this point, filters should be changed.
0.7 1.0	Minimum Maximum	British Standard BS 5726 (1979) for class I microbiological safety cabinets with new unused filters.
0.75 1.0	Minimum Maximum	Howie Code (1978) for class I microbiological safety cabinets. Limits at all times.
0.75	Minimum	Handbook on Radiological Protection in Universities (1966) recommendation for radioactive fume cupboards under working aperture and conditions.
1.0	Minimum	UK Dept of Employment Code of Practice for Health Precautions Against Lead (1973) through any part of the opening of an enclosing hood.
1.6	Minimum	Reported for special units for dissection of infected material.

6.4.3 Special Requirements for Biological Cabinets

A biological safety cabinet (or fume cupboard fitted with a HEPA filter and used for biological work) should be checked by a service engineer with special knowledge of these systems (e.g. a manufacturer's representative) at least every 6 months. Filter changes should only be done by a designated competent person or by the manufacturer's representative. The filters should be fumigated before any operation on them. Protective clothing should be used, the filters sealed into a bag and disposed of as hazardous waste, preferably by incineration.

For a cabinet in daily use, the airflow should be checked weekly by a hand-held device. If it is in infrequent use, then a monthly check is recommended. It is recommended that on the same day of each flow test the cabinet is fumigated and the air grids cleaned of fluff. Combining these routine operations is not necessary, but ensures that some are not omitted.

Fumigation should also be carried out before any work is done on the fans or ductwork or services, and care should be taken that the formaldehyde fumes have been completely cleared before opening a duct. See section 6.4.4.

6.4.4 Fumigation of Cabinets

(The following is adapted from the Howie Report with comments. For a full discussion of procedures see Collins. For further comments on disinfection, see section 10.2.)

If the cabinet manufacturer supplies a fumigation device, this should be used. Commercial packs for fumigation, based on formalin, parformaldehyde or glutaraldehyde are generally good if the instructions are carefully followed.

The amount of formaldehyde required depends on the size of cabinet. A typical small unit has a volume of $0.34\ \text{m}^3$ and the following are amounts for this volume. Larger cabinets should use larger amounts *pro rata* (see caution for method B).

Method A. An electric hotplate is placed in the cabinet. An evaporating dish containing $25\ \text{cm}^3$ of neat formalin (40% formaldehyde) is placed on the hotplate. The front of the cabinet is closed and any joints sealed with tape. The fan should be off. The formalin is boiled off, and the cabinet left overnight. Obviously, it is most convenient if the hotplate can be switched from outside the cabinet.

The following morning, the fan is switched on and the front closure opened a little to allow the fumes to be flushed out. After 5 minutes the front closure is removed altogether and the fan operated for 30 minutes before any further use. (This is a convenient time to test airflows.)

As an alternative to formalin, it is possible to use solid paraformaldehyde in the same way. For this size of cabinet, 4 g is quite adequate. Extra water is *not* added in either case.

Method B. This uses a spontaneous chemical reaction, so the materials should be measured out in advance. A 500 ml beaker containing 25 cm^3 of neat formalin (40% formaldehyde) is placed in the beaker. To this is added 10 g of crystals of potassium permanganate. It is most important that the reagents are added in this order. The front closure is put in place and any joints sealed. A chemical reaction starts within a few minutes and formaldehyde boils off. The cabinet is left with the fan off overnight, then cleared of fumes exactly as in method A.

Some paraformaldehyde may also be used. In this case, 4 g of *dry* paraformaldehyde is mixed with 8 g of potassium permanganate in a large evaporating dish, and 10 cm^3 of water added. The cabinet is promptly sealed.

CAUTION: this reaction should not be scaled up.

The maximum for slightly larger cabinets is 35 cm^3 formalin plus 10 g potassium permanganate, or 5 g paraformaldehyde plus 10 g potassium permanganate plus 15 cm^3 water. For large cabinets, use two or more beakers each containing the individual recipe. *A mixture of 50 cm^3 formalin and 20 g potassium permanganate is likely to explode!*

6.4.5 Organization

Service operations on fume extract systems can be extremely hazardous unless carefully co-ordinated. Users face the hazard of essential services being unexpectedly cut off. Service people face the hazard of chemicals or infective organisms being present.

A permit-to-work system is absolutely essential. Management should consider the special needs of operations on laboratory fume systems which may not be covered by the permit system used for other purposes. A sufficient supply of printed labels (commercially available) is a considerable help. Thus, for example, an electrician can put a label on a fume cupboard (stating it is out of use for electrical maintenance) before isolating the fan.

Local work (i.e. on or around the cabinet or fume cupboard) is usually best carried out during normal working time. The laboratory staff should be instructed to make the unit safe, i.e. remove all apparatus and material and carry out any necessary disinfection or decontamination, before the service person starts work. The presence of the laboratory staff is also useful to answer any questions (e.g. about the trouble which requires repair) or deal with any accident the service person might have with laboratory material.

It is usually best to carry out remote work on a fume system (e.g. repairs to fans on the roof) at a weekend or holiday. This is of course more expensive because of overtime payments, but provides minimum disruption to the laboratory work. In addition, it minimizes the chance that someone will release material into a non-functioning system.

For any work on filters or ductwork, the appropriate fans should be electrically isolated — either by removing the fuses or by tying or locking the isolating switch with a safety tag.

Operations on certain units may need to be done only by specific persons who have received special training and have any required medical tests. This particularly applies to the handling of filters in extract systems handling radioactive substances, dangerous pathogens, carcinogens, or materials causing notifiable industrial diseases. It is likely that anyone working on a duct used for disposal of radioactivity will be required to wear a film badge or dosemeter.

6.4.6 Sub-standard Fume Extracts

Tests on airflow (see section 6.4.1) may show that several fume cupboards or biological cabinets do not meet the modern standards. This can be dealt with in several ways.

Firstly, the units should be clearly labelled with this information. (As a contrary method, only those which are satisfactory may be labelled.)

Secondly, a check should be made for any reason why the units are inefficient. For example, is there a heavily loaded filter which is overdue for changing? Is the airflow greatly improved by opening a door? (In this case as a short-term measure, place a note to the effect that the door must be open, and look at the possibility of fitting a grille or other means of air entry to the room.) Is the fan motor incorrectly wired so as to give the wrong speed or direction of rotation? Are the fan blades clogged with dust and debris? Does the duct have any other openings (e.g. side-arms) in competition with the work unit? Where a fan serves several units, how many can be operated satisfactorily at once?

Thirdly, advice should be sought from a ventilation engineer as to methods of improvement. Many fan systems can be improved by fitting a larger motor. Many fume cupboards can be improved by the fitting of an aerofoil and a rear baffle to give a better directed flow. (At this stage, only advice is required.)

Finally, consideration should be given to limiting the use of particular units. Many older fume cupboards will give an adequate flow when partially closed, but not when fully open. They may be used quite safely if the maximum opening is found which still gives a satisfactory airflow. This can be marked and a notice put on saying that the front screen should not be raised beyond this point when experimental work is in progress.

If there are sufficient units of a high standard for the most hazardous work, then it may not be entirely necessary to upgrade the others, providing the work is carefully segregated. The lower efficiency units should not be used for very hazardous substances, nor for any work which involves a person working at the unit for a prolonged time. However, they may be used as enclosures for prolonged procedures in essentially closed apparatus where human attention is rarely required (e.g. reflux, crystallization, incubation). They may also be used for waste disposal by slow release (evaporation or gas release), leaving the better units available for more active work.

Most laboratories need at least one unit which provides a class I cabinet or high velocity fume cupboard standard of protection. If no units meet this requirement, then at least one should be upgraded or a new unit fitted.

CAUTION: any unit which is inadequate by reason of leakage or an inappropriate exit should not be used at all.

6.4.7 Filters and Absorbers

Extract units can be fitted with three types of air treatment commonly referred to as filters. Firstly there are the depth filters which retain particles (as the name implies) through the depth of the filter material. These have a high capacity for solids before they become clogged, and give a relatively low pressure drop, but by their nature tend to let some of the smaller particles through. They are used to protect other filters (i.e. extend the life before clogging occurs) and may be referred to as 'roughing filters', 'coarse filters' or 'pre-filters'. Secondly there are filters which mainly work on the surface, but give virtually no bleed through of solid particles above a certain size. These are called high efficiency particulate air filters, or HEPA filters for short. They are sometimes referred to as 'absolute' filters, but this is a misnomer as no practical filter can be counted 100% efficient. Note that they filter solid particles: liquid drops (aerosols) may pass through. Thirdly, there are chemical absorbers, which are usually activated carbon or molecular sieve particles. These absorb vapours (and usually aerosols) with an efficiency up to 99%. They are usually fairly effective against dusts, but should not be expected to give protection against biological hazards. They are most effective where the absorber is specific for the work being carried out. Manufacturers offer interchangeable cartridges and sometimes the option of two units in series.

Large-scale industrial extract systems from dusty work may be fitted with cyclones or shaking bag filters. The former collect the majority of the dust before discharge to atmosphere by centrifugal action in a specially shaped chamber. They are commonly sited outside the building and fitted with blow-out panels and vents because of the risk of dust explosions. Large bag filters collecting substantial amounts of relatively coarse dust are mechanically shaken at intervals to dislodge accumulated solids into a collection pan. This extends the time before the filter bags need to be removed (usually for cleaning and replacing). If these systems are used for laboratory extracts, special care should be taken in handling, because of the toxic, reactive or infective material which might become concentrated in the collected solids.

For most laboratory units the maintenance of filters and absorbers simply means replacing them as often as necessary. Modern systems are fitted with indicators or alarms to show when the pressure drop (i.e. resistance to flow) becomes too great. These should not be ignored, but they should not be entirely relied upon. This is for two reasons. In the first place, the indicator may be faulty. In the second place, the filter may develop a hole, in which case it will be unsafe although the airflow seems satisfactory.

CAUTION: some filters have been fitted with manometers to indicate pressure drops. If the liquid dries up or is lost, then air can bypass the filter. A non-volatile inert fluid should be used.

Whether or not a pressure drop indicator is fitted, a significant drop in the working air velocity (as indicated by a built-in meter or a hand-held unit) should be taken as evidence of filter clogging. A reduction to 75% of the velocity measured with a new filter (or to the minimum specified for the work) suggests that replacement is necessary As HEPA filters are expensive, it is usually worth changing the pre-filter alone and testing again. If the flow rate is more than 90% of that for totally clean filters then the HEPA filter need not be replaced on this occasion.

The time interval between filter changes will of course vary with the type and amount of work. However, in view of known instances of mechanical failure and of growth of organisms on filters, it is suggested that no HEPA filter or chemical absorbent filter should be in position for more than two years. Depth filters should be discarded or cleaned (where this is possible) at a similar interval if the work or local regulations do not suggest sooner.

Some self-contained cabinets indicate the end of useful life of their absorbents by emitting a smell. This fact should be generally known in case the user has an impaired sense of smell and other people wonder where the odour is coming from.

All filter changes should be carried out by a technically competent person who has received special instruction wearing gloves and breathing protection. The used filter should be put into a bag immediately. (Some systems allow the fitting of a bag before opening the filter unit to minimize the risk of contamination.) The bagged filter should be disposed of as hazardous waste, preferably by incineration. Special work procedures, measurements and waste disposal arrangements may be required for a filter which has accumulated radioactivity. Filters in biological protection systems should be fumigated before removal. (See section 6.4.4.)

It is important that records are kept of filter changes both for safety and for planning purchase of replacements. It is useful if a label is placed on each individual cabinet or fume cupboard stating when the filter was last replaced.

6.5 VENTING OR EVAPORATING WASTES

6.5.1 Choice of Unit

For the deliberate disposal of waste to atmosphere, the fume cupboard (or direct to a duct via a canopy) is the unit of choice. Biological safety cabinets are not generally constructed of chemical-resistant materials to the same degree, and may not have extracts so carefully placed. It is an obvious point, but occasionally overlooked, that there must be a duct. Units which are self-contained, or exhaust air back to the room, or which (usually 'temporarily') have part of the duct open, missing or broken are quite unsuitable for venting wastes. In addition, type IIA microbiological cabinets or related units with a high degree of recirculation are unsuitable.

For radioactive materials, a particular cupboard and duct will have to be agreed with the inspector, and suitably monitored. It is a good idea to limit the number of cupboards regularly used for waste disposal (as opposed to normal work). These can be specially checked (see section 6.5.2) and any special procedures or apparatus installed.

Type I microbiological cabinets, and (with care) some type IIB cabinets can be used where a fume cupboard is not available, or where the protection afforded by the filters is advisable (i.e. if there is any risk of release of infectious agents along with the waste gas or vapour).

6.5.2 Checking Exhaust Ducts

As a normal safeguard, all fume cupboards and extracted cabinets should have their airflow rate and pattern checked with a hand-held meter in the working space. Details of the correct procedure are available in the British Standards, in many texts on occupational hygiene and from manufacturers of fume cupboards. Note that measurements on a class II cabinet are quite different from those on a class I cabinet or a fume cupboard.

In addition, it is a good idea to have a visible indication of flow in all cases where this is important (i.e. because hazardous materials are in use, or relatively large amounts are being discharged as waste). Some manufacturers supply these as indicators or warning lights. Home-made devices can also be effective. These may be differential manometers fitted to the ducts, moving light objects such as ping-pong balls or even pieces of ribbon attached to the window.

It is strongly recommended that a remote indicator such as a light (or better an alarm) be fitted to signal the failure of a fan, this signal being as close to the chosen cupboard as possible. In the absence of such a device, it will be necessary for the fume cupboard to be more closely watched while waste disposal is in progress.

Before deciding on a particular fume cupboard for regular waste disposal, it should be checked for its ability to disperse a smoke. Some generators can be made or purchased suitable for this purpose, or a sort of firework called a 'smoke candle' or 'smudge pot' can be lit.

CAUTION: always get the manufacturer's exact specification as to the volume rate of release of smoke and the burning time. Check that this is much less than the air volume flow rate for the cupboard. Notify security staff and the local fire service of the exact time and expected duration of the test.

Circulate as many other people as possible with this information. Even so, there is still a chance that someone will set off the fire alarm.

If the weather is peculiarly still, the test should be postponed, purely in the interest of good relations with local residents. Ideally, the test should be carried out under various weather conditions, and the different results noted. Three main checks are required. Firstly to see that there is no noticeable leakage of smoke within the building. Secondly to see that the plume of smoke is carried away from the building. Thirdly to see that it disperses and does not go where it could cause hazard or annoyance. If the smoke test is unsatisfactory on any count, then there should be serious reservations for using the fume cupboard for disposal of hazardous material.

It is difficult to smoke-test cabinets or ducts with filters. The smoke tends to clog the filters, but if they are removed then a true reading is not obtained. Injection of smoke after the filter may be possible in order to check the duct. A more tedious procedure is to release a freon or carbon dioxide or argon or helium into the cabinet and then measure the concentration around the exhaust pipe and building with portable monitors. A portable infra-red analyser is useful for the purpose, but results are obtainable with katharometer and mass spectrometer detectors.

If a suitable non-hazardous gas and a sensitive detector are available, then the above technique is also excellent for pinpointing faults in ductwork.

A simple procedure within the capability of any laboratory is to release a material such as an industrial perfume (also marketed for agricultural use under the name 'deodorizers') or any harmless but strongly (and pleasantly) scented chemical in the fume cupboard or extract system. Any leaks in the duct or any return of outflow fumes back into the building will become readily apparent to anyone with a sense of smell.

6.5.3 Choice of Wastes

(See Appendix C-8 for a classification of commercial gases.)

6.5.3.1 Biological Material

Biologically contaminated dust, spores or aerosols should not be released to an extract system as a means of disposal. At best this means disposal to the filters, at worst a risk of infective fall-out from the exhaust. If it is necessary to dispose of vapour which may contain organisms (e.g. on passing gas through a container or evacuating with a vacuum pump) then it is best to use an in-line filter, and dispose of this via autoclaving and/or incineration. The exit from the filter may be passed into a normal operating microbiological safety cabinet.

Of course, microbiological safety cabinets may be used for treatments (e.g. prolonged boiling under reflux) which might give rise to some free organisms or spores, but organisms should not be deliberately released in a cabinet.

6.5.3.2 Radioactive Material

In general, radioactive dusts should never be discharged to extract systems, as they represent one of the most potent radioactive hazards. If using a vacuum pump or passing gas through a radioactive system, an effective in-line filter should be used, and checked then disposed of as appropriate to the class of radionuclide. Gases and volatile vapours can be discharged within the limits specified by the enforcing agency for the particular duct, but care should be taken not to generate aerosols (which can give rise to dust). A class I or class IIB cabinet may be approved for biological work involving radioactivity.

6.5.3.3 Non-toxic Non-flammable Gases

For practical purposes the following gases may be considered essentially harmless (with the provisos which follow): air, argon, carbon dioxide, helium, krypton, neon, nitrogen, oxygen, xenon. They may be discharged in amounts up to 10% of the known volumetric flow rate of the extract duct. Note that carbon doxide is actually somewhat toxic in higher concentrations: large releases should be avoided, and any regular intake of enhanced levels can lead to stress or other adverse effects.

CAUTION: oxygen can be a severe fire hazard if released in a concentrated form — for example it can cause dust in a duct to ignite.

Cryogenic liquids of the above gases may be permitted to evaporate from a *wide-necked* vessel (to prevent blockage by ice from the air). Gases from pressure vessels may be piped directly into the duct and bled off slowly, preferably via some simple metering device (see section 9.3.4).

CAUTION: rubber or combustible plastic hose should never be used for oxygen, as it may spontaneously ignite.

The following gases may be considered of sufficiently low toxicity to be vented without special precautions, though the volumetric flow rate should be limited to 1%: common freons and halons (CCl_2F_2, $CClF_3$, $CBrF_3$, CF_4, $CHCl_{22}F$, $CHClF_2$, CHF_3, $C_2Cl_2F_4$, C_2ClF_5, C_2F_6) and sulphur hexafluoride. Again, liquids should evaporate from a wide-necked vessel, otherwise a direct tube should be taken into the duct.

CAUTION: some halogenated hydrocarbons are, in fact, rather flammable;

Table 6.3. Flammable halocarbon gases of low to moderate toxicity

Halocarbon no.	Formula	Common chemical name
142B	$C_2H_3ClF_2$	Chlorodifluoroethane
1113	C_2ClF_3	Chlorotrifluoroethylene
152A	$C_2H_4F_2$	Difluoroethane (ethylidene fluoride)
1132A	$C_2H_2F_2$	Difluoroethylene (vinylidene fluoride)
41	CH_3F	Methyl fluoride
1114	C_2F_4	Tetrafluoroethylene (perfluoroethylene)
1140	C_2H_3Cl	Vinyl chloride*
1141	C_2H_3F	Vinyl fluoride

*Low acute toxicity, but carcinogen for occupational exposure. The UK control limit is 3 ppm. It is recommended that the concentration in any extraction duct should be no more than 8 ppm, with the release rate being controlled accordingly.

see Table 6.3. They should not be burnt, because the resulting fumes are more toxic than the original gases.

6.5.3.4 Non-toxic Flammable Gases

The following gases have negligible toxic properties, but are flammable enough to be a considerable fire hazard: acetylene, allene, butadiene, butane, ethane, ethylene, hydrogen, MAPP-gas, methane, propane, propylene. Some halocarbons are flammable; see Table 6.3.

CAUTION: do not underestimate the hazard of small amounts of flammable gases. A gas-air mixture of 100 cm^3, or more, exploding in laboratory glassware can cause serious or even lethal injury.

If there is a large amount to be disposed of, and a suitable burner is available, then the majority may be burnt.

CAUTION: a burner designed for one gas may not work on another. The flame may fail or there may be flash-back in the pipes.

Normally, oxygen should not be used, although this will give a smoky flame in some cases. Many fuel gases may be burnt off in any well-ventilated place where they do not cause a fire hazard.

CAUTION: halogenated gases should not be burnt because they give high toxic combustion products.

Only very small flames (e.g. a bunsen burner) should be used in a fume cupboard. Most biological safety cabinets are unsuitable for flames altogether, owing to the air pattern and their plastic construction.

It is in fact unusual to have a sufficient amount to make burning necesary. Large rented cylinders can usually be returned to the supplier part-full, if notified in advance. For a few litres of gas at normal pressure, or the excess from an experimental purge or gas generation, then there is no advantage in burning, and some risk in maintaining a flame close to the apparatus. It is better to simply vent in these circumstances.

For the special situation of a non-returnable gas container (e.g. some lecture bottles or camping cartridges) where a specific burner is available, the burning off is appropriate. However, this should stop when there is a small pressure still remaining in the container. The last portion should be bled off without burning, into a fume duct.

6.5.3.5 Toxic and Flammable Gases

Some gases, notably carbon monoxide, carbonyl sulphide, coal gas and ethylene oxide, are both highly toxic and highly flammable. Some others are both toxic and liable to spontaneous ignition: arsine, boranes, phosphine, silanes, stibine and the vapours of metal alkyls (e.g. dimethyl zinc). Some toxic gases (e.g. ammonia) are technically flammable, but this is of little relevance to waste disposal on the laboratory scale, as their toxic and corrosive properties are more important. In addition, mixtures will often be used of a toxic gas or vapour in a flammable carrier, e.g. hydrogen cyanide in hydrogen.

It is recommended that with severely toxic agents, some attention be given to chemical methods of treatment (e.g. a suitable absorbent), and only the residue from this treatment be vented.

CAUTION: the hydrides such as arsine, stibine and phosphine are so severely toxic that a short exposure to one part per million (0.0001% by volume) can be very dangerous. Thus ordinary levels of dilution are not sufficient.

Moderate amounts of carbon monoxide, coal gas or similar mixtures can be burnt on a bunsen burner of a type made for this purpose (and not methane gas mixtures). It is technically possible to burn ethylene oxide, but it is usually more practical to vent it very slowly direct to a duct. This is also advised for smaller amounts of carbon monoxide etc.

It is usually inadvisable to burn gases (or gas mixtures) containing sulphur or halogens (bromine, chlorine, fluorine, iodine) — examples are carbonyl sulphide, methylchloride, flammable gases containing halocarbons. The products of combustion of these materials are toxic acids which may be worse than the original material, and may corrode the fume extraction system. Careful dilution with air and slow release to a fume extract system is preferred.

A fire hazard exists when a flammable gas is being diluted with air. For moderate amounts this can be reduced by a preliminary dilution of the gas with nitrogen or carbon dioxide to 10% or less, and placing the outflow in such a position that there is rapid mixing with air. One technique to aid mixing is to pass the gas through an *unlit* bunsen burner, and take a tube from the burner to the extract duct in the fume cupboard. Thus air is drawn in as the gas passes up the burner jet.

A Drechsel bottle may be used as an indicator of the flow of gas. The bottle is filled three-quarters full with water or other suitable liquid, and the gas bubbled through. The bubbles provide a visible check that the flow rate is small. The outlet from the bottle is of course taken up into the fume cupboard extract duct. This is particularly useful for the more hazardous gases. However, water should not be used for extremely soluble gases such as hydrogen chloride or ammonia, as there is a risk of sucking-back.

6.5.3.6 *Corrosive Gases and Vapours*

Some examples of these are acetyl chloride, ammonia, boron trichloride, bromine, chlorine, fluorine, hydrogen bromide, hydrogen chloride, hydrogen fluoride, hydrogen iodide, iodine, mercury, nitrous fumes, ozone, perchloric acid fumes, phosphorus pentafluoride, phosphorus trichloride, silicon tetrachloride, sulphur dichloride, sulphur dioxide, sulphur tetrafluoride, sulphur trioxide, thionyl chloride.

These are all very toxic by inhalation, and most are irritating at low concentrations. For this reason alone, they should only be vented to atmosphere in very small amounts. However, even low level releases repeated often over a number of months or years can cause corrosion of materials of construction of fume cupboards and associated equipment (fans, electric wiring, etc.).

The best general technique is to use some method such as contact with water to absorb the majority, and only vent the residual vapours. The solution can then be chemically treated (see section 9.3.7).

CAUTION: many corrosive vapours and gases react very strongly with water. There should be sufficient flow of water to prevent excessive heating, and precautions should be taken to prevent the water being sucked-back into the gas-filled apparatus.

Where a fume cupboard or extract duct is particularly used for corrosive materials, this should be noted in the maintenance records, and inspections arranged to take this into account.

6.5.3.7 *Liquids*

It should be clear whether the purpose of evaporating a liquid is to dispose of the volatile material or to concentrate an involatile material. For disposal, it is recommended that corrosive liquids (e.g. hydrochloric acid, organic acid chlorides) are not vented but instead treated chemically. However, it is quite reasonable to volatilize traces of irritant material remaining in solutions in this way. Where concentration is intended and a corrosive is to be driven off, then some consideration should be given to a closed system which will collect the evaporated material, e.g. a rotary evaporator. The same applies where significant quantities of single solvents are involved, which may be recycled by this means.

For routine solvents, the main requirements are to keep the quantities small — say 250 cm^3 as an absolute maximum in a fume cupboard at one time, not to have an evaporation going on alongside other work (especially where a flammable solvent is involved), and to make some observations of the evaporation rate.

Further advice on evaporation of liquid waste is given in section 9.3.4.

REFERENCES

[section 6.1.2]

American Conference of Governmental Industrial Hygienists (1984) *TLVs, Threshold Limit Values for Chemical Substances and Physical Agents in the Work Environment and Biological Exposure Indices with Intended Changes for 1984–1985*, ACGIH [annual].

UK Health and Safety Executive (1985) *Guidance Note EH 40: Occupational Exposure Limits 1985*, HMSO [annual].

US Code of Federal Regulations (1972 to date) *29 CFR 1910.1000 OSHA Regulations on Toxic Materials*.

[section 6.4.1]

British Standards Institution (1979) *BS5726: Specification for Microbiological Safety Cabinets*. BSI.

Clarke, R.P. and Goff, M.R. (1981) The Potassium Iodide Method for Determining Protection Factors in Open Fronted Microbiological Safety Cabinets, *J. Appl. Bacteriology*, **51**, 439–460.

[section 6.4.4]

Collins, C.H. (1983) *Laboratory-acquired Infections*, Butterworths.

UK Dept of Health and Social Security (1978) *Code of Practice for the Preservation of Infection in Clinical Laboratories and Post-mortem Rooms [The Howie Report]*, HMSO.

BIBLIOGRAPHY

American Conference of Governmental Industrial Hygienists (1984) *Industrial Ventilation 19th edn*, ACGIH.

British Occupational Hygiene Society Technology Committee (1975) A Guide to the Design and Installation of Laboratory Fume Cupboards, *Ann. Occup. Hyg.*, **18,** 273-291.

British Standards Institution (1959): BS3202 *Recommendations on Laboratory Furniture and Fittings.*

British Standards Institution (1979): BS5726 *Specification for Microbiological Safety Cabinets.*

British Standards Institution (1982): DD80 *(Draft Document) Laboratory Fume Cupboards. Part 1: Safety Requirements and Performance Testing.*

British Standards Institution (1982): DD80 *(Draft Document) Laboratory Fume Cupboards. Part 2: Recommendations for Information to be Exchanged between Purchaser, Vendor and Installer, and Recommendations for Installation.*

British Standards Institution (1982): DD80 *(Draft Document) Laboratory Fume Cupboards. Part 3: Recommendations for Selection, Use and Maintenance.*

Clarke, R.P. (1983) *The Performance, Installation, Testing and Limitations of Microbiological Safety Cabinets*, Science Reviews.

Hughes, D. (1980) *A Literature Survey and Design Study of Fume Cupboards and Fume-dispersal Systems*, Science Reviews.

Laboratory of the Government Chemist (1980) *The Development and Use of Fume Cupboards, Fume Hoods and Ventilated Safety Enclosures for Laboratories (Proc. Symp. London, 22 March 1979)*, Lab. Govt. Chemist UK

Lees. R., and Smith, A.F. (eds) (1984) *Design, Construction and Refurbishment of Laboratories*, Ellis Horwood.

McDermott, H.J. (1976) *Handbook of Ventilation for Contamination Control*, Ann Arbor.

National Sanitation Foundation (1983) *NSF Standard 49: Class 2 (Laminar Flow) Biohazard Cabinetry*, NSF.

UK Department of Education and Science, Architects and Building Group (1982) *Design Note 29: Fume Cupboards in Schools*, UK Dept. Educ. Sci.

UK Health and Safety Executive (1975) *Principles of Local Exhaust Ventilation*, HMSO.

White, P.A.F. (1981) *Protective Air Enclosures in Health Buildings*, Macmillan.

CHAPTER 7

Burning and Incineration

7.1 INTRODUCTION

Burning just means setting fire to something. It is not the same as incineration, which may be described as the efficient combustion of material, usually in the presence of auxiliary fuel. Properly designed incinerators can be effective in destroying many hazarous materials with minimal pollution, especially where the waste is of relatively constant amount and composition.

Burning is often advised as a means of disposal of harmful laboratory materials, either in the open air or in a fume cupboard. It must be stressed that such techniques are generally unsatisfactory and frequently dangerous. They should also be illegal in the UK and USA under anti-pollution legislation.

Althugh incineration is a very important and valuable method of dealing with laboratory waste, it is not an easy option. This chapter concentrates on some of the difficulties, but it is the authors' opinion that more well-managed incineration is desirable. Badly-managed incineration or amateur bonfires would be better avoided.

7.2 OPEN BURNING

7.2.1 Objections

Simple burning is typically carried out in a heap on open ground, i.e. a bonfire, or in a perforated metal container, i.e. a brazier or garden incinerator. So far as the present discussion is concerned, a garden incinerator is not truly an incinerator, merely a well-structured bonfire, which may just be acceptable at an agricultural establishment. Small amounts of waste are sometimes burned on a metal tray or a bed of sand in the open air or in a fume cupboard.

There are several objections to open uncontrolled burning as a method of dealing with harmful material. Firstly, fire has an undeserved reputation for dealing with infected material. In the open air it is most likely that infected particles will be thrown up by the draught of the fire, long before adequate heat sterilization takes place. Even a commercial incinerator may release live spores from the chimney. (However, dispersion at a height causes sufficient dilution and allows the effect of ultra-violet light to minimize the hazard from a domestic refuse incinerator. Possibly sulphur dioxide in the chimney gases also has an effect.) Secondly, if material of any volatility is involved (i.e. most organics and some inorganics) it is virtually certain that a significant portion will simply disperse into the atmosphere. People in the locality may therefore be exposed to harmful amounts of the substance. With some toxic agents, illness can result shortly after. With some others, there may be delayed problems, including sensitization. Thirdly, thermal breakdown can give products that are more harmful than the starting material. For example, polystyrene, a relatively inert plastic, will give off copious fumes of styrene and benzene in the average bonfire. Complex chlorinated materials can give off at best, hydrogen chloride, and, at worst, poisons such as phosgene or dioxin-like substances.

With many substances, even total combustion in an effective incinerator will give rise to harmful or irritant materials being released. For example, phosphorus and many of its compounds are likely to give a smoke of P_2O_5. Chlorine containing waste (such as PVC) almost always results in the production of HCl. These products can be dealt with by a scrubber system in an industrial incinerator, but in the open air will drift about.

Finally, it is unlikely that a sufficient temperature can be generated in simple burning to totally destroy organic molecules and convert them to CO_2, water, etc. As an example, a temperature of 1100 °C would be required, and the material would need to remain at that temperature for at least 2 seconds to be certain of destroying typical pesticides.

7.2.2 Justifications

Although it may be illegal and is usually undesirable, open burning may be technically justified in some situations. (In the USA, there is some evidence that regulatory officials are beginning to accept that open burning is the most practical method of disposal in certain circumstances, such as those which follow.)

7.2.2.1 Fire Pits

A fire pit is usually a firebrick chamber below ground, fitted with a gas or oil burner (usually from above by a forced jet) and often with water sprays, drains, etc. It is typically used for contaminated packaging and sometimes for small amounts of waste chemicals. Some even have forced air ducts and ash removal arrangements.

Where an institution already has a firepit, and has a good record of making sensible use of it (i.e. there is no evidence of nuisance to neighbours or risk to employees), then the authorities may well permit it to continue. However, it is very unlikely that a licence would be granted for a new installation of this nature.

As the facility may be withdrawn at any time, a laboratory is well advised to consider alternatives.

7.2.2.2 Unstable Substances

There are certain materials, namely explosives and some badly deteriorated chemicals, which it would be dangerous to move. The threat to human life then takes precedence over the threat of pollution, and they may be burned on the nearest open space which is clear of people and flammable vapour hazards.

The key here is to place small amounts on a relatively large amount of combustible material (such as wood shavings) so that the hazardous material is all burnt up before the rest of the bonfire finishes. The situation should be as remote as possible and a *reliable* method of lighting used.

With any substantial amount of explosive or chemical which is a potential explosive, then expert advice and help should be sought. This can come from the manufacturers, the fire service, possibly the police or army bomb squads, or a specialist disposal agency such as that of Harwell. See also Chapter 12.

7.2.2.3 Urgent Situations

Circumstances can arise where prompt burning will be less of a hazard than other procedures, especially where these would involve transport. For example, if a spill of flammable liquid or an oxidizing agent, or both, has been mopped up on paper towels, it may well be safest to go outside and burn the material in small amounts, rather than risk it starting a fire in storage. The most senior qualified person available should be consulted, and all possible precautions taken. Clearly this should not be carried out in any zone where all sources of ignition are banned.

CAUTION: note that burning may not mean total destruction, and may expose people to harmful vapours or fumes. See section 7.2.1.

7.3 PUBLIC AND COMMERCIAL INCINERATORS

7.3.1 Refuse Incinerators

Many local authorities operate incinerators to dispose of some of their domestic refuse and trade waste. Clearly, any waste from the laboratory which is no more harmful than domestic refuse is acceptable, and may be collected by the local authority, or delivered by a contractor. However, these incinerators can also be used for more hazardous waste, namely biological and chemical materials. It is unlikely that they would be able to accept radioactives.

To use the system essentially means contacting the local authority and fulfilling their requirements for the presentation of waste. This may mean delivery to a special point, in special bags, on particular days, etc.

A regular service is often operated for certain types of biological waste, e.g. from hospitals, veterinary surgeons, and food industry premises. In addition, the laboratory may be able to negotiate agreement for items such as contaminated packaging or perhaps used samples. It is then the responsibility of the laboratory to ensure that these items are always taken direct to the incinerator, and not to a landfill site or refuse handling station.

Municipal refuse incinerators have been successfully used to dispose of small batches of chemicals. One exercise involved the collection of accumulated waste from schools in the West Midlands. Unidentified chemical packages were added direct to the loading hopper at the rate of one per 7 tonnes of refuse. Similar procedures have been followed elsewhere. The advantage is that the incinerator is large enough to cope with even small explosions (as typically happens with aerosol cans every day) and offers both thermal conversion and considerable dilution into the air and ash. The fact that full identification was not necessary was a useful saving in cost, and was also the safest way to proceed, since there could have been accidents in opening containers for a sample.

It must be stressed that experienced technical staff should be on hand when any kind of hazardous waste is to go in with routine refuse. Under no circumstances should ordinary untrained employees be expected to handle infected or chemical waste. For this reason, authorities sometimes set aside a particular small incinerator for animal carcasses and the like, with specially trained staff and equipment available.

The laboratory should expect to pay more for the disposal of hazardous waste than for ordinary refuse, but the costs are not excessive. It is in everyone's interest that a facility of this nature is not abused by disposal of items other than those agreed, or by not following the correct procedures of notification and transport.

7.3.2 Commercial Incinerators

Ordinary refuse incinerators do not reach a high enough temperature to completely decompose some of the more dangerous chemicals. In addition, there may be no local refuse incinerator or the local authority may not agree to take a category of waste (they are not obliged in any way to accept trade waste). Under these circumstances, it may well be appropriate to send material to a hazardous waste incinerator operated by a private company.

Laboratories are sometimes surprised at the cost of incineration, which may even exceed the cost of purchase. The reason is simple. To ensure complete destruction of a small packet of a highly dangerous material will probably mean its dilution into a tonne or more of combustible matter. This will be burned in a high-temperature high-efficiency unit which must meet stringent pollution standards. In addition, the company is in business to make a profit, which requires a fair charge for each individual task. Safe disposal of hazardous chemicals is a responsible technical business, which needs investment in both skilled manpower and equipment.

In general, the more dangerous or more unusual a waste, the more trouble will be required on both sides. Samples of research carcinogens may need special procedures involving the relevant authorities, whereas ordinary waste solvents can be incinerated at a much lower cost on a routine basis.

Not every commercial incinerator can meet the highest standards required for exceptionally difficult materials. However, in the UK at present time there is excess capacity for incineration of waste.

In certain circumstances (e.g. rare elements) it is possible for the incinerator to return the ash for recovery of valuable elements, or operate its own recovery service via a related company.

7.4 LABORATORY AND INSTITUTIONAL INCINERATORS

7.4.1 General Comments

If a laboratory produces hazardous waste which is suitable for incineration, there are many advantages to possessing its own incinerator. One of the biggest is that the waste disposal process is assured. That is, the laboratory is certain that waste has been destroyed, not somehow lost in transit or disposed of unwisely, and in addition it is not dependent on others. Waste collection and disposal routines can be arranged entirely to suit the institution, and are less affected by outside occurrences such as strikes or the weather.

The main disadvantages are those of cost and management. Cost benefits are often badly calculated and hopelessly optimistic (see section 7.4.4). It is not always appreciated how much management is required in the day-to-day operation of an incinerator service, particularly where the waste producers are not directly responsible for incinerator operation. Wrong selection of unit, or poor arrangements for waste handling are common ways in which a good concept is badly executed.

7.4.2 Principles

Destruction by fire involves two separate processes, though they may occur in the same chamber. The first is pyrolysis, which is the breakdown of material by heat into vapours, ash and char. This can occur in the absence of air. This process is followed by combustion, which is the combination of material with oxygen. For example, if plastic such as polythene is burnt, it melts and degrades to give fumes which combust, forming the flame.

To achieve efficient combustion requires excess air. If there is insufficient (e.g. if an incinerator is overloaded) then pyrolysis and combustion will occur side by side, and some material which has not been combusted will escape as smoke and fumes. The requirements for maximum thermal efficiency (for example, to run a boiler) are different from those for maximum efficiency of combustion. Thus to ensure complete destruction there is a penalty in terms of fuel and heat efficiency.

To convert some resistant material (e.g. polychlorinated biphenyls, some pesticides, some carcinogens) means that all the material must be subjected to a high temperature in the presence of excess air for a certain time period. This may be as high as 1100 °C and as long as 2 seconds, although a temperature of 1000 °C and a time of 0.3 seconds is sufficient for most things. For especially hazardous material, this combination of temperature and residence time may be required by the licensing authority. Tables 7.1 to 7.3 give some official requirements for incinerator performance.

Although a well-balanced flame may have a temperature as high as 2000 °C, the effective operating temperature is lower for several reasons. Firstly, the presence of excess air will give a cooler flame, so this is usually limited if possible. Secondly, the pyrolysis process is endothermic — that is, it absorbs heat. Thus materials which do not have any useful combustion value (such as highly chlorinated chemicals) are mainly converted by pyrolysis, giving significant cooling unless present in only tiny amounts. Thirdly, individual particles may spend different times in the operating zone. For example, even if the average residence time is (say) 1 second, there will be a significant amount of material which spends less than half this time in the operating zone. Finally, the wall of an incinerator may be cooler than the centre gases. In fact, the working limit for common refractories is 1300 °C so the unit is usually designed to operate a safe margin below this.

To achieve complete combustion, there must be good mixing (turbulent flow) in the burning zone. However, this requires relatively high velocities, which means that solid particles can be carried out of the working chamber, and are difficult to remove from the flue gas.

7.4.3 Types of Incinerator

There are a great many commercial designs, each of which can be adapted to an individual user. However, they can all be described in terms of their auxiliary fuel, the type of material they are designed to burn (and its associated feed equipment), the number of chambers used for primary destruction, and the provision of afterburners, scrubbers and other pollution control devices.

A typical small, modern incinerator will have a primary chamber in which the waste is 80 to 95% destroyed, and a second (or even third) chamber to complete the process. See Fig. 7.1. The primary chamber is brought up to

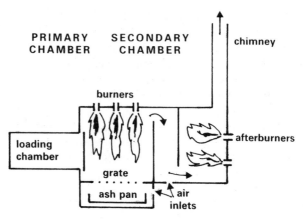

Fig. 7.1. Essential features of a 2-chamber incinerator.

temperature using a commercial ('auxiliary') fuel. Depending on the heat value of the waste, the auxiliary fuel is reduced or cut off altogether, as the waste burns, auxiliary fuel being added automatically to maintain the temperature as required. Smoke and gases pass to secondary chambers which are supplied with additional air and kept at a high temperature with more fuel in burners (sometimes called the afterburners).

   Most incinerator designers try to achieve as much combustion as possible in the primary chamber, and use the afterburners to treat any material which escapes. However, some designs limit the amount of air in the primary

Table 7.1. Recommendations for biological waste incinerators, based on the U.K. Code of Practice.*

Combustion chamber temperature (pyrolysis units may operate at lower temperatures)	Minimum	850 °C
Afterburner chamber temperature	Minimum	800 °C
Residence time (of gases in the zone above 800 °C)	Minimum	1.0 s
Flue gas exit temperature	Minimum	150 °C
Black smoke emission per any 8 hour period	Maximum	10 min

In addition, the emission of dust, grit and acid gases should not cause pollution,† and odours should not cause offence. Gas cleaning apparatus may be necessary to achieve these requirements, and should be fitted to all units of a capacity greater than 750 kg per hour.

*U.K. Dept of the Environment (1983) Waste Management Paper No. 25: Clinical Waste, HMSO.
†Based on the Clean Air Acts 1968, and subsequent regulations.

Table 7.2. U.S. Environmental Protection Agency requirements for hazardous (chemical) waste incinerators*

Efficiency of destruction of Principal Organic Hazardous Constituent(s) (POHC(s))	Minimum	99.99%
Solid particles in flue gas emission	Maximum	180 mg m^{-3}
HCl in flue gas emission	Maximum	1.8 kg hr^{-1}
or		
Efficiency of removal of HCl from flue gas	Minimum	99%

*Federal Regulations 46 FR 7666, 23 January 1981, and 47 FR 27516, 24 June 1982.

Table 7.3. Recommendations for incinerators for hazardous and refractory organic waste, based on the US proposed standards for halogenated waste*

Combustion temperature	Minimum	1200 °C
Residence time in combustion zone	Minimum	2 s
Excess oxygen	Minimum	3%
Ratio in flue gas $CO_2/(CO_2 + CO)$	Minimum	99.9%

*Federal Regulations 43 FR 59008, 18 Dec 1978.

chamber so that combustion is incomplete. This is known as 'starved air combustion', and the unit is often referred to as a 'pyrolysis incinerator'. Greater thermal efficiency and reduced smoke emission are claimed.

Most small laboratory incinerators have the chambers as distinct separate constructions. For large commercial units, the chambers may in fact be zones within one chamber. Conversely there may be six or more chambers in 'multiple hearth' incinerators. These large specialized units are outside the scope of this book.

It is unusual but possible for a laboratory to burn only liquid waste. This is normally done in an incinerator which uses an auxiliary fuel to provide a hot mixture of air and flue gas, into which the liquid is sprayed by a nozzle called an 'atomizer'. More commonly, an incinerator is primarily designed to burn solid waste, and may or may not be fitted with atomizers for liquid waste. This can be a useful addition if the volume of liquid waste justifies it.

Very large-scale units normally operate continuously, with automatic feed and ash removal. However, the most usual arrangement for a laboratory unit is for it to be brought up to temperature then loaded with a series of charges of waste during the day. The incinerator is allowed to complete its final cycle ('burn down') then cool overnight. In the morning the ash is removed prior to start-up. Automatic systems to aid addition of waste or removal of ash are often a significant extra cost. However, 'ashing out' is not a pleasant task, and operators welcome any mechanical aids.

For modest-size units, where complete automation is impracticable, there are two common methods to protect the operator when loading in a fresh charge of waste. In the better of these, a loading hopper is filled and closed. A ram mechanism then delivers the waste to the firing zone, keeping both the waste and the combustion area isolated from the operator. This is the most expensive option, but is essential for the most hazardous waste.

The old direct access door is quite unsuitable for laboratory waste (or waste of any hazard at all) so an intermediate method has been developed. This uses a curtain of clean air between the firing zone and the outside when the loading door is open. If correctly operating, this curtain will effectively prevent gases and particles from escaping during normal operations. However, a small explosion may cause material to be ejected through the air curtain. It is also possible for a flashback to occur if the waste on the loading platform gives off flammable vapours.

7.4.4 Economic Aspects

Incinerators are sometimes justified on the basis of supposed cost savings, particularly where energy recovery is included. However, many users find the original forecasts were optimistic, or based on unrealistic assumptions. Even large industrial users sometimes find it difficult to achieve the real energy savings they supposed, so a small laboratory is advised to treat with caution any suggestion that a large investment can be recovered in a short time.

The basic fact is that for much (but not all) laboratory waste, incineration is a sound method of destruction, with many important advantages which are difficult to value. For example, on-site incineration reduces the risk of infection in transit, and provides an assured disposal service for difficult waste, which is not dependent on outside agencies. The main cost saving is in reduced payments outside for disposal of the same waste. However, note that there will still be the ash to dispose of. For typical mixed solid, this amounts to 40% of the weight and 20% of the volume of the original material. Moreover, the ash from incineration of chemically contaminated waste may well have to be disposed of to land as hazardous waste. (In the USA, the Environmental Protection Agency assumes such ash to be hazarous unless there is proof to the contrary.)

All incinerators will require a supply of auxiliary fuel, which must be paid for. The only possible exception is an incinerator for waste oil of high calorific value. However, such waste can normally be sold or given away. (If the oil is contaminated with harmful material, then the afterburner will require clean fuel.) In addition, the operation of an incinerator will occupy an important part of at least one person's working day.

To count against these costs, it is sometimes possible to make use of the incinerator heat to provide hot water or steam on site. It is for this purpose that erroneous calculations are often made. If an institution such as a hospital or large research establishment has a regular large supply of combustible waste (paper, and plastics other than PVC) then heat recovery may well be justified. The difficulties are as follows. Firstly, the economic calculations are often based on a fully loaded unit working 24 hours a day. Hence an incinerator working fully loaded for the normal working day will perhaps only be making a contribution for 6 hours (allowing for start-up and burn-down) and will take four times as long to recover the cost of the heat recovery boiler. Furthermore, during start-up and burn-down there is a daily cost in auxiliary fuel which is ignored in the continuous model calculation.

In practice, laboratory incinerators often run only part-loaded for much of the day. A 75% load may give a 50% loss in recoverable heat. Another source of losses is the period between charges, when the incinerator maintains its temperature by burning auxiliary fuel. This is an expense for any incinerator, but represents a greater penalty where a boiler is fitted, because heat recovery boilers have about half the thermal efficiency of a direct-fired boiler. That is, 2 litres of fuel will have to be burnt in an incinerator to raise the steam which could come from one litre in a conventional boiler. (This is because of the need to ensure complete combustion rather than maximum thermal efficiency: the basic design criteria for a steam boiler and an incinerator are very different.)

A common economic error arises from poor forecasting of either the supply of waste or the demand for heat. If the unit is sized to cope with the maximum waste supply, then it will usually be working at part-load and thus seriously reduced efficiency. Conversely, if there is no demand for the heat at the time it is produced, then the energy saving is purely an illusion.

It is recommended that a careful measurement be made of the amount and approximate calorific value of waste over a reasonably long period, and the smallest possible incinerator with heat recovery boiler be priced to cope with the minimum assured supply of waste (or possibly half the maximum supply). If such an incinerator plus heat recovery system does indeed save money compared with simple incineration or removal by contractor, then installation may be considered. A second incinerator without heat recovery can be installed to cope with the remaining waste. The heat recovery unit is run for as many hours as possible at optimum load, and the other unit is run for as many hours as necessary to cope with the supply. Very little extra manpower should be required for the second unit.

The combustion of waste typically produces corrosive gases and often ash or smoke which can cause increased costs for maintenance on the boiler, compared with a conventional fuel boiler. The cost of the boiler may therefore (justifiably) be greater than for a simple boiler of the same heat duty. Some combinations of incinerator and boiler are considerably better than others in projected life and maintenance requirements. It is a good idea to contact someone who has already installed a similar unit from the same manufacturer to verify the claims of the supplier.

For hazardous waste in the USA, the Environmental Protection Agency may demand a 'test burn' to ensure that the material is destroyed with 99.99% efficiency. To fulfill the necessary requirements, this trial can be very expensive, but may be worthwhile if there are important amounts of difficult materials regularly produced in a facility.

7.4.5 Operation and Management

7.4.5.1 The Operator

It is absolutely essential that the operator receives formal training in two aspects of the work. This training is best performed by competent persons in time set aside for it. Casual instruction on the job is not adequate. Firstly, the operator must be fully trained in the operation and service requirements of the incinerators, and in any emergency procedures. Secondly, the operator must receive sufficient good training to make him or her aware of the hazards of the waste (physical, chemical, biological or radioactive) and competent to protect himself from them. Training may come from the incinerator supplier, the laboratory institution, or (more usually) a combination of both.

It is sensible to have at least two people on site trained to operate the incinerator, and the training programme may be extended to cover site engineers or maintenance personnel at very little extra cost.

In addition, the operator must have clear instruction as to the aims of the incineration programme, and the different requirements of the units. For example, one unit may be reserved for relatively harmless waste and operated for maximum results from a heat recovery boiler, whereas another may be for destruction of specific small loads of hazardous waste, carefully diluted in routine waste.

The operator should be provided with protective clothing for normal operation, and also equipment to deal with spillages (e.g. from torn or split bags). This should include tongs to pick up potentially harmful items, and some extra bags and closures.

Where infected material may be present, the operator should be provided with disinfectant sprays, and clear instruction in their use. The temporary storage area should be regularly disinfected, and the operator's work practice with clothing etc. should take into account the risk of infection.

7.4.5.2 The Waste

Great care must be taken in the collection of waste for incineration. Clearly-labelled receptacles must be provided, and waste producers made aware of the need for segregation. For example, aerosol cans should be rigorously excluded, as they can cause serious damage to a small unit. Most incinerators do not tolerate glass very well, because it melts and may clog burners or ash collectors. However, some units are available which will accept glass if required, though this is usually by lower temperature operation.

The waste should be collected in closed burnable containers (i.e. plastic sacks, paper bags, cardboard boxes) of such a size that they can be put into the incinerator loading hopper without opening. It is preferable to use a double container system for harmful waste (e.g. a sealed bag inside a sealed bag, or a sealed bag inside a cardboard box).

Considerable thought should be given to the delivery of waste and storage near the incinerator so that this does not produce a fire hazard or a health risk to the operator. If at all possible there should be some element of segregation, such as separate bays for different classes. This both aids the operator in routine operation and is helpful in identifying spillages. It is often worth while arranging for certain wastes to be delivered at a specific time so that they can be immediately incinerated with no intermediate storage.

Non-combustible items, such as most metals and ceramics, will be retained in the ash. Sharp objects can be a serious hazard to the operator, and large ones may lodge in inconvenient places. Depending on the design of incinerator, it may be possible to put boxes of disposable sharps into re-usable metal trays, which can be more easily recovered from the ash.

Waste of low fuel value (e.g. pathological waste) should normally be placed in with a larger amount of combustible material, as far as possible. Some incinerators can burn 100% plastics, but in general it is wise to limit plastics to 15% of the load. As circulation of air is important, waste packaged for an incinerator should not be baled or compressed. (This can also cause jamming of the loading mechanism as the bale releases.)

Aqueous liquids can be incinerated in a standard solids unit, by soaking them onto sawdust or another combustible material. For hazardous substances, this can be done by filling a plastic jar about 10% with the liquid, then filling with the absorbent. The jar is closed with a combustible cap, taped and sealed into a plastic bag to make it safe to handle. Oils or combustible liquids of high flashpoint can also be incinerated in this way, but vermiculite should be used in place of sawdust.

It is possible to incinerate small amounts of volatile or toxic organics in a solids incinerator by the following technique. The material is soaked onto granular activated charcoal and mixed with a large excess of charcoal in a polythene bottle or jar. This ensures that there is minimal vapour release at room temperature, owing to the strong absorptive power of the charcoal. The jar is then taped closed and may be burnt in with normal waste. This technique can be adapted to deal with material collected up from a spillage, but has a limited application owing to the cost of chemical charcoal.

CAUTION: highly flammable liquids should not be directly incinerated

except in special facilities, as they may explode or flash back as they are being loaded. Exceptionally, small amounts may be made into a 2% solution in oil, then burnt via an atomizer or absorbed on vermiculite.

7.4.5.3 Management

The main task of management is to see that initial good practices do not deteriorate with time. The most usual problems come from inappropriate materials being put in waste containers for the incinerator. Occurrences should be publicized, and complaints from the operator dealt with seriously.

Other problems come from incorrect loading of the unit, usually to save time. The unit will in general perform inadequately if it is given the first charge before it is fully up to temperature, if it is loaded with too much at one time (e.g. by filling the loading hopper with dense material), or if there is a poor balance between combustible and non-combustible waste.

Damage can be done to the materials of construction by corrosive gases in the presence of water. Thus it is essential that nothing but auxiliary fuel is burnt until the unit is up to operating temperature. The most crucial part is the cool end of a waste heat recovery unit, which should be above 260 °C. The most common source of these gases is PVC, which should preferably be limited to 0.5% of the load as far as practicable (e.g. by not sending PVC items for incineration, or by not sending them all at once).

Damage can also be done by ordinary atmospheric moisture if a unit is not used for prolonged periods. The manufacturer's advice should be followed for shutdowns of more than one week.

7.4.6 Future Prospects

Incineration is currently receiving intense study by commercial organizations and government agencies. A very likely development is small-scale units for hazardous materials, based on fluidized bed or molten salt technology. At least two commercial projects are well advanced.

Further study should in the near future enable manufacturers of modest-size units to produce specifications which will satisfy the US Evironmental Protection Agency as to the efficacy of certain designs for particular classes of hazardous waste. This should make 'off the shelf' technology for hazardous waste destruction more feasible.

In the UK there have in recent years been moves away from large-scale refuse incinerators owned by local authorities, but there seems a resurgence in smaller incinerators directly used by institutional waste producers. In particular, it is becoming policy for all hospitals of any size to have (or share) an incinerator for clinical waste.

In the USA, small incinerators have been permitted to operate with less stringent requirements for pollution control equipment, and are thus less costly to install and simpler to operate than larger units. However, with all anti-pollution legislation, local interpretations vary, and each case has to be considered on its merits.

7.5 OTHER COMBUSTION TECHNIQUES

Devices other than incinerators have been used, particularly for chemical residues. The following notes are not exhaustive, but indicate some options which have been used. Each situation should be carefully examined, both for the material to be destroyed and for the nature of the facilities. Great caution should be exercised to prevent there being any fire risk or danger to health. (See section 7.2.1.) Sections 6.5.3.4 and 6.5.3.5 deal with disposal of flammable gases. Section 9.3.5 is on burning of chemicals.

7.5.1 Fuel Boilers

It is possible to burn combustible liquids in an ordinary steam boiler in two ways. Either the liquid can be sprayed into the firing zone by a special nozzle, or the liquid can be added to the fuel supply. Where there is a regular supply of waste liquid of predictable composition, then a spray nozzle is advised. The nozzle will require water cooling, a feed pump and a small storage vessel. It is usually a good technique to mix the waste 50% with the liquid fuel, if they are miscible.

Liquid may also be injected into the feed pipe to the boiler burner, providing the boiler uses liquid fuel and the waste is compatible. Many waste oils can literally be added to the fuel tanks of an oil-fired boiler, with no discernible effects, providing the amounts are small and there are no sludges or particles which might clog nozzles.

In both the UK and the USA, the burning of combustible wastes in a steam boiler is not counted as a waste disposal operation, and is therefore exempted from many requirements of licensing and control. However, this only applies if the waste has a genuine fuel value, i.e. it has a heat of combustion greater than 8000 Btu/lb (18.6 MJ kg^{-1}). This excludes many halogenated compounds.

Non-combustible wastes may be destroyed in a fuel boiler, but the unit then becomes a waste disposal device and is subject to controls and licensing. However, this can be an acceptable low-cost way of treating some aqueous liquids which are unsuitable for the sewer. These are fed into the combustion zone by a spray nozzle.

Advice should be sought from the manufacturer or other competent person as to the feasibility of waste disposal via a particular boiler. Care should be taken that the waste does not cause problems of smoke emission, corrosion, burner clogging or any danger to health of the operator.

Modern solid fuel boilers will generally tolerate only a very small amount of material other than the principal fuel. They should not be used for the indiscriminate disposal of plastic and paper. Spray injection of liquids is often possible, as is addition of occasional items through a hatch, but serious disturbance to the boiler and danger to the operator can result if the operation is not tightly controlled (see section 7.4.5).

7.5.2 Laboratory Furnaces

It is obvious that standard furnaces and combustion devices may be used to convert certain types of waste to ash. This may be important where precious metal is to be recovered, but it can also be a method of reducing the bulk of wastes such as filter papers.

It is recommended that an analytical chemist familiar with combustion techniques be consulted. Many organic compounds will volatilize, as will many metal-organics (which are generally a severe health hazard). The volatility of certain inorganic compounds is surprisingly high.

At the very least it is vital to have good local extract ventilation — for example a hood over the furnace. The best arrangement gives a slow passage of air through the working zone and out to an extract duct. In general it is best to operate the furnace at 100–140 °C for a while in order to dry the material before taking it up to the combustion temperature of 400–850 °C. This reduces spattering.

For materials which volatilize (i.e. vaporize or give off fumes by thermal degradation) a quartz tube contained in a series of tubular furnaces may be used. One section of the tube is packed with ceramic and heated to a suitable combustion temperature. The waste is placed in a colder portion of the tube. The waste is then heated up, and a slow flow of air passed through so that fumes from the waste pass through the combustion zone.

7.5.3 Atomic Absorption Spectrophotometers

The flame of an atomic absorption spectrophotometer is an expensive but occasionally suitable method for disposal of small amounts of liquid waste (or tiny amounts of soluble solid). It can only be applied where there is already an adequate extract system for the flame fumes (see Chapter 6). The method is suitable for both aqueous and organic liquids, but dilution may be sometimes be advisable to get a better flame characteristic or to limit corrosion of the burner.

The sample is drawn up into an oxidizing flame (air in excess over stoichiometric) by means of the standard nebulizer tube. Where very harmful materials are involved, it is advisable to set the flame conditions using a similar but less harmful liquid (e.g. the pure solvent without the solute).

As a typical nebulizer only passes 40% or less of the liquid into the flame, it is necessary to collect the drainage from the nebulizer chamber into a suitable vessel, and offer this into the nebulizer tube. Repeated passages of pure solvent and combustion of the drainage will be required to clean out the nebulizer chamber.

Precautions should be taken so that the operator is not exposed to fumes from the waste, particularly the drainage, which is often warm.

7.5.4 Closed Combustion Devices

Some laboratories have access to oxygen combustion bombs or other closed and efficient devices used for analytical measurements. It is perfectly possible to use these for destruction of very small amounts of harmful material, providing some thought is given to the products of combustion, which must be safely flushed out of the device.

7.5.5 Individual Designs of Combusters

A number of devices were described in the literature in the past for destruction of particular kinds of waste. Although certain of these home-made devices may have worked well, there is little evidence of measurement to show that their emissions were acceptable by today's standards. In addition, some aspects of the design and operating procedures could give cause for concern.

There is no design which these authors can recommend. Laboratories intending to construct their own devices should be warned that combustion design is a specialist art, which should not be undertaken lightly, especially under recent safety and pollution legislation.

7.5.6 Wet Combustion

The term 'wet combustion' is used by chemists for a strong chemical oxidation (usually aided by heat) which is almost totally effective in converting organic compounds to water, carbon dioxide, nitrogen and sulphur dioxide.

CAUTION: the chemicals used are extremely dangerous. This sort of procedure should not be attempted by anyone unless he or she is fully skilled in similar operations with chemicals.

References to wet combustion are given in section 9.3.7.1. Some comments on chemical oxidation are given in section 9.3.7.4. Wet combustion techniques for the destruction of carcinogens are referenced in section 12.1.2.

There is an industrial technique in which a flame is submerged in a vessel of liquid. This has some application for incinerating chemicals present as a dilute solution in water (from a chemical process, for example). It is unlikely that it could be adapted for ordinary laboratory use.

BIBLIOGRAPHY

American National Standards Institute (1977): ANSI/NFPA 82-1977, *Incinerators, Waste and Linen Handling Systems.*

British Standards Institution (1964): BS3813 Part 1, *Specification for Incinerators for Waste from Trade and Residential Premises.*

British Standards Institution (1973): BS3107, *Specification for Small Incinerators.*

British Standards Institution (1973): BS3316, *Specification for Large Incinerators for the Destruction of Hospital Waste.*

Corey, R.C. (ed.) *Principles and Practices of Incineration*, Wiley–Interscience.

Edwards, B.H., Paullin, J.N. and Coghlan-Jordan, K. (eds.) (1983) *Emerging Technologies for the Control of Hazardous Wastes*, Noyes Data Corp.

Gorsuch, T.T., (1970) *Destruction of Organic Matter: Intl Series Monographs Analyt. Chem. Vol. 39*, Pergamon.

Rosenhaff, M.E. (1974) *An Evaluation of Requirements for Pathological Incineration Facilities*, Lab. Animal Sci., **24**, 905–909.

Sittig, M. (1979) *Incineration of Industrial Hazardous Wastes and Sludges*, Noyes Data Corp.

US Federal Register 43 FR 59008-59009, Dec. 18, 1978 250.45-1, *Incineration.*

CHAPTER 8

Outside Contractors

8.1 INTRODUCTION

Most laboratories make use of an outside contractor for some or most of their waste. Clearly, specialists are required for the removal of hazardous or difficult materials, and it is important that the contractor is both competent and reliable. The laboratory management should take positive steps to be certain that the service is adequate and fulfils the legal requirements (see Chapter 2). Care must be taken in selection of a contractor, and there should be occasional checks to ensure that unwise changes have not been made in the service.

A contractor may legitimately change the location or the method of disposal (usually for economic reasons) but the laboratory should be aware of the change and should positively agree to it. For example, if waste which is normally incinerated is instead taken to landfill, the laboratory should ascertain that this procedure is environmentally acceptable. As incineration is usually the more expensive option, the laboratory should insist that any waste which truly requires incineration receives this treatment, since the price is based on the more costly procedure.

Changes of management and ownership are not uncommon in the waste disposal industry. It is advisable for the laboratory to note any alteration in address or letterhead details which may indicate such a change. Sometimes a better service results: sometimes it does not. Often a new owner will wish to alter the waste disposal point (e.g. to another facility owned by the same group). It is the responsibility of the laboratory to confirm that the new disposal method is appropriate to the waste. (See Chapter 2, especially sections 2.3.2 and 2.5, for legal obligations.)

8.2 SPECIAL SERVICES

Outside contractors can offer many services apart from simple removal of waste. A select list is given in Table 8.1. Large organizations may be able to offer these services in addition to waste collection, but there are likewise many smaller reputable agencies which give only one specialist service. Repair and refurbishment of broken equipment can be a valuable alternative to waste disposal which can sometimes be carried out by very small firms of only one or two people but with certain special skills and equipment.

Table 8.1. Some specialist services offered by outside contractors

Air filter servicing
Asbestos removal
Drain unblocking
Fume cupboard testing
Fumigation
Mercury cleaning
Perchloric acid decontamination
Radioactive decontamination
Repair of broken glassware
Repair of electric and electronic apparatus
Refurbishing valves and pumps
Second-hand equipment purchase
Silver recovery
Solvent cleaning

8.3 SELECTION

Personal recommendation is the best means of finding a good contractor, so it is worth asking colleagues in other laboratories for their experiences. Control agencies do not normally recommend as such, but are often able to give a list of local companies licensed for the particular work (where this is applicable) or known to have the necessary facilities.

There are relevant trade associations who can provide lists of members. However, this is not an absolute guarantee of reliability, and good firms may exist outside the trade association. Occasionally suppliers are able to give the name of a contractor experienced in handling waste from their products. Failing these sources, advertisements in trade journals and entries in telephone directories may be consulted.

Note that a waste removal operation is in three parts, and it is possible for the contractor to sub-contract one or all of them. Firstly the waste has to be collected (which should be by people properly equipped and trained for the particular waste). Secondly the waste has to be transported (which must comply with transport regulations, and in the USA must be licensed for hazardous waste). Thirdly the waste has to be deposited or destroyed (on suitably equipped premises, which have to be licensed if the waste is hazardous). (See Chapter 2.)

When negotiating with a disposal contractor, the laboratory should find out which portions of the disposal process will actually be performed by the contractor, and which will be (or sometimes may be) sub-contracted. If sub-contractors are to be used, then there should be some evidence of their competence in dealing with the particular waste.

Quite often, the contractor is essentially a transport operator who deposits waste to sites and facilities he does not himself own. There is no fundamental harm in this: it can even give some element of choice, but it does mean that the contractor cannot speak on behalf of the disposal site. Where hazardous waste is involved, it may be necessary to make a separate contact with the company owning the disposal points.

The most crucial aspect is for the contractor to convince the laboratory that he personally has the necessary expertise and can get access to the necessary facilities to safely and legally dispose of the waste. Where large amounts or regular arisings of especially hazardous waste are involved, it is a very good idea for a representative of the laboratory to inspect the disposal operation to be sure that it is adequate for the purpose.

Clearly, it is good practice to get quotes for service from more than one contractor, but price alone should not be the deciding factor. For example, some companies will offer other services such as packaging, analysis and technical advice on waste disposal prior to removal. A contractor selected for routine waste may be chosen for his ability to offer an occasional exceptional service, such as the disposal of a small quantity of extremely hazardous substance once a year, or an out-of-hours emergency service.

Finally, the selection of a contractor will depend on personal judgment as to reliability, which can best be formed by observing his work. Are the operatives properly dressed for the work, and do they seem to know what they are doing? Can the representative discuss legal and technical requirements? Do the prices seem fair (too cheap may be suspicious)? Do you think they will stay in business?

8.4 THE CONTRACT

Most disposal contractors have a set of 'standard conditions and terms' which are produced at the conclusion of negotiations. Instead, these should be requested at the outset and time taken to read them and get advice. By and large these are usually reasonable, but there is no need to accept outrageous waivers, such as disclaimers for any and all damage caused by the contractor's vehicles and employees, nor excessive financial penalty clauses should you find the service unsatisfactory and wish to terminate. If these points are raised at the beginning, then the laboratory is in a much stronger position to have the conditions altered (or go elsewhere).

Conversely, it should be appreciated that the contractor is in business and himself requires some protection. As the rewards from laboratory waste are usually small, a laboratory manager who tries to totally rewrite normal contracts may not find a firm willing to deal with him or her at all!

A price advantage is usually offered for long-term contracts and for regular arrangements. However, it is unwise to commit oneself for more than a year initially. Only when a regular and satisfactory routine has been established should longer terms be considered.

In law, contractual obligations are not limited to the conditions given in the formal contract. It is possible for the laboratory to make special provisions without needing to alter the main document. The best way of doing this is in a letter of acceptance, by means of a phrase such as 'the contract is accepted on the understanding that....'. For example, it may be wise to specifically mention that wastes in certain coloured bag must always go direct to an incinerator without any intermediate handling, or to say that the disposal site or process will not be changed without the prior agreement of the laboratory. Unless a letter to the contrary is received, the contractor will normally be taken to have accepted these special conditions.

The letter of acceptance is also a good place to record any verbal promises which may have been made by the company's representatives — such as the provision of an out-of-hours emergency service, or an undertaking to vary the time or amount of collection without penalty.

Of course, the laboratory has obligations to the contractor. These are principally to pay for services rendered and also to inform the contractor as necessary for a safe and effective service. A common failing is not to tell the contractor when services are not required (e.g. during a holiday). If this is not done, the contractor may legitimately make a charge for a wasted journey. Much more serious is a failure to inform him of special hazards, which may both break the contract and render the laboratory liable to be sued for damages.

8.5 LOCAL GOVERNMENT SERVICES

In most areas the local government authority operates a refuse disposal service which will accept a limited range of trade waste. Some laboratories find that they can arrange for nearly all their waste to meet the requirements for the local refuse collection service. The ability of local authorities to cater for specialized waste varies enormously. Some even have units which specially deal with laboratory waste. Many offer a service to veterinary practices which can be useful to laboratories with biological wastes. Sometimes the authority will permit certain categories of hazardous waste to be deposited (with prior agreement) at its landfill sites or incinerators, but does not itself arrange collection. This can be a useful facility for very small amounts of waste which can perhaps be brought in by laboratory staff.

In accepting waste from a laboratory, the local authority is effectively a commercial contractor rather than a service to citizens, and the mutual legal obligations are on this basis. In general, local authority arrangements for collection and charges are less negotiable than they are with a purely commercial enterprise, since they are fixed by committees representing local citizens rather than by a company official.

8.6 EMERGENCY SERVICES

In a situation of extreme emergency (for legal or safety reasons) it may be necessary to have an exceptional service for which an exceptional payment will be required. This may range from a fairly routine collection but at an unusual time (e.g. a Sunday or other holiday), perhaps because an order has been served upon the laboratory requiring the removal of accumulated waste, to a life-or-death situation such as decontaminating after an explosion.

Obviously, such circumstances are a poor time to negotiate a good price or to compare tenders. However, it is important to get a firm statement of the method of pricing as soon as possible, so that the expenses can be controlled to some extent. For example, there may be an additional rate payable after a certain number of hours. Keeping specialist personnel may not be justified after the crisis has passed and the situation can be dealt with by laboratory staff. Sometimes it may be cheaper for the laboratory to borrow or hire necesary equipment which otherwise might have to be transported in some distance by the contractor at a consequently higher cost.

When a contractor is retained on a regular basis, his prices for urgent or emergency services should be requested and put on file so that they can be referred to when necessary.

The Fire Service have special expertise in dealing with emergency situations involving hazardous substances, and will normally be called in when there is a serious risk to life or property. However, it should be noted that their services are not necessarily free. A charge is particularly likely to be levied if they are called in to deal with something which could in fact have been handled by a commercial contractor.

8.7 USA PRACTICE

An example of a draft contract (with comments) under US law is given by: Fischer, K. E. (1985) Contracts to Dispose of Laboratory Waste, *J. Chem. Educ.*, **62**, no. 4, A118–A122.

CHAPTER 9

Chemicals

9.1 INTRODUCTION

Unwanted chemical substances can be difficult, expensive and time-consuming to dispose of safely and legally. For this reason it is common (bad) practice to defer the problem by simply storing them. This results in disposal at a later date becoming more expensive and quite possibly more dangerous. Many materials can become dangerously unstable with age (see Appendix C-6), and information about mixtures may become lost.

In a perfectly managed laboratory, the time to dispose of waste is immediately the material becomes unwanted. The time to decide on a waste disposal procedure is *before* the chemicals are bought, when the experimental work is being planned. This particularly applies for pilot-scale work, or where there are practical restrictions (e.g. a mobile laboratory). It may sometimes be worth considering alternative chemicals or even alternative procedures. On a chemical plant, the cost of disposal of byproducts is considered an added cost to the raw materials, a procedure which could well be taken up by many laboratories.

Of course, the perfect laboratory does not exist, but in a well-run institution, staff will be encouraged to deal responsibly with chemicals, and there will be systems of labelling, housekeeping and regular audits to minimize the problems of human oversights.

It is recommended that there should be a clear policy against unnecessary storage of chemical substances. The buying of chemicals should be monitored and compared with stores' stock. (It often happens that people buy materials which are already available. If they are unwilling to use up the stock because of its age or condition, then there is no point in retaining stock which will never be used.) However, the administration must be made aware that there can be small but important differences between grades or manufacturers of chemicals, and this judgement should be left to the responsible scientist.

It is now general good practice that all chemicals (and kits etc. containing chemicals) should be stamped with the date upon delivery. Manufacturers are now also beginning to indicate the likely shelf life, which may be several years or a few months. (For example, ethers are now given a shelf life of 6 months unopened, one month after opening, because of the spontaneous formation of peroxides. Some biochemical assay kits lose their guaranteed calibration in as little as two months.)

Bench or other reagents not in the manufacturer's original container can be controlled by regular annual changes. An alternative is to affix a coloured self-adhesive dot when the bottle is filled, using a new colour for each year or academic term. It should be a house rule that *all* special mixtures (including intermediates, by-products, washings, etc.) which are stored overnight or longer should have the initials of the person concerned. This cannot be considered onerous, but is extremely helpful in deciding what to discard, and methods of disposal.

9.2 GENERAL PRINCIPLES

The following points are suggested for safe disposal.

9.2.1 Knowledge

The person organizing the disposal should be generally knowledgeable about chemicals, and be as well informed as possible about the materials involved. Time spent in talking to people in order to accurately identify a particular substance (i.e. finding out who used it and for what) can be the most important part of the disposal process. The important properties of the substances (toxicity, flammability, reactions, etc.) should be got from as reliable a source as possible. Beware of people with limited technical knowledge who do not actually know, but are willing to make an inaccurate guess.

The person carrying out the disposal must be certain as to the procedures to be adopted, in as much detail as necessary. No important matter (e.g. to be carried out in a fume cupboard, in the presence of alkali, in small amounts etc.) should be assumed to be known.

9.2.2 Safety

Safety should be the overriding consideration. If the necessary skills and facilities are really not available, then it is better to pay a specialist than to risk an accident. Equally, a senior person in the organization may have to spend time overseeing the waste disposal process, if he or she has some relevant specialist knowledge.

Removal of hazardous material, or its destruction on site, will almost certainly require some disruption to work (if only to ensure a clear gangway) which should be organized in advance.

Procedures should be so arranged to expose the minimum number of people to the minimum amount of risk. This can be achieved to a large extent by the choice of time and place of disposal (but of course should not leave anyone in isolation with hazardous material). All procedures should be limited to the competence of the person concerned.

9.2.3 The Law

(See also Chapter 2.) If the substance is the subject of any special legislation (e.g. a dutiable substance, a listed poison, a controlled drug, a radioactive material, etc.) then the legal formalities must be followed.

Even if not the subject of specific laws, chemicals from laboratories are generally considered hazardous, and the law creates limitations on the way in which they are transported and where they are deposited. Particularly in the USA, it can be very difficult to dispose of unidentified chemicals legally.

The very act of waste disposal may technically change the status of the laboratory. Disposal on site, or even the storage of substantial quantities may require licensing. Usage of existing equipment (e.g. a paper incinerator used to burn chemicals) may be an offence if not specifically permitted for this purpose.

See sections 2.5 and 2.6, and Appendices C-1 and C-2 for discussion and lists of UK 'Special Waste' and USA 'Hazardous Waste'. See Appendices C-3.1 and C-3.2 for lists of UK and USA 'Controlled Substances' (carcinogens). Some of the formalities for disposal to land or to drains are mentioned in the appropriate sections which follow (sections 9.3.2 and 9.3.3).

9.2.4 The Condition of the Material

It is vitally important to realize that chemicals of the same general description may require totally different handling. In extreme cases, a chemical originally supplied as a stable substance may deteriorate so as to become shock sensitive. Deaths have occurred through opening or moving such materials. The following check list is suggested before the disposal process is attempted.

(1) Is the material accurately identified?
(2) Is the description complete? (For example, a bottle of a solvent may contain some sodium wire to keep it dry.)
(3) Is the material known to deteriorate so that it may be dangerous to move or open? (See Appendices C-4, C-5, C-6.)

(4) Is the container secure?
(5) Is there any visible evidence of deterioration to a hazardous condition,
     e.g. visible crystals in solvents (may be explosive peroxides), contami-
     nation around the closure (may be friction sensitive), drying out or
     absorption of moisture, encrustations, colour change?
(6) From the age and observation, is the material likely to be in a usable or
     recoverable state?
(7) What are the principal hazards of transport/disposal?
(8) What are the principal hazards of keeping it?
(9) What are the practical options for disposal?

9.2.5 Options for Disposal

The following represents a rough order of preference so far as a typical
laboratory is concerned. Some are applicable to a wide range of chemicals,
some to only a few. Any decision to reprocess or chemically convert should
be taken with due regard to the limitations of the laboratory concerned and
the facilities available.

(1) Use for the intended purpose. (Only if the substance is not likely to
     deteriorate or pose a hazard in storage. See Appendix C-6.)
(2) Find another use. (For example, use a high quality chemical where a
     less pure grade would suffice, in order to finish off a part-used bottle.
     Some wastes can be used to neutralize other wastes — see section
     9.3.7. For secondary uses for solvents, see section 14.6.2.)
(3) Sell the material. (Mainly applies to expensive metals — see section
     14.4.)
(4) Give the material away. (Another laboratory in the same organization
     or in the same business may be glad of a small amount for trial
     purposes. Schools can often make use of standard chemicals, providing
     they are within the syllabus and not contaminated with harmful agents.
     Some waste disposal authorities and some education authorities make
     a specific effort to obtain unwanted chemicals for school experiments.
     Outdated pharmaceuticals can often be used by university pharmacy
     departments for research or training.)
(6) Purify for re-use. (See Chapter 14.)
(7) Dispose into normal refuse (Very few chemicals. See section 9.3.1 and
     Appendix C-7.)
(8) Dispose into drains. (Only if suitable, or can be made suitable. See
     section 9.3.3., Tables 9.1, 9.2, 9.3.)
(9) Dispose via a contractor. (See Chapter 8.)
(10) Evaporate or vent. (Small quantities of volatile liquids or gases. See
     sections 6.5 and 9.3.4.)
(11) Burn. (Only if the procedure is both safe and effective. See Chapter 7
     and section 9.3.5.)
(12) Chemically convert. (And deal with residues. See sections 9.3.6 and

9.3.7 for general principles and sections 9.2.6 and 9.3.8 for some specific procedures.)

(13) Dispose to a licensed landfill site. (Usually via a contractor. See sections 2.3.2, 2.5 and 2.6 for comments on the law. See Chapter 8 for comments on contractors. See section 9.3.2 for general practice. See Appendices C-1, C-2 and C-7 for relevant lists of chemicals.)

9.2.6 Recipes for Disposal

Specific methods for disposal are described in a few books, notably Gaston's *The Care, Handling and Disposal of Dangerous Chemicals*, the International Technical Information Institute's *Toxic and Hazardous Industrial Chemicals Safety*, the Manufacturing Chemists Association's *Guide for Safety in the Chemical Laboratory*, the National Research Council's *Prudent Practices for Disposal of Chemicals from Laboratories*, and Strauss's *Handbook for Chemical Technicians*. In addition, some manufacturers (notably Aldrich and MCB Manufacturing Chemists Inc.) give specific disposal advice in their catalogues (derived from the Manufacturing Chemists Association guide). Wall charts by BDH Chemicals, Cambrian Chemicals, and J. T. Baker Chemical Co. give methods of spillage disposal. See section 12.1 for a list of publications giving specific methods for chemical carcinogens.

It must be recognized that the recipes given in these publications usually apply to the unadulterated chemical as originally manufactured. The method may need to be modified in the case of mixtures or old or dirty chemicals. Moreover, many methods advised a few years ago (also in chemistry textbooks and in articles or letters in journals) would now be illegal.

Usually, chemical procedures only convert a waste into a less hazardous form which itself requires disposal. The procedures must be treated as suggestions rather than as guaranteed methods, and may require changes. However, it is often inadvisable to change the procedure by scaling up, as this may produce too fierce a chemical reaction. If the recipe is for 100 g, it may be possible to treat 200 g in one batch, but it is likely to be safer to repeat the process several times rather than trying to treat 1000 g at once. Where a particularly dangerous material is involved, it is advisable to carry out a test run with a very small amount.

Rather than repeat a list of recipes for specific subjects, this book gives some general techniques and their problems, which may be adapted (or used in conjunction with recipes from the references or other sources) for waste including mixtures. Section 9.3.8 gives more specific information for certain substances which seem to cause concern or difficulty in a number of laboratories. Chapter 14 gives techniques for recycling chemicals and solvents. Chapter 15 includes information on spillage disposal.

Note that in the USA, the NRC's *Prudent Practices for Disposal of Chemicals from Laboratories* may be considered the equivalent of a code of practice, but does not of course cover all chemicals or mixtures.

9.2.7 Packaging

The enforcement agencies regard empty chemical containers as chemical waste, which must be disposed of according to the relevant regulations. Containers which have been properly decontaminated, however, may be treated as ordinary waste. The method of decontamination must be appropriate to the chemical, but an accepted standard is 3 good rinses with an effective solvent. (Water is an effective solvent for very many substances.) To prevent misunderstandings, it is advisable to remove the labels, or clearly mark the containers 'Empty'.

Many containers can be returned to the supplier (see section 3.11.3.)

Small amounts of paper or plastic items can often be incinerated along with other waste, but the agreement of the enforcement agency must be sought for controlled substances (e.g. controlled carcinogens, radioactives). In addition, care must be taken to exclude substances which are severely toxic, and significant amounts of materials which are explosive, flammable or oxidizing agents. Most incinerators do not tolerate glass or PVC except in very small amounts. (See Chapter 7.)

If the above options are not available, then the packaging must be put in another container for safe transport to a suitable disposal point.

9.3 METHODS

9.3.1 Disposal to Ordinary Refuse

The institution will normally have arrangements with the local authority or a private contractor for the removal of office, washroom, kitchen waste, etc. as ordinary trade refuse. Care must be taken that hazardous chemicals are not inadvertently included, because of the risk of fire or injury. In particular, chemical-soaked paper towels, used adsorbents, active metals (such as finely divided catalysts) and oxidizing agents (including neutral salts such as sodium nitrate) have caused problems in the past.

It is obvious that certain unwanted materials from a laboratory are no more hazardous than ordinary refuse. A selection is given in Appendix C-7. To this list can be added many minerals and edible biochemicals (such as sugars and dried foods or food components) which are free of heavy metals and do not have a physiological effect on human beings. (For example, caffeine — a stimulant — and castor bean meal — an allergen — are not acceptable.)

It is best if the materials are contained in plastic bags or plastic jars so that there is little chance of material being released in transport. This is to prevent any possible difficulty of identification, and fears that some dangerous chemical may be involved.

Many laboratories routinely deal in essentially harmless materials such as many food items and some inert minerals. It is as well to inform the authority or contractor that there will be regular inclusions of (for example) dried banana, cocoa powder, crushed concrete, plastic granules or whatever.

With the agreement of the removal operator, it is also possible to dispose of slightly more active material in this fashion. The usual arrangement is that not more than one jar (500 g or 1 kg) may be included in a standard sack of rubbish. This is normally restricted to non-corrosive non-oxidizing salts such as sodium sulphate, and solid organic compounds of low toxicity such as citric acid. Obviously, no poison, no liquid, no explosive or fire hazard substance can be contemplated.

The principal advantage of this procedure is that it permits removal of limited amounts of material in their original containers without recourse to more complex formal procedures which are not justified by the nature of the material. However, it is essential that the waste collecting agency agrees, and that it is not extended to substances which could present any hazard to the collector — even though laboratory staff consider it easy to handle. As a rule of thumb, nothing should be included with any cautionary sign or warning by the manufacturer, or which is listed in any of the texts on laboratory safety.

9.3.2 Disposal to Landfill

Ordinary burial is no longer considered satisfactory, though it is suggested in some publications which are still available. In the USA and Europe, sites for disposal of chemicals have to be licensed, and operated in a specific fashion. The site licence is quite specific as to the type of waste permitted, and it is *not* for the producer to assert that the waste is acceptable. For this reason, specialist contractors may use more than one site, plus perhaps an incinerator or other facility. In this case, the contractor will normally specify how the waste should be presented, and may himself repackage it to meet the regulations.

The manner of disposal to landfill in the UK is quite different from the practice in the USA, and will be dealt with separately. This fact should be borne in mind when referring to texts advising disposal methods which may advocate a procedure which is inappropriate to the other country.

9.3.2.1 *United Kingdom Landfill Practice*

The commonest method of disposal of small quantities of hazardous material is by mixing them into domestic refuse. This policy is known as 'dilute and disperse'. A limited number of sites for industrial hazardous waste practice a containment technique in which containers of laboratory chemicals may be sealed within concrete, clay or other impervious material, which is known as 'encapsulation'.

For land disposal, it is completely unacceptable to have any material such as sodium metal which could start a fire, or anything which is explosive or aids fire (a concentrated oxidizing agent such as sodium nitrate). However, combustible and (within limits) flammable substances are dealt with. Heavy metals and other poisons should be in an insoluble form (by chemical pre-treatment if necessary). The licensing authority normally specifies that the container should permit the material to come into contact with the site environment, which means that some containers must be pierced. However, the transport arrangements must be secure enough to prevent human contact. Local methods vary, but may require items to be packaged in sacks with other refuse or in an inert absorbent in a lidded drum which is opened at the site, or in cardboard boxes or other containers which may be expected to fail in a short period.

The procedure at the site is either to spread the waste at the 'working face', i.e. the line at which refuse is currently being tipped, so that it is promptly covered, or to excavate a trench in existing refuse and pour or tip the material in, then cover over. This latter method is particularly used for liquids, for radioactives, and substances posing a greater-than-ordinary toxic hazard such as cadmium compounds.

It is important that the correct notification procedure under the Control of Pollution Act 1974 is followed. Details can be obtained from the local waste disposal authority, who will also advise on transport and labelling requirements. (See also Chapter 2 and Appendix C-1.)

Encapsulation normally involves ordinary closed laboratory packages surrounded by an absorbent mineral in a steel drum. The drum may be filled with a sealing agent (such as a polymer or slow-setting concrete) or may be placed in a larger drum, surrounded by a sealing substance. This drum is then buried at some depth in an impervious substance (clay or something like concrete) in a so-called 'sealed site' which means that chemicals are unlikely to escape into the general groundwater.

9.3.2.2 United States Landfill Practice

The intention of US practice is to place compatible materials together and totally secure them from the environment at a relatively small number of locations — a policy of 'segregate, concentrate and contain'. One method for industrial wastes is via borehole to a great depth into a suitable geological formation. However, this 'deep well injection' is unlikely to be used for the relatively small amounts of materials from laboratories.

The usual method involves burial in drums, surrounded by lime, fly-ash or other buffer material in a secure site, i.e. one which has little or no interaction with groundwater, and covering with an impervious clay layer.

A peculiarity of the system is that chemical liquids are legally 'solid waste', and strong oxidizing agents are counted as 'ignitables'. The container to be buried may be referred to as a 'lab pack' which commonly means a 55 gallon drum containing many actual laboratory packages.

It is completely unacceptable to include any material such as sodium metal which could start a fire, or anything which is explosive or aids fire (a concentrated oxidizing agent such as sodium nitrate). Flammable solvents are accepted provided the flashpoint (of the mixture supplied) is greater than 60 °F, providing they are not peroxidizable (e.g. ethers), and providing they are not liable to exothermic polymerization (e.g. styrene). Water-reactive chemicals (e.g. calcium carbide) are prohibited.

In the past, scrap containers were often used for final transport of waste. Nowadays, reconditioned drums may be used, but the container not only has to be secure, it has also to conform to specific Department of Transport (DOT) requirements. This can even apply to materials which are not regulated by the Environmental Protection Agency. The DOT approves other containers such as boxes, but these may or may not be convenient for the individual site. Some companies will make up lab pack drums from smaller amounts collected from different sources.

Each drum can contain many bottles, jars or other containers. Each must be secure and individually labelled. The individual containers must only have substances in the same EPA hazard class, and which are non-reactive if mixed. The containers are separated by an approved absorbent (typically vermiculite) sufficient to absorb all the contents should any or all of the containers fail.

The drum must be labelled correctly, and accompanied by a manifest listing the contents. Spot checks may be made by the site operator or regulatory authorities, and there may be financial or other penalties if the tested contents do not agree with the manifest description.

A recent commercial innovation for small or occasional producers is called 'Tox Box' (trade mark of Lab Safety Supply, Janesville, WI) which is a 4 litre glass wide-mouthed bottle filled with an absorbent, and protected by an approved packaging. Liquids are soaked into the absorbent, solids put in place of it. The purchaser simultaneously buys the cost of disposal and the necessary paperwork.

Advice on the legal requirements and local licensed contractors is available from the local office of the Environmental Protection Agency, or the State Waste Disposal Agency, where there is one. Local regulations may vary. (See Chapter 2 and Appendix C-2.)

9.3.2.3 Identification and Mixing of Wastes for Landfill

Most chemical wastes are not pure substances, but mixtures of a somewhat indeterminate composition, yet the regulations seem to demand descriptions of great precision. Furthermore, the landfill facility may misunderstand (through an understandably cautious attitude) laboratory descriptions: for example supposing 'arsenic-free' to imply the presence of arsenic, or 'neutralized acid' to belong in the acid category. On the other hand, laboratories have in the past (before current legislation) dumped materials with descriptions as vague as 'strong acid' or even 'laboratory waste chemicals'.

It is suggested that a list is offered for inspection by the contractor or facility, before packing up for disposal and any queries dealt with at this stage. Where materials of a very similar nature arise (or can be made to arise by chemical treatment, e.g. precipitation of insoluble metal compounds) then they may be put together on the manifest or even in the same jars to reduce the volume and simplify the paperwork. For example:'Mixed chlorinated solvents — carbon tetrachloride, chloroform, trichloroethylene: 6 × 500 ml bottles' or 'Insoluble heavy metal sulphides — copper, lead, mercury: 2 × 500 kg'.

The purpose of the labels is to inform, and this means giving the necessary information in as direct and unambiguous fashion as possible. A familiar description is a help, and in the USA, a DOT or EPA description should be selected where applicable. (See Appendix C-2 for the EPA list.)

9.3.3 Disposal to Drains

9.3.3.1 General Principles

(See also Chapter 5.) In general, the effluent from a laboratory should be made as harmless as possible as soon as possible. In the past it was considered that laboratory drains would accept untreated strong chemicals, whose effect would be mitigated by dilution in large tanks, often containing marble chips to reduce the acidity. Even if such 'neutralizing' tanks are fitted, this is nowadays completely bad practice. The only effect of the drains which can be expected is one of moderate dilution.

It is first of all necessary to know where the laboratory drains lead. A small laboratory sewer joining the main drains of a large non-chemical-using institution may expect more dilution than if many laboratories are involved, or the institution itself produces chemical waste. In the latter instance it is probable that there will be an on-site treatment plant, and the laboratory drainage must have regard to its nature. In particular, note that what seem trivial amounts of many chemicals can severely upset a biological treatment plant of typical works size, and even heavy loads of non-toxic organics (e.g. sucrose) can cause disturbance.

All material put down laboratory sinks and sluices should be completely soluble in water, except for non-toxic biological material which is ground fine enough to be freely carried by the water flow. Even small amounts of insoluble chemicals should be avoided as far as possible, because of the chance that they will accumulate. It is possible for flammable solvents to collect in dead high points in the system, with a risk of fire during maintenance. Other materials will generally collect in traps or other low points. It has been known for explosives (used in safe amounts in the laboratory) to accumulate to a dangerous level in a trap. There is also a considerable risk with materials such as solid sulphides which can give off toxic gases in contact with acid which may later be poured down the same sink.

As far as human action allows, chemicals should be compatible with other materials they may meet in the drains. Thus they should not affect the drainage system itself, and should not give hazardous reactions with other known or likely effluents. That is, they should not react dangerously with the process effluent (if the laboratory drains join the works drains) or common chemicals or unusual ones known to be in use in the laboratories. Laboratory drains are usually highly interconnected, and it may be necessary to protect oneself against bad practice by someone in an adjacent room.

Some possible hazardous reactions in drains are given in Table 9.1.

Reactive chemicals should be substantially neutralized. All chemicals should be dilute. Table 9.2 gives the sort of range of chemical properties which may be accepted by a public sewer system. It is suggested that materials put into laboratory sinks or sluices should not exceed these values by a factor of more than 10 for ordinary laboratory quantities. If very tiny amounts are involved, a larger dilution effect may be assumed, but on the other hand there is very little trouble involved in dilution. Where very large quantities are involved (e.g. in a pilot plant) then the actual limits set down by the authority should be adhered to. Note that with acidity or alkalinity, a difference of one pH unit is a factor of 10.

The specification of Biological or Chemical Oxygen Demand is not always understood, but in general is a measure of the amount of biologically degradable material to be processed by the sewage works. This is included as a rule-of-thumb guide in Table 9.3, which is intended to be a working limit for typical materials, put down a sink and assuming a dilution factor of 10 in the nearby drains.

9.3.3.2 Practical Considerations

A sense of proportion (in both the figurative and the literal senses) is the main requirement for safe and sensible use of the drains. Wash-waters containing trace quantities of almost every common chemical are daily discharged without any real danger or bother, because the materials concerned are well dispersed. Any competent scientist, technician or laboratory manager should be able to carry out the simple calculations required to estimate the drains concentration. If the amounts are large (e.g. disposal of whole bottles of old stock) or the disposal is regular (e.g. automatic machines, teaching classes, daily or weekly routine procedures) then a calculation should be carried out.

Note that over a long period, small amounts of some chemicals can have an adverse effect on the drain material. Dilute solvents or oxidizing agents may cause cracking of plastic pipes. Dilute acids, ammonia or some complexing agents will gradually eat away metal sink fittings, joints or pipes of copper, brass or chrome. Sodium azide is a traditional preservative for water and some biochemical solutions, but reacts with lead and copper to give an explosive compound, which has accumulated to give significant explosions in drains of several hospitals. (See section 5.4.5.)

Table 9.3 gives a rough guide to the sort of concentrations of common materials which may in general be safely poured into the sink waste pipe in one litre batches. For larger amounts or continuous disposal, refer to Table 9.2 initially, then actual authority limits.

The drains are a practical route for only very tiny amounts of toxic agents, e.g. the residues from washing or neutralization, but in conjunction with chemical treatment provide a safe means of dealing with much laboratory waste. It may be possible to extend this facility for a specific limited problem by discussion with the drainage authority. For example, permission is sometimes given for standard amounts of unwanted pharmaceuticals to be directly disposed to the foul sewer, rather than risk them getting mislaid elsewhere. Similarly, after a spillage where rinsing to drain is the most practical option, the authority may agree to accept on one occasion the particular chemical load.

Only cold water should be used to dilute chemicals, and any disposal should be followed by sufficient clean water to rinse out the sink and immediate drains. It should be made clear to new staff what materials may be passed to drain, and what are prohibited, as they may have some misapprehension from previous work or teaching experience. This is regrettably a common source of error.

As a basic policy, all laboratory workers should be instructed that chemical solutions must be diluted 10 times before discarding to sink, even after neutralization, and that they should then be poured slowly into a stream of running water across the base of the sink. Acids and alkalis should be approximately neutralized, and mild reactions should be used to remove the majority of oxidizing or reducing power of solutions. It will be helpful to have plastic beakers or bowls available in certain sinks to assist the dilution procedure.

The above will be adequate for many laboratories only using small quantities of chemicals and with a plentiful use of water. In a more strongly chemical orientated laboratory, it will usually be necessary to ensure that more precise controls are in force and ensure personal responsibility for individual disposals. The dilution process should be treated as a chemical procedure, and make use of standard apparatus such as a dropping funnel to feed solution slowly into running water. Individual laboratory workers should leave a note (or preferably have permanent cards with their names and the words 'disposal in progress') where they leave material being run to drain. The laboratory manager should consider any long-term hazard, the chance of different workers putting incompatible materials in proximity, and the likely amount of any problem material to be disposed of in this way. If a significant amount of poison is disposed to sewer, it is at least advisable to consider an alternative procedure.

9.3.3.3 Some Dilution Procedures

For the majority of chemicals, dilution to a safe level is merely a continuation of ordinary laboratory practice. However, the following practical methods have been found useful in certain circumstances.

(1) Solids may be drawn into solution by placing them on a filter paper in a

Table 9.1. Possible hazardous reactions in drains

The following is a selection of chemical reactions with hazardous consequences. The combinations are chosen because cases are known where the reaction has occurred in a drainage system, or where comparable incidents are known and it is likely that mixing could occur in a laboratory drain by ordinary errors or poor practice. It does not include the very many possible hazardous reactions from gross mixing of concentrated chemicals.

The combinations given are those with a good chance of a hazardous reaction. However, not all combinations will give the reported reaction under all circumstances, and this list should not be taken as a criticism of correctly controlled practice which may involve a particular combination. The purpose of the list is to indicate material combinations it is prudent to avoid in the drains, and which may not be obvious to laboratory workers.

Note that traps and catchpots often contain mercury from broken thermometers, metal dross and some solids such as heavy metal precipitates. Sections (b) and (c) are short-term hazards, but the dangers given in section (a) can sometimes take years to build up.

(a) Reactions giving explosive deposits

Acetylide	Brass	
Azide	Cadmium	Metals, oxides
Carbide	Copper	or salts
Fulminate	+ Lead	
Picrate	Mercury	
Styphnate	Silver	
	Zinc	
Ammonia	Bleach	
Ammonium salts	+ Hypochlorite	
	Iodine	
	Periodate	
Ammoniacal silver	Allowed to dry out in an	
solutions	unused sink trap, or as	
Organic peroxides	+ deposits from a leaking joint	
Perchlorates		
Phosphorus		
Picric acid		

(b) Reactions giving highly flammable or explosive gases or vapours

Barium	As:
Calcium	Amalgams

Cold water	+	Lithium	Amides
		Potassium	Carbides
		Sodium	Hydrides
		Strontium	Metals
Hot water	+	Flammable solvents	
Mineral acid	+	Most metals	
		Azides	
		Phosphides	
		Aluminium powder	
Strong alkali	+	Devarda's alloy	
		Zinc amalgam	
		Zinc dust	

(c) Reactions giving highly toxic, irritant or offensive gases and vapours

Acids	Salts of: azide, cyanide, phosphide, selenide, sulphide telluride Salts of: bisulphite (Disulphite), dithionite (hydrosulphite), sulphite, thiosulphate Antimony, arsenic, germanium, selenium, tellurium: metals, alloys or some compounds of these elements in the presence of an acid-soluble metal, e.g. aluminium, iron, zinc
Hydrochloric acid	Above, also Bleaching powder (hypochlorite) Cleansers (some) for sinks and toilets Formalin Manganese dioxide Sterilizing tablets or powders (some kinds)
Alkalis, e.g. NaOH	Ammonium salts Methylamine
Bleach (liquid), i.e. hypochlorite	Ammonium salts Bleaching powder Cleansers (some) for sinks and toilets Sterilizing tablets or powders (some)
Hot water	Volatile poisons, e.g. acrylaldehyde, acrylonitrile, ethanthiol, formaldehyde, methylamine

Table 9.2. Quality of water discharged to sewer

The following are typical of limits which might be laid down by a local drainage authority for materials discharged to its sewers. Actual limits vary.

Factor	Limits
Temperature	0 to 45 °C
pH	6 to 11
Oil, grease, tar	50 g m^{-3}*
Undissolved solids	400 g m^{-3}
Chemical Oxygen Demand	600 g m^{-3}
Total iron	250 g m^{-3}
Total heavy metals (Cd, Cr, Cu, Hg, Ni, Pb, Zn)	30 g m^{-3}
Total dissolved heavy metals (as above)	10 g m^{-3}
Dissolved Cd, Hg, Cr(VI)	1 g m^{-3} each
Sulphate	2000 g m^{-3}
Ammonia	100 g m^{-3}
Chlorine (or hypochlorite)	50 g m^{-3}
Formaldehyde	20 g m^{-3}
Sulphur dioxide (or sulphite)	10 g m^{-3}
Cyanide	10 g m^{-3}
Sulphide	5 g m^{-3}

*Nil discharge demanded by some authorities.

    Buchner funnel and rinsing through. If the Buchner funnel is placed on a Buchner flask and suction applied by a water-jet pump, then further dilution to drain occurs directly.

(2) A Buchner flask can be used as a dilution vessel by attaching a rubber hose to the side-arm and passing tap water in, so that the flask overflows in the sink The solution to be diluted is placed in a separating funnel with the outlet submerged in the Buchner flask.

(3) Another useful dilution vessel is a plastic bowl in the sink, fed by a hose from the tap to just below the water surface. A home-made clip or clamp should be used to keep the hose in place. Material can be continuously fed to the bowl, or periodic additions of solid or liquid can be made.

(4) Many laboratories have peristaltic pumps. These are ideal for the continuous controlled feeding of waste solutions to a sink (via an overflowing beaker or as in (2) or (3) above).

(5) A sink with an overflow stand-pipe can itself be used as a dilution vessel. A rubber hose should be taken from the tap to below the water surface

Table 9.3. Concentrations of chemicals to be poured down a laboratory drain

The following is a guide to the maximum quantity of chemical dissolved in one litre of cold water which might reasonably be poured down a laboratory drain, where no other hazards or regulations or work procedures prevent.

General nature of chemical	Amount per litre
Insoluble, light, inert material (clay, foodstuffs, paper pulp, etc.)	1 g
Soluble harmless organics (sugars, some sodium salts, etc.)	10 g
Neutral soluble inorganics of low toxicity ($NaCl$, Al_2SO_4, $Fe(NO_3)_3$, etc.)	10 g
Soluble salts of heavy metals	0.1 g
Unreacted sulphide, cyanide, chromate	0.05 g
Soluble organic or inorganic cleaning agents and/or disinfectants (except pure phenol)	10 g
Phenol	0.01 g
Soluble phenol or cresol derivatives	0.1 g
Water miscible solvents of low toxicity (acetone, ethanol, etc.)	5 g
Acid and alkaline substances	pH 4 to 12

to avoid splashing. To deal with jars of solid, the jar can be opened under water and flushed out with the hose still under water. If desired, an alkaline environment can be maintained by periodic additions of sodium carbonate (for slightly acid substances or to aid hydrolysis or solubility).

(6) As an alternative to (5) a standard overflow stand-pipe may be modified by cutting a notch in the side, so that water runs through this rather than over the top. An ordinary funnel can then be placed in the top, and the waste poured into this funnel. See Fig. 9.1.

9.3.3.4 Disposal from Machines

Some laboratory machines produce a liquid waste which may be suitable for direct disposal to drain. Permanent connections will normally be made for washing machines and steam autoclaves. However, some analytical devices, such as autoanalysers, liquid chromatographs, atomic absorption spectrophotometers, may also produce a liquid waste stream of low volume. Of course, if the liquid phase is an immiscible solvent, then disposal to drains is inappropriate.

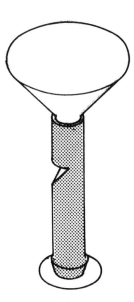

Fig. 9.1. An overflow stand-pipe for a sink drain, modified by having a notch cut in the side. This makes a convenient holder for a funnel to allow suitable liquids to be diluted and flushed away. The notch must be big enough to permit a reasonable flow of water from the sink.

It is recommended that flexible pipes from a machine are taken directly into the drain, rather than dripping into a sink or drain cup. Usually very small diameter tubing is involved, so this is not difficult. The tube should be passed at least 15 cm into the drain pipe in the downstream (i.e. direction of flow) direction via a small, vertical tube fixed to the horizontal portion of a drain. A drain cup (as used for returns from condensers) may be used if convenient. If these arrangements are not possible, the tube may be passed into a sink drain so that the end is in the drain trap. Some laboratories have floor-mounted glass drain receivers with several ports for sink drains. The waste from the machine may conveniently be installed via an unused port. If no more permanent arrangement can be made, the waste line from the machine can be run into a plastic beaker in a sink, which also receives steady flow of water from a tap, thereby diluting and overflowing.

In every case there must be a continuous flow of clean water from upstream of the connection point. A notice can be placed by the machine instructing users to open a certain tap before starting work, and the tap itself should be labelled so that it is not inadvertently turned off. This simple dilution is generally all that is required, on account of the very small amounts (say less than 5 cm^3 per minute) of liquid involved. However, some partial neutralization of acids may be achieved by placing some pieces of limestone or marble in a drain receiver, or some chalk in a sieve through which the upstream tap water supply is fed.

If the machine is sometimes used for liquids such as immiscible organic solvents, then there must be an alternative receiver available, and the work so organized that the waste line is changed over when required.

9.3.4 Evaporation and Venting

For gases and volatile materials (say boiling point below 150 °C) then discharge to atmosphere is a perfectly acceptable method, providing the amounts are small, and the process is controlled so as to produce no fire or toxic hazard. Sometimes it does not completely dispose of all components, but leaves a residue which must be separately treated.

The aim should always be to keep the concentration of any toxic substance below the Occupational Exposure Limit (UK), Threshold Limit Value (USA), Permissible Exposure Limit (USA), or other human exposure limit, if one exists, and this should govern the release rate. In the case of gases, the flow rate should be controlled and measured (at least approximately) by a simple rising float tube flow meter or a soap bubble meter (which can be made from an old burette). See Fig. 9.2. For liquids, a small amount should be exposed and the time for evaporation noted, to give an idea of the rate. Note that the evaporation is essentially proportional to the area exposed: twice the amount in the same evaporating basin will take about twice as long.

For aqueous liquids which are unsuitable for disposal into the drains, evaporation can be a way of concentrating prior to removal or treatment as a solid, with the bonus of removing traces of volatiles such as ammonia or formaldehyde. A plastic washing-up bowl in a fume cupboard can be used for this purpose, with occasional stirring to break up any crust of solid. A considerable quantity of liquid can be reduced to a small amount of solid with very little trouble by this method. Usually a slurry is left. If desired, this may be dried out by warming in a safely vented oven, or by mixing in a compatible drying agent such as plaster of paris. As far as possible, the liquid should be neutralized (acid/base, oxidizing/reducing components) and only compatible liquids should be mixed.

Organic solvents may also be evaporated off in a fume cupboard, but no more than 250 cm^3 should be present in a fume cupboard at one time. This should be reduced if there are particularly dangerous components in the mixture. Some laboratories with a small use of solvents are able to deal with all their solvent waste in this fashion, instead of accumulating it. Care should be taken with any residues which remain: these may be chemically dangerous. With certain solvents (notably ethers), repeated evaporation from the same vessel may cause accumulation of explosive peroxides to a dangerous level. (See sections 9.3.7.5 and 14.6.7, Tables 14.1 and 14.4, Appendix C-6.)

Heating is not generally advised: the aim is to evaporate rather than boil off. Uncontrolled heat can cause a serious fire hazard with flammable solvents. For solvents with a boiling point above 100 °C, warming with hot

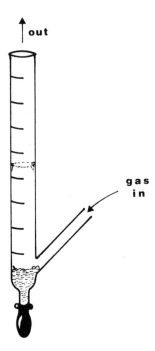

Fig. 9.2. Soap bubble flow meter. This can be purchased or made from an old
burette or glass tubing. It is filled with soap or detergent solution to the level
shown. To operate, the teat is compressed so that the incoming gas blows a bubble
film. The movement of the film between the two graduations is timed to give a flow
rate. A sharp point can be made from wire or even a paperclip to pierce the bubble
film at the top of the tube.

water may be used. For solvents with a boiling point above 200 °C, the
method is generally not practical. For most materials, it is preferable to use
(a) a longer time (b) a better draught (c) a greater surface, rather than heat.

It is possible to evaporate volatile material outside providing certain
precautions are taken. Usually an iron tray, or a tray of sand is used in a
place where few people have access (commonly the roof). It is important
that the area is genuinely secure and account is taken of maintenance staff,
window cleaners, sunbathers, and others. Particular care must be taken to
check that the procedure is not anywhere near an air intake of any form. It is
preferable if the evaporating tray is protected from direct rain by some
simple cover, but otherwise well exposed.

No more than 500 cm^3 should be put in one tray, and there should be no
more than 2 trays in use at once. It may be convenient to have a small metal
bin or cabinet in which to store waste awaiting treatment, but the amounts
should not be permitted to accumulate to more than, say, two week's work.
Fire fighting equipment should be nearby.

The person who places material on an outside tray must be a responsible
one, who has received specific training in the handling of the materials and
in extinguishing fires. Even for the most routine of materials (e.g. acetone),

eye protection and an apron must be worn. For more hazardous substances, an appropriate cartridge respirator may be advisable. Someone else should always know when this chore is being carried out, and if at all possible there should be another person in the near vicinity in case of accident. Non-fuming aqueous solutions may be concentrated initially by the use of a steam bath, but this should be stopped as soon as crystals start to form or the viscosity is noticeably high (e.g. formation of large, slow-moving bubbles). If heating is continued further, there is a danger of superheating and explosive boiling occurring.

Evaporation should not be carried out simultaneously with practical work in the cupboard, except possibly where only water is being evaporated off. Some laboratories routinely evaporate overnight. If this is done, it must be realized that the fan could possibly fail, and the work procedure organized to take this into account.

To vent gases from cylinders (or low-boiling liquids from ampoules warmed in tap water) a pipe should be taken right into the throat of the fume extract duct and secured by a reliable mechanical means (e.g. a clip or wire). Gases should not be vented-off overnight. See section 9.4 for details of gas cylinders. See sections 6.5.3.3 to 6.5.3.6 and Appendix C-8.

Small spillages mopped up in paper towels, sand or mineral absorbent may be evaporated-off in a fume cupboard providing this gives no fire, corrosion or toxic hazard. They may need to be restrained from blowing away: for example, a desiccator cage may be used. When dried out, it is preferable that the towels or solid are put in plastic bags and disposed of carefully. They may still be highly combustible or corrosive due to residues.

Where a tray of sand is used, the sand must be routinely removed and safely disposed of to landfill or incineration. Tarry residues on a tray likewise need periodic removal.

Where more volatile liquids are involved, it can be a useful practice to cool them down in a refrigerator or cold-room (providing the units are internally spark-proofed) or in an outside safe store. This makes the pouring out process safer for the person involved.

With all venting processes, only a small amount of smelly, irritant or corrosive vapours can be released. The excessive release of smells or irritants is one of the most common cause of complaints against laboratories. (See Chapter 6.) In addition, corrosive vapours may damage ducts, fans or machinery even where the amounts are small, but exposure is prolonged and repeated.

See section 6.5 for a general discussion of venting or evaporating different classes of waste.

9.3.5 Burning

Burning is not generally recommended for on-site disposal of waste chemicals, unless there are very special facilities. The advice given in some texts to burn materials in a fume cupboard can pose a serious risk of fire and may deposit corrosive or otherwise hazardous materials on the cupboard and duct interior. Attempts to dispose of chemicals via a boiler or incinerator intended for waste paper can cause serious accidents. Open air burning can give extremely hazardous fumes.

Burning may be justified if the laboratory is part of a site which has chemical destruction facilities (e.g. certain chemical or paint works, munitions manufacture, etc.) which are appropriate to the chemical. Open air burning may sometimes be used as an emergency procedure for materials which are too unstable to handle or transport. See Chapter 7 for a more detailed discussion.

9.3.6 Separation

9.3.6.1 General Comments

Many laboratory techniques exist for separating chemicals. They can be a very effective method of minimizing waste disposal problems. In general, faced with a mixed waste, the responsible person should ask if there is any easy technique available which would change the character of the waste in a useful way. This particularly applies to large volumes of dilute solutions of harmful material. Removal of the harmful material into a more concentrated and/or inert form may literally reduce the size of the problem.

Note that total separation is not always necessary. For example, take a pilot laboratory process producing at pH 4.5 (slightly acid) washings at the rate of 200 litres per hour and containing about 1000 g m^{-3} of heavy metals. To bring this down to a concentration acceptable to the sewers (see Table 9.2) could be achieved simply by diluting with 100 times the volume of water, which is a large amount. However, a precipitation process (say addition of sodium hydroxide to pH 9) which was 98% efficient would bring the metals content down to 20 g m^{-3}, which would only require one volume of water. (The practical efficiency would be lower than the equilibrium solubilities suggest, owing to loss of fine particles through settling tanks or filters, and other factors.)

The following notes are not intended to be exhaustive nor definitive, but to indicate some of the possibilities which are available. Details will of course vary with the individual mixture to be treated, and the laboratory facilities.

9.3.6.2 Solvent Stripping

In general, evaporation using a rotary evaporator (or moving film evaporator) is preferred to distillation. The process is gentler and can more easily be left unattended. It is possible for some solutes to carry over, but this occurs less often. Organic and aqueous solutions can be divided into relatively clean solvent (mixture, possibly) and a concentrate. If desired, either fraction can be distilled afterwards to get purer components. The advantage of making an initial cut is that smaller apparatus is required to distill the (possibly harmful) concentrate and more efficient separation can be achieved.

It is sometimes useful to take up a material (e.g. a viscous liquid) into a solvent for treatment, then strip off the solvent afterwards by this means. See also, section 14.6.

9.3.6.3 Adsorbents

A carefully selected adsorbent can be extraordinarily effective in removing low concentrations of harmful substances from solution. For example, many biologically active chemicals are quantitatively removed from solution by stirring with activated charcoal. The charcoal may then be carefully dried (if necessary) and incinerated. (See sections 7.4 and 7.5.)

Ion-exchange resins (or even ion-exchange paper) may be used to remove metal ions from simple solution in water and some other ionizing solvents. They can be less effective where the metal is complexed, or where its ionization is suppressed (e.g. for elements such as mercury with some covalent character). Ion-exchange paper has the advantage that it is easily ashed. Adjustment of pH may be necessary. Ion-exchange resins are occasionally very successful in adsorbing ionizable organic species, including some pharmaceuticals, particularly if the solvent conditions can be adjusted.

Activated alumina is well-known for the removal of hydroperoxides from ethers. Acid or base activated material can be used to adsorb a variety of other solutes from a range of solvents, as is shown by its use in liquid chromatography. In fact any agent which is useful in liquid chromatography may have an application for removal of solutes. The relevant literature should be consulted.

9.3.6.4 Precipitation

Solid can be precipitated from solution by a variety of methods. These are described in textbooks on inorganic and organic wet analysis. Suggestions may be sought from anyone familiar with gravimetric analysis or the preparation of organic derivatives. However the following points are relevant to precipitations for waste separation.

CAUTION: great care should be taken where the material to be precipitated is a potential explosive, for example peroxides, nitro-organics or azides. Sometimes there may be combinations of substances in the precipitate which create unexpected unstable compounds. Common examples are the azide, styphnate, perchlorate, thiocyanate, and ammonium ions in the presence of iodine, bromine, chlorine or their generating salts (e.g. hypochlorite).

Most commonly, precipitation is achieved by the addition of a reagent which forms an insoluble compound or complex with the important solute. For solid formation to be near 100%, it is necessary to have an excess of the reagent. However, a large excess should be avoided for two reasons. Firstly the reagent may itself be harmful, e.g. the use of sulphide to precipitate heavy metals. Secondly, an excess of reagent may often redissolve some of the precipitate, e.g. the hydroxides or carbonates of many metals will dissolve if the solution is made too alkaline with excess hydroxide or carbonate.

A large particle size is helpful for easy separation by settling or filtration. This is helped by slow addition with stirring. A very fine precipitate will sometimes aggregate if the suspension is kept hot (or even refluxed) with stirring for some time. So-called 'filter aids' are fine particles of paper or mineral to which precipitate can adhere and thus agglomerate. They are often most effective if added before precipitation, as they provide nuclei for solid formation.

Precipitation can also be achieved by altering the solvent condition. For example, many weakly acid organic compounds are soluble in aqueous base, but come out of solution if brought to a slightly acid pH. (Excess acid should be avoided as it may cause hydrolysis.) Organic solutes are often brought out of water solution by the addition of inorganic ions, such as sodium chloride ('salting out'). Inorganic solutes may sometimes be displaced from aqueous solution by mixing in an organic solvent such as acetone or methanol.

In general, a more perfect separation can be obtained from solutions in water than in organic solvents. However, where very difficult or expensive materials are involved, the separation from organic solvent may often be improved by diluting the solution with a larger volume of another more appropriate solvent, or by the use of a very specific derivatizing reagent.

9.3.6.5 Extraction

Liquid-liquid extraction can be tremendously effective in many cases where there is a relatively small amount of a difficult substance in a large volume of liquid. Essentially, the process consists of shaking the liquid in the presence of another immiscible liquid, which absorbs the substance. The two liquids are allowed to settle into two layers, and dealt with separately.

Chemists and many other scientific workers will be familiar with this technique using a separating funnel. Others should take advice on the correct method. In general, several small extractions are better than one large one — for example three extractions with 100 cm^3 liquid are more efficient than one extract with 300 cm^3 liquid.

There are available devices which are more or less continuous and/or automatic for use if the procedure is likely to be repeated often.

A common example is where organic solvents contain dissolved acid substances, which could corrode solvent cans or interfere with recovery processes. These substances can often be removed by extraction with water containing sodium carbonate. If necessary, the organic solvent can be dried with sodium sulphate after the water has been separated off.

It is likewise possible to remove peroxides or other oxidizing agents from solvents by treatment with aqueous solutions of reducing agents, such as ferrous sulphate. In general, if the water contains a chemical which will react with the substance to be removed to give a water-soluble product, then the extraction will be favoured. For example, weak organic acids may be insoluble in water, but can be converted to their soluble sodium salts by sodium carbonate. Hence they will be extracted into water containing sodium carbonate.

Water containing small amounts of organic (and some other substances) may give them up to an immiscible organic solvent. The advantage of this procedure is often that a much smaller volume of waste results (e.g. 1 litre of water being extracted 3 times with 10 cm^3 solvent, giving a final hazardous waste volume of 30 cm^3). The resulting solution may be more suitable for incineration, or other disposal method.

In general, substances come to equilibrium in this procedure according to their solubilities and the amount of the liquid phases. For example, if a certain substance is 10 times as soluble in water as in hexane, and equal volumes are used, then the result will be 91% of the substance in the water and 9% in the hexane. However, if volumes of water and hexane were used in the ratio 1 to 10, then there would be an equal amount of the substance in both phases.

If solubility data is available, the effect of an extraction can be calculated as follows. Let s_A be the solubility of the substance in liquid A. Let s_B be the solubility in liquid B. Let the volumes of the two liquids be V_A and V_B. Let the mass of substance in liquid A after extraction (i.e. at equilibrium) be m^A, and the mass in liquid B be m_B. Then the relative amounts of substance in the two liquid phases are as follows:

$$\frac{m_A}{m_B} = \frac{s_A V_A}{s_B V_B}$$

Note that this is only valid if there is sufficient volume of extracting liquid for the substance to dissolve and remain below the solubility limit s_B.

It should be appreciated that any liquid–liquid extraction will cause the original liquid to be slightly contaminated by the extract liquid to the extent of their mutual solubility. This may occasionally affect subsequent procedure for disposal or recovery.

9.3.7 Chemical Conversion

9.3.7.1 General Principles

In principle all hazardous chemicals can be converted by chemical reaction to less hazardous ones. The only limitation is the skill and knowledge of the user, and the facilities (including time and money) available. However, it is quite possible for the treatment process to be unacceptably hazardous, or even to give rise to compounds with new hazards, for example producing volatile poison by-products.

It is therefore essential that any chemical procedure is vetted by an experienced chemist before being applied to a mixed laboratory waste. The chemist should consider all the probable components of the mixture (including contamination), not just the one which is to be treated. He or she should review the likely products, and any possibility of side reactions, including catalysis or inhibition by small amounts of active compounds. This latter has been the cause of many troubles in waste disposal, both in giving unexpected reactions (such as explosions) and in making some reactions ineffective (e.g. by poisoning the catalyst) so that the procedure is completed without full conversion of the hazardous substance.

It is clear that chemical destruction in practice is more difficult than is often suggested. Recipes given in many publications for the treatment of a pure substance may be used as a starting point, but are not absolutely guaranteed for mixtures. In fact, the disposal process is exactly as demanding as any other chemical operation with the substances concerned, and may require as much attention as a synthesis or analytical method.

The problem can be reduced by good management of wastes. For example, wastes should not be mixed unnecessarily, should be kept free of contamination, properly stored and labelled, and in general treated like chemicals rather than rubbish prior to treatment. Procedures for the separation of certain components can often greatly simplify the problem. (See section 9.3.6.)

The above warnings should not be taken as complete discouragement. For relatively simple mixtures, for regular waste disposal, or for highly hazardous substances, then a chemical reaction can often be a safe and effective method, which will greatly reduce the risk of accidents in disposal outside, and incidentally reduce the paperwork required for outside disposal of hazardous waste. It is therefore well worth spending a reasonable amount of time and effort in developing a procedure.

It is often the case that a very simple reaction is obvious for treatment of one component. For example, solvents containing traces of corrosives or catalysts may have these neutralized before storage, incineration or recovery. This gives large benefits for a small effort and is therefore worth while.

Extended and complex procedures are usually only justified for compounds of high toxicity. As the laboratory should be equipped to handle these substances, it is proper for them to be converted before the waste is sent elsewhere. In such situations, the waste should be analysed before and after treatment to check (at least approximately) the amount of hazardous substance remaining. For a lengthy treatment (such as prolonged boiling under reflux) samples can be withdrawn from time to time to monitor the progress of the reaction. This ensures that sufficient treatment is given, but the procedure is not prolonged unnecessarily.

In the case of highly toxic substances, it may be necessary to use powerful and hazardous reagents to ensure full conversion. However, it would be foolish to have a disposal procedure of high hazard to dispose of a substance with a small hazard. Thus more gentle reagents are preferred whenever possible. The cost and availability within the laboratory should also be considered.

It often happens that one laboratory waste can be used to treat another, at least in part. Obvious examples are acids and bases, oxidizing and reducing agents. The 'spent' regent from one procedure may still retain sufficient activity to constitute a gentle reagent to treat another waste. Good management will be able to schedule waste disposal procedures to take advantage of these arisings. If the wastes come from routine procedures, then the process can be tested and developed, with particular apparatus set aside as for any regular operation.

CAUTION: when mixing wastes, attention should be paid to the possibility of reactions due to minor components.

Waste disposal reactions can often be made more convenient by adaptations of ordinary devices. For example where it is wished to react a solid with a liquid and a gas is evolved, then a Kipps' apparatus (or an equivalent gas generator) may possibly be used. The rate of reaction is controlled by adjusting the rate at which the gas is allowed to escape (normally via a tube into a fume duct). An extraction thimble can be a useful method of permitting solid-liquid contact where the reaction is less vigorous. The references for section 9.3.7.1, given at the end of this chapter, review and describe various methods and types of apparatus which might be adapted.

9.3.7.2 Water

A great many wastes can be treated to advantage with water. Obviously, those chemicals which react powerfully with water can have this particular hazard removed by simply carrying out the reaction under controlled conditions. However, water can be beneficial for other materials.

Hydrogen peroxide (30% or more), perchloric acid (60% or more), and nitric acid (70% or more) are all very powerful oxidizing agents which can cause immediate fires in contact with many combustible materials (including organic solvents). However, the oxidizing power is considerably reduced if these materials are diluted to concentrations below those given in brackets above. Therefore diluting these chemicals down to a safe margin below these concentrations can greatly reduce the danger in storage or handling.

CAUTION: even a dilute solution can give rise to a high fire risk if it is spilled onto an absorbent combustible, such as paper or wood, and allowed to dry out.

Ordinary sulphuric acid (96 to 98%) generates a great deal of heat when in contact with water, which can cause a steam explosion. Controlled addition of water to a final concentration of 80% can greatly reduce this hazard.

CAUTION: the essential factor in dilution of sulphuric acid is control by a skilled person under protected conditions. Liquid water should never be poured directly into sulphuric acid stronger than 80%, or into other very concentrated acids.

Taking sulphuric acid as an example, the following are methods generally applicable to substances which react very powerfully with water, and where it is desired to carry out this reaction in order to overcome this hazard.

(a) The material can be added to a large excess of water in a well-stirred container. For small amounts it is convenient to use a dropping funnel to control the rate of addition, and a thermometer to indicate the temperature. The addition of 100 g of sulphuric acid to 1000 g of water will given an overall temperature rise of about 20 °C, but it could be much greater locally if the rate of addition is too quick.

(b) The material can be added to a lesser amount of water in a stirred and externally cooled container — for example a flask in a stirred ice-water bath. In this case control of addition is absolutely essential, as is a thermometer. Unless the cooling bath is refrigerated, then additions of ice will be necessary. Addition should be stopped whenever the temperature rises to a pre-determined limit (say 50 °C). Good stirring, without vortexing or splashing, is vital.

(c) The material can be added to crushed or flake ice. The heat absorbed by melting ice greatly reduces the heating effect of dilution of the suphuric acid or other material. After the first cautious additions have produced a slurry of ice and liquid, this should be stirred for further additions. External cooling can be used in addition.

(d) Addition of water can be through the vapour phase. This generally takes some time, perhaps several days, so that heat is easily dissipated. As sulphuric acid has virtually no vapour pressure, it is the water vapour which will move. The vessel containing the sulphuric acid is connected by an overhead tube to a source of water vapour — such as the general atmosphere (slowest), a vented vessel of cold water (faster), or a vented vessel of warm water (faster). Care should be taken that the apparatus cannot be knocked or disturbed so that liquid water travels down the vapour tube.

    If the sulphuric acid vessel is graduated, then the progress of absorption can be noted from the increase in volume.

    Air-sensitive material can be treated with water by bubbling nitrogen through water, then through an empty vessel to remove spray, then through the material to be treated. For example, this has been used to decompose organometallic compounds such as Grignard reagents and alkyl lithiums.

Where the substance to be reacted is itself volatile then its vapour can travel to the water. Examples are some reactive alkyl halides and some inorganic liquid halides. This should not be confused with an involatile substance in a volatile solvent. One procedure is simply to connect up two vessels via an overhead tube, with water in one and the substance in the other. A liquid seal should be used to allow air into the system (to prevent a vacuum) without permitting vapour to escape. Alternatively, a gentle stream of nitrogen can be bubbled through the liquid, then through a trap to prevent suck-back, then through at least two vessels containing water to absorb the vapour carried over.

(e) Water can be added as a dilute solution in an inert solvent. For example, to convert SO_3 in fuming sulphuric acid (oleum) to H_2SO_4, water is added as a 2 to 4% solution in H_2SO_4. That is, ordinary 96 to 98% H_2SO_4 is slowly mixed in with cooling and stirring, giving a final mixture which is virtually 100% H_2SO_4. To dilute this, 80 to 90% H_2SO_4 can be used, also with stirring and cooling.

To decompose organic substances (and others in organic solvents) water can be used as a low concentration in an organic solvent (e.g. rectified ethanol, tetrahydrofuran, dichloromethane). Where only small amounts are involved then the water present in the ordinary manufactured grade (i.e. not specially dried) may be sufficient. The solvent must be inert to the mixtures concerned.

9.3.7.3 Hydrolysis

Hydrolysis is the conversion of chemical compounds effectively by the addition of water to the molecule, aided by heat or chemical agents — usually base or acid. That is, more vigorous conditions are required than simply adding water.

Surprisingly, many materials such as acid anhydrides are not immediately converted by cold water. To completely convert acetic anhydride will require several hours refluxing or the addition of sodium hydroxide. In the past many toxic compounds were buried with the expectation that hydrolysis would convert them to less toxic forms over a matter of months or years. This is no longer permitted, so more efficient methods have to be carried out in the laboratory.

A great many (though by no means all) toxic organic compounds containing active halogen atoms (chlorine, bromine, iodine, sometimes fluorine) can be fairly easily hydrolysed with concentrated (10 to 15%) NaOH. As a rule of thumb, if the compound is lachrymatory (i.e. strongly irritating to eyes and breathing) then this procedure is likely to work. The literature should be consulted for the particular compound concerned. The material is stirred with the NaOH solution and left overnight. The next day it should still be alkaline (test with phenolpthalein indicator) or further NaOH is needed. The products may then be disposed of via drains, landfill or incinerator as appropriate. Heating or refluxing may be necessary for more resistant compounds to ensure conversion.

CAUTION: the presence of strong base (e.g. NaOH) can cause explosive polymerization of hydrogen cyanide, acrylonitrile, plastic monomers and many organic compounds containing double bonds, such as vinyl derivatives and dienes.

A solution of KOH in ethanol provides a stronger attack on more resistant compounds such as nitriles (e.g. acrylonitrile), epoxides, alkyl halides (but not fluorides) and dialkyl sulphates.

The procedure is to make up a 25% solution of solid KOH in rectified ethanol or industrial methylated spirit. A 20% molar excess is required. The dissolution typically causes heating up to 60 °C, but if a stock solution is used it should be warmed to this temperature. The material is added slowly with stirring so as to maintain a gentle reflux (heating or cooling may be required). Solids should be dissolved in ethanol or a compatible solvent for addition. The stirring and reflux should be maintained for 2 hours, then the mixture added to an equal volume of water, cooled and neutralized. Solids are removed and the liquid diluted to drain.

Acid hydrolysis is more suitable for some compounds, notably carbox-amides (e.g. formamides, peptide derivatives). These are typically conver-ted by refluxing for 5 hours with concentrated (36%) hydrochloric acid.

9.3.7.4 Oxidation

Oxidation applied to waste disposal is sometimes misunderstood, and can actually increase the hazards. It is vital to realize the exact intention of the oxidation reaction and ensure that it is achieved. A variety of reagents and procedures are available to give different degrees of oxidation: reference should be made to standard chemistry texts. A mild oxidation may be used to convert one compound to another of less difficult nature (less toxicity, different solubility, lower volatility, etc.), but care should be taken that it does not proceed too far to give other products. For example, a nitration may give rise to explosive polynitro compounds, and many aromatic compounds can undergo oxidative coupling to give complex substances of significant acute toxicity and cancer-causing potential. However, with compounds which are readily oxidized in relatively simple mixtures, this can be extremely effective.

Aqueous solutions of hypochlorite (in 25% molar excess) can be used to destroy mercaptans, hydrazine and some hydrazine derivatives, and inorganic cyanides. (Note that it is in general not effective for organic cyanides, i.e. nitriles.) About 10% of the substance should be added with stirring, and the mixture warmed to initiate the reaction if it does not start spontaneously (i.e. if the mixture remains cold). The remainder is added slowly over about an hour so that the temperature remains below 60 °C. Cooling with an ice bath may be necessary. A further 2 hours stirring is allowed for completion of the reaction, and the reaction mixture can be diluted with water and flushed away to drain.

The main feature of the above procedure is to prevent the oxidation becoming a runaway reaction by making the addition rate roughly equal to the rate of conversion and ceasing addition as the temperature starts to rise. An exactly similar procedure can be used for other compounds with their appropriate oxidants. Table 9.4 gives some combinations of substances and oxidants, and appropriate modifications to the technique. Note that it is sometimes advisable to add the oxidant to the solution of the substance, and sometimes necessary to make the mixture acid or alkaline, but in all cases, addition of chemicals should be gradual with stirring so as not to exceed the specified temperature.

For organic compounds which contain available double bonds, these may be oxidized along with the target group, so there should be sufficient oxidant to allow for this. For example, acrylaldehyde $CH_2=CHCHO$ will require twice as much $KMnO_4$ (4 moles instead of 2 moles) as propionaldehyde CH_3CH_2CHO.

The two agents used for strong oxidation of waste are nitric acid and chromic acid. These are traditional cleaning agents because they can convert many organic materials, often to carbon (char) and water and carbon dioxide. However, they have been involved in many explosions owing to the violence of the reaction. Mixtures with organic solvents (e.g. ethanol or acetic acid) can be particularly dangerous.

It is recommended that these reagents are not used for general cleaning purposes. They may be used where appropriate by a skilled and knowledge-able person for a specific chemical reaction. No general recipe is given here, as the prime chemical literature should be consulted. However, it is noted that chromic acid can be appropriate for decontamination (as opposed to routine cleaning) of laboratory ware, particularly when severe carcinogens have been used or when a special type of clean surface is required. Chromic acid, used carefully, is one of the most reliable methods for destroying a wide range of organic carcinogens. See section 12.1.

9.3.7.5 Reduction

As with oxidation, it is important to realize the exact purpose of a reducing reaction, ensure that it is achieved, and carry out the procedure slowly with temperature control to prevent a runaway reaction.

A typical procedure is to have the material to be treated in aqueous solution or suspension at a concentration of no more than 5% and in acid solution (pH 1 to 2). A solution of ferrous sulphate or of sodium bisulphite is added gradually with stirring. There should be at least a 25% excess of the reducing agent available. When about 10% of the reducing agent has been added there should have been a rise in temperature or other indication

Table 9.4. Oxidizing conditions for selected groups of substances

Material	Oxidant	Maximum temperature °C	Variation in procedure from that given on page 156
Carbon disulphide	Hypochlorite	30	
Hydrazines	Hypochlorite	60	
Mercaptans	Hypochlorite	60	
Inorganic cyanides	Hypochlorite	10	In alkaline solution, slowly add ice-cold hypochlorite, leave overnight after procedure
Hydrogen cyanide	Hypochlorite	10	In 3 volumes of ice-water, add one mole equivalent of 10% NaOH then oxidant, leave overnight after procedure
Metal carbonyls	Hypochlorite	60	Under nitrogen atmosphere, carbonyl added in inert solvent
Alkyl, aryl boron, phosphorus, arsenic	Hypochlorite	60	Some mixtures give solid products
Nitriles (organic cyanides)	Hypochlorite	60	First treat with molar excess (25%) NaOH in ethanol: stir for 1 hour then evaporate to near dryness. Make a slurry with water and add to hypochlorite as in the procedure. Leave overnight afterwards before flushing away

organo mercury compounds	Hypochlorite	70	Prolonged treatment may be necessary owing to low solubility. Product is solid oxide
Aldehydes	Neutral $KMnO_4$	70	Aldehyde in water then addition of oxidant
Oxalic acid, oxalates	Acid $KMnO_4$	70	Oxalate in water with sulphuric acid 1 to 2 molar warm to 60 °C, add aqueous $KMnO_4$ dropwise until known equivalence and permanent pink colour. Dark brown manganese (IV) oxide may precipitate
Phenols	30% H_2O_2	60	Phenol in water + about half its weight of $FeSO_4.5H_2O$. Adjust pH to 5 to 6. Addition of oxidant as in procedure

of a chemical reaction. If not, the mixture may be warmed, or a little sulphuric acid added to initiate the reaction. Further addition is controlled so that the temperature does not exceed 70 °C.

The above procedure can be used, for example, to convert highly toxic chromium(VII), e.g. dichromate, to the much less toxic chromium(III). Sodium bisulphite is the preferred reagent for this, and is in fact effective against the widest range of oxidants. However, ferrous sulphate is preferred for peroxides.

A common problem is the destruction of small amounts of peroxides in organic solvants such as ethers. If the solvent is immiscible with water, it can be shaken with a mixture of 25% $FeSO_4.5H_2O$ in 50% aqueous H_2SO_4. For one 2.5 litre bottle, 50 cm^3 of this mixture should be sufficient, given periodic shaking for about 30 minutes.

CAUTION: before attempting this procedure, check that the solvent does not contain any sodium wire or pellets of sodium hydroxide (often added for drying purposes)

The success of the procedure can be confirmed by testing as follows. The solvent (10 cm^3) is shaken with 10% freshly prepared KI solution (1 cm^3) with a drop of starch solution. Hydroperoxides give an obvious blue colour. Test strips using a similar principle are available from Merck (Darmstadt).

Note that ferrous sulphate mainly reduces hydroperoxides, which are the most prevalent products of air oxidation of solvents. If other peroxides are known to be present (e.g. if they were used as reagents) this procedure should not be relied upon.

Trickling the solvent through a column of basic activated alumina is effective in removing most peroxides, and is suitable for solvents (e.g. tetrahydrofuran) which have significant miscibility with water. The alumina should be incinerated or treated with acid ferrous sulphate as above.

CAUTION: solvents which contain peroxides have exploded with lethal effects. No attempt at treatment should be made where crystalline material is visible in the solvent. If a bottle is several years old, or if there is any visible solid around the cap, then no attempt should be made to open it. Both cases should be treated as likely to explode if subject to shock. See sections 12.4.3 (13) and 14.6.7, Tables 14.1 and 14.4, and Appendix C-6.

9.3.7.6 Acid–Base Neutralization

This is probably the most common chemical treatment for waste, and should present no problem to the experienced laboratory worker. However, the following points may be noted for the benefit of good management.

First and foremost the process should be mild. Under no circumstances should a strong base (e.g. solid NaOH) be treated with a strong acid (e.g. concentrated H_2SO_4). Instead, any strong acid or base should be diluted before treatment (see section 9.3.7.2). For solid agents (e.g. NaOH, P_2O_5) it is important to allow plenty of time for dissolution, and to take care that local overheating does not occur — this can cause breakage of the vessel or splashing.

The final product should be as near neutral as is reasonably practicable: it is worth while adding an indicator if one is available. The mixed indicator methyl red/thymol blue is particularly useful: it shows red for acid, blue for base and yellow (or colourless) for the pH range which is generally acceptable for discharge to the sewers. So-called 'universal' indicator will give yellow to green in the same range. Other indicators must be selected by a knowledgeable person with regard to the neutralization.

To achieve a balanced neutrality, it is best if the acid and base are matched in general character. For example, ammonia is a weak base which is brought to neutrality by acetic acid, a weak acid. However, sodium hydroxide is a strong base which is better neutralized by (say) hydrochloric acid, a strong acid.

For general use, sodium carbonate solution is suitable for most acids, and sodium hydrogen sulphate solution is suitable for most bases. Others can be used. (See section 15.2.3.) The choice may be affected by the final disposal method intended. For example, neutralizing a sulphuric acid waste with calcium hydroxide or calcium carbonate will produce a solid product, hydrated calcium sulphate, which is suitable for landfill, but inconvenient for disposal to sewers. (See section 9.3.6.4.)

To neutralize solvents, they can be treated with an aqueous solution of acid or base, providing they do not mix with water. For low levels of acid in aqueous or non-aqueous liquids it is sometimes convenient to use a solid base, such as chalk or barium hydroxide. (See section 9.3.6.5.) Ion exchange is possible, but not usually economic unless the amounts are small or the substance is particularly hazardous.

9.3.8 Particular Substances

Although not intended as a recipe book, the following notes are intended to assist in dealing with some common problems.

9.3.8.1 Sodium, Potassium, Lithium, Calcium, Magnesium Metals

These metals have much in common, but some extremely important differences are vital in waste disposal. The condition of the metal is also important, i.e. whether it is in lumps, wire or a dispersion, whether it is in good condition or badly aged, and what other materials are present.

The metals will all burn, and can be incinerated. However, the products of combustion are very corrosive — the mixture of waste or the incinerator design must be such as to deal with this fact. Containers of suspensions or solid lumps or wire under hydrocarbon may be put directly into a municipal incinerator, if the operating authority agrees. Lumps and residues from chemical procedures should be placed in a bottle filled with dry paraffin or mineral oil prior to incineration. This method is not feasible for industrial blocks and is not recommended for amounts more than 100 g or lumps larger than 10 g.

Larger amounts of the metals (except for magnesium) are best destroyed by water in a place where no-one can be injured. This is safer than the use of large amounts of alcohols, which would be a major fire risk, and gives off less corrosive fumes than burning. Methods which have been used successfully are:

(a) burial of laboratory sticks in very dry sand, so that ground moisture would gradually affect them;
(b) weathering in a pile of sand and soda ash in an open place but protected from direct rain by a non-combustible roof;
(c) dropping into a large hole part-filled with water, using a long pole. In one case the hole was specially dug and in another case it was rainwater collection in a very large excavation scheduled for chemical waste disposal

For any of the above methods the greatest care should be taken to protect people. Canals, rivers, ponds or natural watercourse should not be used. The disposer must be absolutely certain that people (including trespassers) will not interfere with any burial or weathering material.

It is in fact possible to recover pure sodium and similar metals from degraded stock. See section 14.5. The following methods are for simple disposal.

Sodium can be converted to sodium ethoxide by simply putting small pieces into absolute alcohol. If coated with oxide or in larger lumps it can be cut up with a knife in a porcelain dish filled with paraffin or mineral oil. This reaction is more gentle than that with water, but also gives off hydrogen. For sodium wire this procedure may be adequate, but lumps of sodium may take several hours to dissolve. The use of 95% ethanol (i.e. rectified alcohol) is quicker but gives an unacceptable fire risk.

A better technique is to use a round-bottom flask which is continuously purged with a trickle of nitrogen to reduce the fire risk and fitted with a reflux condenser, dropping funnel stirrer and thermometer. The sodium is put into the flask (as wire, dispersion or cut-up lumps). It is preferable to have some liquid present in the form of an inert but not too viscous hydrocarbon, such as light paraffin. This aids mixing and heat dispersion in the early stages. Ethanol (95%) is slowly dripped in with stirring. A total of 13 cm^3 ethanol is required per gram of sodium. The rate of addition is controlled to give gentle reflux. Heating may be necessary to prolong the reflux until all the sodium has dissolved. When this has happened, water is added slowly in a volume equal to that of the ethanol and at a rate which gives only mild refluxing. (This hydrolyses the sodium ethoxide to sodium hydroxide.) The resulting solution is cooled, neutralized with hydrochloric acid (approximately 5 mol dm^3) and flushed to drain.

Lithium can be destroyed by the same method as for sodium, but requires rather more ethanol (30 cm^3 per gram). However, its reaction with cold water is much less violent than that of sodium. Some workers routinely dispose of small amounts of lithium wire by dropping them into a stirred beaker of ice-water in a fume cupboard. This should not be done for powdered lithium or suspensions in solvent, which are better treated with 95% ethanol. Lithium is much harder to cut up than sodium.

Potassium is the most difficult of the metals to dispose of safely for two reasons. Firstly its reaction with water and alcohols is much more vigorous. Secondly it forms an oxide crust on exposure to air or on prolonged storage. Unlike sodium and lithium this crust is violently explosive.

CAUTION: potassium metal showing a yellow coating should not be cut up (even under hydrocarbon) as an explosion may result. Explosions can result when the yellow oxide comes in contact with organic solvents such as kerosene or xylene. Old stock bottles are best disposed of by incineration as above.

Very small scraps of potassium may be decomposed by dropping into a large open vessel of cold water in a fume cupboard. The effect is dramatic as the released hydrogen catches fire, but as water is not combustible there is no further danger. Very tiny pieces are particularly likely to start fires if put into alcohols.

Potassium dispersions are uncommon, but may be destroyed by a 50% mixture of *dry* ethyl acetate with toluene or xylene. The mixture should preferably be left overnight after the potassium has disappeared, then extracted with water and neutralized. (See sections 9.3.6.4 and 9.3.7.6.)

Lumps of potassium may be treated in the same way, or in a round-bottom flask as for sodium, with the following modifications. The pieces are put into the nitrogen-purged flask and covered with glycerol (98+ %). A slow decomposition results, which may be speeded up after a couple of hours by dripping in 95% ethanol with stirring and maintaining the nitrogen atmosphere.

Potasium which is heavily encrusted may be decomposed in the same vessel by using tertiary butanol and allowing a long time, probably overnight with a reflux condenser, a nitrogen purge but no heating. It is the authors' opinion that it is better to allow the time than to try to speed up the process with heat or other more active alcohols.

Sodium-potassium alloys should be regarded as being at least as active as potassium and be disposed of accordingly.

Calcium metal is less vigorous in its reaction with water, and can normally be decomposed by stirring into a large amount of cold water in a fume cupboard. However, hydrogen is liberated and heat is generated. The reaction is sometimes slow to start which can lead to the mistake of adding too much calcium to the vessel. Fine powder is particularly dangerous: it can even ignite in air. Small particles of calcium can be made safer to handle by dusting them with fine powdered calcium carbonate (chalk). Suspensions in solvent should be diluted with a compatible high-boiling solvent then agitated for a prolonged period with water. The water layer may be removed and replaced several times. (See sections 9.3.6.5 and 9.3.7.6.)

Magnesium is most commonly used in the form of ribbon or turnings. These may be cautiously added to a beaker of 5% hydrochloric acid, with stirring and cooling if necessary, in a fume cupboard. As with all these metal dissolutions, hydrogen gas is given off. Magnesium powder is very liable to ignite if moistened. It should be dealt with as in the following section.

9.3.8.2 Fine Metal Powders and Catalysts

Many metals in the form of a dust are pyrophoric — that is, they are liable to catch fire spontaneously in air. A common example is Raney nickel, which is used as a catalyst. Other metals which are not pyrophoric when purchased may become so during use, notably as hydrogenation catalysts in organic synthesis. Obviously, metals which are known to be pyrophoric should only be handled under an inert atmosphere. Less obviously, catalysts which have been filtered out of reaction mixtures should not be allowed to dry out, and must never be discarded into a waste bin with paper etc.

One method is to keep the metal wet with water, and pour a slurry out onto a non-combustible tray so that the powder is spread thinly. This should be left for a couple of days in a place where it will not be disturbed and where a fire would just burn out. In practice it is unlikely that a fire will occur; more usually the metal oxidizes slowly to a non-pyrophoric form.

CAUTION: do not place the tray in any position where powder could be blown away, particularly near air intakes or machinery which would be adversely affected by metal dust.

An alternative for many metals such as nickel is to dissolve the powder in dilute (5%) hydrochloric acid. The metal is put in a beaker with sufficient water so that it can be stirred easily. Dilute hydrochloric acid is added dropwise with stirring, pausing any time the reaction becomes too vigorous. It may be necessary to decant off the top solution after a while. Hydrogen is released, so the process should be carried out in a fume cupboard away from ignition sources. If a small amount is involved the solution may be flushed away to drain, otherwise the metal can be precipitated out for recovery or disposal.

Some laboratories have established procedures where used catalysts are treated with aqueous thioacetamide or hydrogen sulphide to 'poison' them and render them non-pyrophoric. However there is a toxic hazard from both reagents, and thioacetamide is a cancer-suspect agent (actually, so is nickel).

Precious metal catalysts do not in general dissolve in hydrochloric acid, but can be poisoned. They should of course be recovered — see Chapter 14.

9.3.8.3 Phosphorus (Red and White)

The best method of disposal of laboratory quantities is for the unopened jars to be placed (one per charge) into a large refuse incinerator. Open air burning should be with a large pile of combustible material (e.g. wood shavings) to ensure that all particles are thoroughly burnt. The fumes produced are very harmful (though they pose no long-term hazard to the environment) and the greatest care should be taken to avoid the exposure of people or animals to the fumes.

White phosphorus can be converted to phosphoric acid by aqueous copper(II) sulphate (1 mol dm^{-3}). One litre of the reagent will treat 6 g of phosphorus. The process takes about a week (with occasional stirring) and produces a black precipitate of copper phosphide as a by-product, mixed with copper metal. This should be filtered off and treated with 500 cm^3 of 5% sodium hypochlorite for an hour with stirring.

CAUTION: copper phosphide can give rise to severely toxic phosphine gas. The precipitate should not be allowed to dry out. All operations should be in a fume cupboard.

Red phosphorus can be converted to phosphoric acid by prolonged (5 to 10 hours) refluxing with potassium chlorate. For 5 g of phosphorus, use 33 g $KClO_3$ in 2 litres of 0.5 mol dm^{-3} H_2SO_4. The excess chlorate can be reduced by 14 g of sodium bisulphite when cool, then the mixture may be neutralized with NaOH or Na_2CO_3 and flushed to drain.

9.3.8.4 Metal Hydrides and Amides

Alkali metal hydrides and related materials are commonly used for drying solvents or certain kinds of reduction. Disposal problems can be minimized by not using too much for a particular purpose, so that there is only a small excess to be dealt with. Larger amounts, i.e. unwanted reagent jars, are best dealt with by incineration in a large municipal incinerator, one container at a time, or by a specialist disposal firm.

The usual waste is a residue in solvent. NaH, KH, LiH, CaH_2 and $LiAlH_4$ can be converted to the hydroxides by gentle exposure to water, e.g. by the passage of nitrogen containing water vapour through the vessel. This also safely flushes away the hydrogen which is formed.

For gram quantities or where a more rapid procedure is required, the hydride should be suspended in an inert solvent (dioxan is effective, but should be added cautiously as it may itself contain water). The mixture is stirred and cooled while ethanol is added dropwise. (Some workers find certain other alcohols best for certain hydrides or certain waste situations.) When hydrogen bubbles cease to appear then water is cautiously added until the volume of mixture is doubled. If no other components prevent this, the water layer can be flushed to drain and the organic layer (if any) disposed of as solvent. This procedure may be used for the above hydrides and amides.

CAUTION: in the presence of heavy metals such as silver, copper or lead, amides may be explosive. Old samples of sodium or potassium amide can develop an orange-beige coating of explosive oxide. The above procedure may still be followed, but lumps should not be cut up and especial care should be taken.

In anhydrous liquid ammonia, sodium amide can be converted to sodium chloride by ammonium chloride. This is actually an acid-base neutralization and can be monitored by phenolphthalein indicator. The equivalent reaction occurs for potassium, lithium and calcium amides.

$$NaNH_2 + NH_4Cl = NaCl + 2\,NH_3$$

Sodium borohydride $NaBH_4$ is not a drying agent, but a selective reducing agent in both organic and inorganic chemistry. It is almost indefinitely stable in alkaline solution but can be decomposed by heat or by acidification with the liberation of hydrogen gas. The slightly alkaline aqueous solution can be oxidized by ferric ions or many other oxidizing agents.

CAUTION: acidification or heating in the presence of compounds of arsenic, tin, germanium, antimony, bismuth, selenium, or tellurium may give rise to volatile hydrides of these elements. The hydrides are extremely toxic — more so than hydrogen cyanide. Strong acid may give toxic borane gas.

The commercial solution is 12% in aqueous sodium hydroxide. It should be diluted with 4 times its volume of water, then carefully acidified with stirring to allow gentle evolution of hydrogen. Acetic acid is suitable, and a nitrogen purged vessel is recommended to reduce the fire hazard.

The solid may be dissolved in alkaline water and treated as above. It may also be dissolved in neutral water (without hydrogen evolution) and decomposed by boiling. A concentration of no more than 3% should be used for this. The solid can also be burnt. It can be ignited by a match and burns quietly with very little in the way of harmful fumes, but the glassy residue may cause problems in a small incinerator.

9.3.8.5 Cyanides, Acetonitrile, Acrylonitrile

Cyanide is well known as a rapid poison, and in many forms can be rapidly absorbed into the body via the lungs, eyes and skin, as well as by mouth. There are important differences between certain forms, which are relevant to waste disposal.

All cyanide reagents can be destroyed by combustion. For contaminated packaging, incineration is the method of choice. If facilities are available for liquid incineration, then combustion of solutions in water or organic solvent (preferably at a concentration no more than 5%) as appropriate is a good technique. Attention should be paid to any likely residues such as metal salts in the products of combustion: these may need to be disposed of separately.

Potassium cyanide and similar salts readily dissolve in water to give the free cyanide ion, which can be easily complexed or oxidized. Where only tiny quantities are involved (as in many procedures) a basin of dilute neutral ferrous sulphate may be used for rinsing gloves and apparatus. This converts cyanide to the ferrocyanide complex which is much less toxic and volatile, and may usually be flushed to drain without further treatment.

Larger amounts (a gram or more) are better oxidized by hypochlorite to cyanate. See section 9.3.7.4. Kilogram amounts should be dealt with by a professional waste disposal service.

CAUTION: great care should be taken that cyanide salts do not come into contact with acid, which liberates highly toxic hydrogen cyanide gas, except as part of a properly designed and controlled system where this factor has been taken into account.

Mercury cyanide and many organic cyanides do not give rise to the free cyanide ion, and therefore treatment with ferrous sulphate is ineffective, and treatment with hypochlorite may be very slow. Prolonged hydrolysis with NaOH or alcoholic KOH may be necessary to liberate free cyanide for conversion by these techniques. See sections 9.3.7.3 and 9.3.7.4. Mercuric cyanide is best treated in solution with excess potassium iodide (which gives free cyanide) then hypochlorite. If there is sufficient to warrant recovery of the mercury, it can alternatively be precipitated by sulphide, and the liquid treated as for cyanide. As this procedure involves three severe poisons, a fume cupboard and great care are absolutely essential.

Hydrogen cyanide itself may be used as a vapour or a liquid (boiling point 26 °C). The commercial product contains 0.1% H_2SO_4 as a stabilizer, and it is often dried with concentrated H_2SO_4, which may therefore be present in the apparatus. In the absence of these stabilizers, it polymerizes. Contact with a strong base (e.g. NaOH or NaCN) can cause an explosion. It should therefore be diluted with (or absorbed into) 3 or more volumes of water, preferably near to 0 °C, which is then made alkaline with a molar equivalent of NaOH (allowing for any H_2SO_4 present). The resulting solution of sodium cyanide is then treated with excess hypochlorite with cooling and stirring. See section 9.3.7.4.

CAUTION: heavy metal cyanides such as mercury or silver can be friction and impact sensitive explosives. Traces of these compounds in HCN can initiate explosions.

Acetonitrile CH_3CN is unfortunately liable to explosive polymerization initiated by H_2SO_4, so care should be taken when it is to be destroyed in the presence of HCN (see above). Solutions in solvents are best destroyed by incineration. Otherwise it can be hydrolysed by KOH or NaOH (see section 9.3.7.4) then treated with hypochlorite.

Acrylonitrile $CH_2=CH.CN$ is normally stabilized with a little ammonia. Like HCN it is liable to explode in contact with strong base. It is more sensitive than acetonitrile to the presence of acid, which can cause explosions, and care should be taken with mixtures involving HCN (see above). Otherwise it can be hydrolysed by KOH or NaOH (see section 9.3.7.4) then treated with hypochlorite.

9.3.9 Exceptionally Hazardous Substances

For substances of extreme toxicity, such as dioxin, ricin or plutonium, the waste disposal procedure must be built into the experimental work. Furthermore, it should be performed within the contained working area (e.g. a glove box). The method should be determined and practised before the substance is brought in or made in the laboratory.

Where a laboratory discovers that it has a substance of extreme toxicity or explosive potential, it is vital that contact is made with a genuine expert in that substance, before any attempt is made to remove or destroy it. An incomplete conversion process may be more dangerous than simply keeping the material.

Some notes are given on carcinogens, teratogens and cytotoxic agents in section 12.1. Some notes are given on explosives in section 12.2. Some notes are given on mercury in section 14.4.5.

9.3.10 Recycling and Re-use

See Chapter 14, particularly sections 14.4, 14.5 and 14.6.

9.4 GAS CYLINDERS

9.4.1 General Comments

Materials supplied in pressurized cylinders have the physical hazard of pressure in addition to their chemical properties. In addition, release into the laboratory atmosphere gives the greatest danger of poisoning or fire. It is a legal requirement that gas cylinders are periodically tested in an approved manner. Retention of a cylinder beyond its permitted life is a serious offence, and would make the institution guilty of negligence in the event of an accident. (See, also, section 3.11.1 for economic arguments against retention of unwanted cylinders.) Normally this testing is carried out by the gas supply company, when cylinders are returned near their test date.

Appendix C-8 gives a classification of materials supplied by cylinders, *for the purpose of venting only.* Users should have adequate data for safe usage of any gases they handle. Section 15.3.6 gives advice for gas cylinders involved in fires.

9.4.2 Disposable Lecture Bottles and Canisters

A few materials, notably camping gas (butane or propane) and sulphur dioxide, are provided in single-use containers. These should not be opened until required, and should be used up as soon as possible thereafter. Camping gas canisters in particular are likely to dissipate their contents once the seal is broken, and will not store indefinitely.

If no further use is likely within a month, the canister can be vented off or burnt off, as suggested in sections 6.5.3.3 to 6.5.3.6, and in Appendix C-8. A single, empty canister is usually acceptable in a load of trade waste. If there is any doubt, or many canisters are to be disposed of, then a notification as Special Waste (UK) or Hazardous Waste (USA) may be required.

CAUTION: stringent precautions should be taken to ensure that used canisters (even empty ones) are not put in refuse for in-house incineration, baling or maceration.

9.4.3 Rented Lecture Bottles

Many users of small 'lecture bottles' are under the misapprehension that the cylinder is disposable. Many others mistakenly assume that a price quoted in a supplier's catalogue for a gaseous chemical is the total charge. In fact, the usual practice is for the catalogue price to refer to the cylinder contents only. An additional charge is attached for the purchase of the cylinder. (This is usually much greater than the cost of the gas.) If the cylinder is returned within 18 months, then a proportion of the cylinder price is returned. Refunds are not usually made after this time.

The best practice is to use the cylinder contents within 6 months and return the cylinder to the supplier. This both saves money and reduces the waste disposal problem. The cylinder need not be empty, but it should be secure and labelled 'part-full'.

Suppliers can usually be persuaded to accept their cylinders back, even if several years old, though of course no refund will be made. This is the best method of disposal.

Failing this, the cylinder should be emptied by venting safely (see sections 6.5.3.3 to 6.5.3.6 and Appendix C-8). The empty cylinder should be disposed to landfill accepting chemical wastes. It should not be put into scrap metal for recycling.

Old cylinders may be corroded or have the valves jammed.

CAUTION: excessive force should not be used to undo the valve, as it might break off. Heat should never be applied, nor should penetrating oil, which can undergo an extremely dangerous reaction with some gases.

In these circumstances, the valve should be brushed clean and inspected to see if it is open or closed. If closed (or if in doubt) the cylinder can be weighed to ascertain how full it is (liquid can sometimes be heard inside). The local waste disposal authority may be able to help, but it will usually be necessary to pay a specialist hazardous waste contractor to arrange safe disposal.

9.4.4 Large Gas Cylinders

Most industrial-size gas cylinders are rented from a gas supply company. It is important that they are returned to the supplier when no longer needed, even if part-full. (See section 3.11.1.)

If an institution has its own cylinders, then it is responsible for ensuring that these are tested at the legally prescribed intervals (which varies for the cylinder usage). If a cylinder is discovered which is beyond the testing date, then this cylinder cannot be used and cannot be tested. It must be discarded.

If the institution has the facilities, then any material in the cylinder may be transferred to another suitable cylinder. The residue can be vented off. See sections 6.5.3.3 to 6.5.3.6 and Appendix A. The gas supply company will usually be able to advise on disposal of the empty cylinder, and may even carry out disposal. However, a gas supplier will generally refuse to handle an out-dated cylinder which is part-full, unless that cylinder is owned by the company.

If a laboratory finds itself the owner of a part-full cylinder which is jammed or corroded shut, then it will be necessary to employ a specialist hazardous waste disposal contractor. The same applies for an out-dated cylinder where the laboratory does not have the facilities to safely dispose of the contents.

9.4.5 Leaking Gas Cylinders

The usual cause of a leaking cylinder valve is excessive force applied when closing it. If a valve does not seal, then it should be opened at least a quarter, and some gas vented off (in a safe manner). This may dislodge grit and permit the valve to close properly again, when it is gently closed. If the leak persists, then act as follows.

A small 'lecture bottle' cylinder may be placed in a fume cupboard. A large cylinder will normally need to be outside in a well-ventilated area away from people.

CAUTION: even a non-toxic gas such as nitrogen can cause rapid unconsciousness or even death by asphyxia, e.g. in the confined space made by

a few items of furniture. If it is necessary to transport a leaking cylinder by lift, under no circumstances should it be accompanied by a human being.

A cartridge respirator may give some protection against irritating fumes, but only an air-supplied breathing apparatus will properly protect against high concentrations of gas.

In the case of rented cylinders, the supplier should be contacted as soon as possible for assistance. If the leak is not too bad, it is often possible for the cylinder to be taken away by the usual delivery vehicle. However, a charge may be made if an extra journey is involved.

The supplier is often able to send an expert who can deal with more serious leaks. If the situation is more urgent, then the fire service should be called.

CAUTION: under no circumstances should laboratory staff attempt to repair or adjust the cylinder valve itself, or its fixing to the cylinder.

Re-usable cylinders owned by the institution (e.g. breathing apparatus, medical gases) which start to leak should be emptied and submitted to the supplier (or other specialist company) for refurbishment.

REFERENCES

[section 9.2.6]

Gaston, P.J. (1964) *The Care, Handling and Disposal of Dangerous Chemicals,* Northern Publishers (Aberdeen).
International Technical Information Institute (1978) *Toxic and Hazardous Industrial Chemicals Safety,* Int. Tech. Inf. Inst. (Japan).
Manufacturing Chemists Association (1972) *Guide for Safety in the Chemical Laboratory 2nd edn,* Van Nostrand Reinhold.
National Research Council (US) (1983) *Prudent Practices for Disposal of Chemicals from Laboratories,* National Academy Press.
Strauss, H.J. (1976) *Handbook for Chemical Technicians,* Reinhold.

[section 9.3.6.5]

Linke, W.F. (1958 and 1965) *Solubilities of Inorganic and Metal Organic Compounds 4th edn, Vol. 1,* Van Nostrand; *Vol. 2,* American Chemical Society.
National Research Council (US) (1926–1929) *International Critical Tables Vols. I–VI,* McGraw-Hill.
Sorensen, J.M. and Arlt, W. (1979–1980) *Liquid Liquid Equilibrium Data Collection (Chemistry Data Series Vol. V),* Parts 1, 2, 3. Dechema.
Stephen, H. and Stephen, T. (1963–1964) *Solubilities of Inorganic and Organic Compounds Vol. 1, Parts 1 and 2; Vol. 2, Parts 1 and 2. Vol. 3* (ed. Silcock, H.L.) (1979) *Parts 1, 2, 3,* Pergamon.

Timmermaus, J. (1959–1960) *The Physico-Chemical Constants of Binary Systems in Concentrated Solutions Vols. 1–4,* Interscience.

[section 9.3.7.1]

Analyst (1960) Methods for the Destruction of Organic Matter, **85,** 643–656.
Gorsuch, T.T. (1970) *Destruction of Organic Matter, Intl. Series Monographs on Analyt. Chem. Vol. 39,* Pergamon.

BIBLIOGRAPHY

Disposal Procedures

Armour, M.A., Browne, L.M. and Weir, G.L. (1982) *Hazardous Chemicals: Information and Disposal Guide,* Dept of Chemistry, University of Alberta, Canada.
Gaston, P.J. (1964) *The Care, Handling and Disposal of Dangerous Chemicals,* Northern Publishers (Aberdeen).
International Technical Information Institute (1978) *Toxic and Hazardous Industrial Chemicals Safety,* Int. Tech. Inf. Inst. (Japan).
Manufacturing Chemists Association (1972) *Guide for Safety in the Chemical Laboratory 2nd edn,* Van Nostrand Reinhold.
National Research Council (US) (1983) *Prudent Practices for Disposal of Chemicals from Laboratories,* National Academy Press.
Scottish Schools Science Equipment Research Centre (1979) *Hazardous Chemicals — A Manual for Schools and Colleges,* Oliver & Boyd.
Strauss, H.J. (1976) *Handbook for Chemical Technicians,* McGraw-Hill.

Reactive Hazards of Mixtures

Bretherick, L. (1979) *Handbook of Reactive Chemical Hazards 2nd edn,* Butterworths.
Bretherick, L. (ed.) (1981) *Hazards in the Chemical Laboratory 3rd edn,* Royal Society of Chemistry.
National Fire Protection Association (US) (1975) *Manual of Hazardous Chemical Reactions. NFPA publication no. 491M,* NFPA.

General Chemical Hazards

(The above publications also give details of toxic and other hazards. A limited amount of chemical hazard information is included in many textbooks and handbooks. The following are mainly data compilations.)

Dutch Association of Safety Experts (1980) *Handling Chemicals Safely 2nd edn,* Dutch Chemical Industry Association & Dutch Safety Institute.
Fairchild, E.J. (1978) *Agricultural Chemicals and Pesticides,* Castle House.

Mackinson, F.W., Stricoff, R.S. and Partridge, L.J. (eds.) (1978) *NIOSH/OSHA Pocket Guide to Chemical Hazards. DHEW(NIOSH) Publication no. 78–210,* US Govt Printing Office.

Meidl, J.H. (1972) *Hazardous Materials Handbook,* Glencoe.

National Fire Protection Association (US) (1973) *Hazardous Chemicals Data. NFPA publication no. 49.*

National Institute of Occupational Safety & Hygiene (US) (annual) *Register of Toxic Effects of Chemical Substances,* (NIOSH), US Govt Printing Office.

National Research Council (US) (1981) *Prudent Practices for Handling Hazardous Chemicals in Laboratories,* National Academy Press.

Royal Society of Chemistry (1983) *The Agrochemicals Handbook,* Royal Society of Chemistry.

Sax, N.I. (1979) *Dangerous Properties of Industrial Materials 5th edn* Van Nostrand Reinhold.

Shugar, G.J., Shugar, R.A. and Bauman, L. (1973) *Chemical Technician's Ready Reference Handbook,* McGraw-Hill.

Sittig, M. (1979) *Hazardous and Toxic Effects of Industrial Chemicals,* Noyes Data Corp.

Steere, N.V. (1971) *Handbook of Laboratory Safety 2nd edn,* Chemical Rubber Co.

UK Health & Safety Executive (1985) *Occupational Exposure Limits 1985: Guidance Note EH 40,* HMSO.

Weiss, G. (1980) *Hazardous Chemicals Data Book,* Noyes Data Corp.

Windholz, M. (ed.) (1976) *The Merck Index. 9th edn,* Merck.

CHAPTER 10

Biological Materials

10.1 INTRODUCTION

Medical and biological laboratories have as their overriding interest the prevention of infection or the exposure of human beings to powerful chemical agents. However, in concentrating on these hazards, it is easy to overlook other dangers which can be aggravated by inadequate waste disposal practice.

Firstly, these laboratories are often a greater fire hazard than is realized. It is a common error to keep large quantities of paper and plastic disposal items within the laboratory, with the result that a small fire rapidly becomes a very large one. Furthermore, flammable solvents are often in use for cleaning, staining or as anaesthetics. Particular care should be taken that solvent-soaked swabs are not deposited in paper bins. Even the ether-impregnated fur of a dead rat has been known to cause an explosion in a refrigerator, resulting in a major fire.

Secondly, very sharp objects and fragile glass (such as pasteur pipettes) are routinely handled by people protected only by very thin gloves. Serious injuries can occur, even from small non-infected items. The disposal technique should ensure that sharps are easily put out of reach and never handled except in an appropriate container thereafter. (See section 10.6.)

Where extremely dangerous materials are involved, then extreme methods of containment and disposal may be required. As far as possible, material should be rendered inactive within the experimental area (e.g. glove box), or if not then within the containment area (specially sealed

174

rooms). Adequate waste disposal and methods of dealing with spillages should be considered at the design stage.

The following chapter deals with the more general types of risk, since specialized laboratories should have the necessary knowledge to implement methods for the particular organisms they handle. (See section 10.3.)

10.2 STERILIZATION, DISINFECTION AND HYGIENE

10.2.1 Definitions

Sterilization means the total killing of all living organisms, including spores, and is much more difficult to achieve than is commonly supposed. Disinfection means the inactivation (generally by chemical means) of troublesome organisms, and in practice is only a temporary means of keeping infection away. Good hygiene implies careful separation of organisms from substrates on which they can proliferate (including human beings) and regular effective cleaning to remove both substrates and any organisms which may have been deposited, before harmful accumulation occurs.

The most important point in infection control is to recognize that procedures can be ineffective. A common mistake is to assume that anything coming out of a sterilizer is necessarily sterilized. (Infection may remain for several reasons: see section 10.2.2). Similarly, items put into a bath marked 'disinfectant' may not actually be disinfected (e.g. if the active agent has lost its strength). Likewise, a cleaning procedure may actually be a way of spreading infection (by sweeping dust into the air, or spreading organisms on a dirty cloth).

In practice, good hygiene and sanitary procedures will remove most of the risks from waste. Sterilization or disinfection may be considered as worthwhile additional safeguards, depending on the level of hazard.

10.2.2 Sterilization

It is difficult to guarantee total sterility: even burning is not always totally effective (see Chapter 7). However, for practical purposes it can generally be achieved by dry heat, by steam, incineration, gamma radiation or some of the more severe chemicals. The most usual chemicals are the gases formaldehyde or ethylene oxide, and the liquids formalin, glutaraldehyde, hypochlorite and chlorite solutions.

In all cases the key to success is good penetration by the sterilizing agent. This is particularly important for large objects (e.g. bags of waste, carcasses) where the centre may be relatively untouched after an hour or more's exposure on the outside.

Viruses are relatively impervious to chemical disinfectants, though their characteristics vary considerably. Phenolic agents have virtually no effect, and ordinary strength hypochlorite will only deactivate a limited number. Chlorite mixtures may be more effective. Formaldehyde, glutaraldehyde and ethylene oxide are fairly reliable.

However, particular viruses may be susceptible to particular chemicals. Where a broad spectrum of viruses may be present, then heat is the most reliable method. Exposure to 60 °C for 20 minutes will destroy the majority, though it is reported that a few of the most resistant may survive dry heat up to 170 °C under certain conditions. Wet heat at 80 °C for 30 minutes may normally be considered reliable.

Other microbes (bacteria, fungi, protozoa, slime moulds, algae) are generally susceptible to disinfectants and to heat. The exception is the spore state of fungi and bacteria. In particular, some bacterial spores are very resistant indeed, and present the major problem in practical achievement of sterility.

In the presence of water, all active microbes are killed by a temperature of 80 °C maintained for 10 minutes. (The exception being a very few organisms adapted to hot springs, which are rarely found in laboratories.) This will also kill most (but not all) fungal spores, but will leave most bacterial spores still viable. In boiling water at 100 °C, an exposure of 20 hours is necessary to guarantee inactivation of the most resistant spores. The effect of different temperatures and dry heat is discussed hereafter.

Chemical disinfectants vary in their toxicity to different organisms. Where properly matched to the organism they can be extremely effective. However, of the commonly used materials, only formalin, glutaraldehyde and chlorite are generally effective against resistant bacterial spores.

10.2.2.1 Dry Heat

Dry heat has the advantage of requiring simpler equipment than steam sterilization, and not being dependent on the penetration of steam. The only limitation is that certain materials (e.g. wood, plastic) degrade at the temperatures required. This means they are not re-usable, and the fumes may affect the oven adversely.

It is vital that time is allowed for items to heat up to the necessary temperature. This can be aided by arranging items to allow circulation of air, and by using a fan-assisted oven. It is generally important that infected material is not exposed on a free surface from which it might release spores on drying, especially if forced circulation is used.

One hour's heating at 100 °C will kill all non-sporing organisms; at 120 °C it will kill fungal spores; and at 160 °C it will kill bacterial spores. Roughly double the time is needed for 10° lower temperature, roughly half for 10° higher.

For very small objects such as slides or filter papers, an infra-red lamp may be used. Ten minutes exposure is usually sufficient. Radiant ovens are used industrially and may possibly be useful for sterilizing the outside of containers.

It is possible that microwave ovens will come into use for sterilizing, since they penetrate tissue very well. However, this penetration is not total, and larger objects (e.g. carcasses) will tend to cook from the outside. Also note that a certain bulk of material containing water is required, so that dried smears on glassware may not be sterilized. Certainly, it would be possible for a laboratory which deals with foodstuffs to cook waste in a microwave oven.

CAUTION: under no circumstances should metal objects be included, even small ones, in waste put into a microwave oven as this can damage the oven. Some plastics are unsuitable for use in a microwave oven.

An alternative to an oven is to place a container in an oil bath. The container should either be vented or be strong enough to contain the internal pressure. An 8 hour immersion at 120 °C or a 3 hour immersion at 140 °C will normally suffice. A temperature of 180 °C for 30 minutes will deactivate even the most resistant spores and viruses.

10.2.2.2 Boiling Water

Where facilities are limited, boiling water can be used quite effectively for a range of organisms and materials. A minimum boiling time of an hour is recommended, unless spores or resistant pathogens are likely to be a problem, in which case several hours (up to 20 in the case of very resistant spores) may be required.

The use of a 2% solution of sodium carbonate greatly enhances the killing power of boiling water, reducing the minimum time required to about 10 minutes. Note the alkali will attack some metals, and may be neutralized by organic acids if present in any amount.

As an alternative to prolonged boiling, the technique of Tyndallization can be used. In this, the materials are boiled in a covered container for 30 minutes then allowed to cool and stand for 24 hours. The container is again brought to the boil for 30 minutes, allowed to cool and stand overnight and then boiled again. This process kills live cells then gives time for spores to develop to a point at which they are vulnerable.

10.2.2.3 Steam Autoclaves

The active agent in autoclave treatment is hot, wet steam. The pressure is usually above atmospheric to increase the boiling point and hence the temperature, but low pressure units (vacuum autoclaves) are used for heat-sensitive materials to give true steam (not water vapour) at 80 °C or even less. Low pressure steam is usually used in conjunction with formaldehyde or ethylene oxide.

With small, externally heated units (or if faults develop on certain other units) it may be possible to vaporize all the water and for the temperature to exceed the boiling point. In such a situation, the steam behaves as a dry gas, and the process is no more effective than heating in a dry oven at the same temperature (see section 10.2.2.1). For this reason, items in an autoclave which has been allowed to run dry must be considered non-sterile.

For similar reasons, it is important that all the air be displaced from the unit. Sterilizing time should really be taken from the moment the autoclave and its contents are completely filled with pure, wet steam. Vacuum autoclaves remove air very quickly. Large autoclaves, in which air is removed by downward displacement, take somewhat longer than average. This accounts for the differences in operating times recommended by the manufacturers for different models.

CAUTION: a primary cause of incomplete sterilization is the failure to follow, or even consult, the manufacturer's instructions for an individual autoclave. A suitable pressure/temperature/time cycle for one machine may not be appropriate on another unit.

Autoclaving may be used (a) as a final treatment following disinfection and cleaning (b) as a preliminary treatment prior to removal for incineration (c) as the only treatment against infective hazards. These require different standards of sterility. Likewise, the type and nature of organisms involved effect the time and trouble which should be taken.

Where treatment is (a), a final safeguard, then it will normally be possible to arrange items so that steam can penetrate, e.g. in a bucket and with containers open (having previously been opened under disinfectant).

Where treatment is (b), preliminary, then it may be inadvisable to handle the material. One effective method is the use of autoclavable plastic bags with self-opening and self-closing rings. These expand on heating to permit steam to enter and close during the cooling part of the cycle. For safety, this bag should be placed inside another strong bag with more conventional closure immediately upon removal from the autoclave.

Disposable plastic items (particularly Petri dishes) tend to partially melt and fuse together in the autoclave. Whether a sealed bag, bucket or self-opening bag is used, it is likely that some non-sterile material will remain in the middle of any bulk of waste of this type, unless the process is maintained for several hours. However, absolute sterility throughout is not normally necessary. Providing the outer layers are sterile and there is as little disturbance of the contents as possible, there will be no real infective risk in transport (e.g. to an incinerator). The items should be put into a suitably strong outer sack or container to minimize the risk of spillage. (See section 3.4.)

Where treatment is (c), the only control of infective hazards, it is vital to ensure that it is successful. Where organisms of low infective potential are involved, and where they are known to be heat-liable, then it is mainly necessary to ensure that items are not packed too tightly together, that the autoclave operating instructions are followed, and that staff observe good hygienic practice. It is worth while taking an occasional sample for culture to ensure that the process is successful.

Wire baskets are a traditional and highly effective container for bottled wastes. However, they will not contain a spillage, and therefore are prohibited by the Howie code (see section 10.3) for laboratories handling pathogenic micro-organisms, infectious agents or pathological material. The Howie code specifies 'Material to be autoclaved must be placed in containers with solid bottoms and sides', but does not give dimensions. It is in fact important that these containers are relatively shallow — no more than 250 mm deep — or there will not be adequate air displacement. Autoclavable polypropylene trays are commercially available, and are very convenient for this purpose. They may also be used to support autoclavable bags, which may otherwise spill or split in use.

Polypropylene jars with snap-on lids are used in some institutions for transporting waste to an autoclave. The lid, of course, gives extra security in transport, but it is essential that it is removed during autoclaving. The lid must be left in the autoclave so that it, too, is sterilized.

If 'sharps' are to be autoclaved, they must be in a container which will remain puncture-proof during and after the process. Polypropylene, polycarbonate, or metal containers are generally better than cardboard for this purpose. (See section 10.6 and Appendix B-2.)

The Howie code further requires that the operation of the autoclave be checked. This is done by an initial test with a dummy full-load of non-hazardous material approximating in character to that which is to be treated. Thermocouples are placed within the load and at the bottom of the chamber. The time taken for the thermocouples to reach a satisfactory temperature (usually 121 °C) is measured. This is then used to specify a proper work cycle for future use. It is recommended that thermocouple tests are repeated (e.g. monthly) to ensure that the machine is continuing to function correctly.

Autoclave tape is so convenient and inexpensive that it should be used every time. A piece is placed at the mouth of the container, *not* in the load. A failure to change certainly indicates a failure of the autoclave process. However, a change in the autoclave tape does not guarantee sterility. A more reliable indicator (though not absolute) is the tube or indicator strip which changes colour when exposed to sufficient prolonged heat, and which can be placed in the load. Commercial items are available in various forms.

It is possible to carry out a biological check on the autoclave function by the use of commercial sachets which are placed within the load. These contain *Bacillus stearothermophilus* on filter paper. After autoclaving, the sachet is opened and part of the contents placed in nutrient broth and incubated. Turbidity indicates incomplete sterilization, but may take 24 to 48 hours to develop.

The drain and vent valves on autoclaves can present a hazard since they may release non-sterile material. Where human pathogens are involved it is vital that the autoclave is not allowed to discharge into the general laboratory atmosphere or an open drain (such as a sink or floor channel). (See Chapter 5.) Where radioactive substances are in use these routes for discharge of radioactivity must be taken into account. (See Chapter 11.)

Loading of autoclaves should be such as to permit good steam circulation, and to prevent materials running out of packages or containers into the chamber. Agar (and similar gels) can present a particular danger by blocking drains. Where material has leaked into the chamber, it should be removed from the floor and walls as soon as possible.

CAUTION: autoclaves present hazards of high temperature and high pressure, and require full compliance with the manufacturer's safety instructions.

The following are some examples of hazardous situations due to incorrect use which have been reported.

Incorrect autoclave. There are many machines available for special sterilizing duty, such as for surgical dressings, bottled fluids, bedpans, etc. In general, they are *not* suitable for typical mixed laboratory waste, though cases are known where they have been supplied for this purpose.

Chemicals. Chemicals should not be autoclaved (except as part of a specialist manufacturing facility). In particular, bottles of solvent may explode or catch fire, and corrosives (including bleach) may damage the pressure regulator or safety valves.

Sealed Bottles. It is preferable for bottle caps to be loosened before autoclaving. Where it is necessary for them to be processed sealed, extra time should be allowed (at least an hour) for them to cool down before removal, as they have been known to explode when taken out hot.

10.2.3 Disinfection and Cleansing

Correctly used, proprietary and non-proprietary chemical agents can be very effective in controlling the risk of infection. As with dry and wet heat, it is necessary for the agent to penetrate at sufficient strength and for a long enough time to achieve maximum effectiveness. Penetration is aided in some proprietary formulations by combining the disinfectant with a detergent. As with autoclave treatment, it may sometimes be acceptable to have only the exterior of some items treated, as a purely temporary safeguard against contamination.

There are two main methods of use: either items are put into a bath of disinfectant, or the disinfectant is applied to the items (e.g. a bench or wall or large object). In the latter case it can be difficult to ensure total contact except by regular repetitions of the treatment. Hand or powered sprays usually help, but a proper procedure must be followed so that the spray does not present a direct hazard to people or animals.

Where baths of disinfectant are used, it is essential that they are maintained at the correct strength and discarded frequently. As a general rule, items should be in the bath for at least 30 minutes before they can be considered treated. In practice this means that no item should be removed until 30 minutes after the *most recent* addition to the bath. Closed containers should preferably be immersed before opening, to reduce the risk of exposure to the worker.

The Howie Report (see section 10.3) recommends a written disinfectant policy, stating which disinfectants are used for what purpose and the working concentrations to be used. This can easily be added to the laboratory safety policy. The disinfectant policy should contain a clear statement of the storage time and in-use time for baths or swabbing solutions.

The Howie Report says that 'the use–dilution [should be] based on in-use tests for each' (para 18 1 i). This has been misunderstood in some places. It does *not* require the laboratory to carry out any of the so-called 'standard' tests by which disinfectants are rated. These may be left to the manufacturers and specialist establishments. Most laboratories can simply use the manufacturer's specifications or published information to establish an effective and economic concentration. 'In-use' tests are then used to check that the procedure is working. In essence, a sample is taken of material after disinfection (a swab from a surface, a sample of solid, or 1 cm^3 of liquid from a bath at the end of its life). Residual disinfectant is deactivated by dilution (e.g. by 10 volumes of quarter-strength Ringer's solution for typical phenolics) or neutralization (e.g. by 0.5% sodium thiosulphate for hypochlorite). Specimens are then cultured (nutrient agar at 37 °C and at ambient, or as appropriate to the organism). Any growth within 72 hours indicates that the disinfection has not been complete.

Appendix B-5 gives details of some commercial disinfectants.

10.2.3.1 Cleaning Agents

Although no cleaning agent can be counted a sterilant, they are all to some extent disinfecting (by lysing cell walls). In addition, they can greatly enhance the effect of other agents by the removal of dirt and by aiding penetration. Regular removal of dirt (on which micro-organisms would otherwise breed) is of course essential basic hygiene. To maintain a low level of contamination in biologically active rooms, it is common practice to wash down with a cleaning agent and with a disinfectant on alternating occasions.

Soaps have only limited bactericidal properties, which can be enhanced by the addition of a compatible disinfectant — the main use of such formulations being for cleaning the skin. Detergents are more effective, though their main use is really to remove dirt and aid the penetration of more active chemicals. Quaternary ammonium compounds (often referred to as 'quats') are surface active compounds which are true disinfectants which can be used for surface cleaning and other purposes. As they are non-toxic and non-irritant to humans and most mammals, they are popular in food laboratories and in animal rooms. They are generally non-corrosive and can even be applied as an aerosol. They have, however, a limited activity range and are readily deactivated by protein, soap, non-ionic detergents, some plastics and some other natural materials.

Ampholytic detergents are chemical substances which combine a detergent group on one molecule. They are generally used for washing down floors and walls where it is acceptable and advantageous to leave an inhibiting residue.

A range of proprietary alkaline cleansers have been developed for laboratory use (see Appendix B-4). If correctly used (i.e. not overloaded and kept in contact for the right period) then they may often be used in place of a disinfectant for decontaminating laboratory ware. The high pH and lysing effect will kill or inactivate many micro-organisms, and the powerful cleaning effect ensures that biological material is removed from surfaces. A short, hot soak or a cold overnight soak is generally recommended, then a rinse with sterile water (best with a trace of hydrochloric acid). For critical work the items can of course be autoclaved or treated with an appropriate disinfectant before use. The soak baths should be discarded weekly, or earlier, if there is obvious accumulation of dirt, or if the pH has fallen (tested daily with a paper test strip), or if particularly active or difficult organisms may be present.

Acid cleaning agents are occasionally used (usually in washing machines) where certain types of scale occur.

Cleaning baths usually work better if items are *not* autoclaved beforehand (since this can bake on protein) but this will sometimes be a necesary precaution if there is a risk of more harmful organisms being present. A combination of cleaning agent and phenolic (commercially available) may be appropriate in such circumstances.

Note that cleansers intended for use in soak baths have very different properties from detergents as used for hand or machine washing. The two should not be confused, even though they may come from the same manufacturer and have similar trade names.

CAUTION: in the past, so-called cleaning mixtures have included powerful oxidising agents, notably chromic acid or nitric acid mixtures. These materials can react explosively with organic material, and have even been known to explode in storage. They do not generally perform better than other safer cleansers. If the use of such an agent is considered essential, it is recommended that the item is first cleaned as far as possible by a surface active agent then well rinsed before treatment with the oxidizing agent. It is especially dangerous to use a solvent in conjunction with the oxidizing agent.

10.2.3.2 Alcohol

Ethanol is an effective and widely used agent for disinfection of surfaces. It is most effective at a concentration of about 70%. Stronger mixtures, including 100%, are actually *less* effective. So-called 'absolute alcohol' is the most expensive starting material, and not usually necessary. Industrial methylated spirit (often known as IMS) or surgical spirit is usually acceptable, at a much lower cost. When made up with water, it is effective against most wet vegetative microbes. When made up with 0.1 molar HCl or NaOH it will kill the majority of bacterial spores (but the solution is correspondingly more corrosive and irritant).

Propan-2-ol (commonly known as isopropyl alcohol or IPA) may also be used as a substitute for ethanol, but its optimum concentration is 80%. Both alcohols are used as solvents for biocides which are not readily water-soluble, e.g. to deposit a longer-acting chemical on a surface.

It is unusual to use alcohol for bulk disinfection or in a soak bath, both because of the cost and because of the fire risk. However, it is frequently used to swab surfaces or as a spray. Swabs should be discarded into covered containers, and especial care should be taken to avoid sources of ignition when sprays are used, and for some time after.

10.2.3.3 Phenols and Cresols

For many people, disinfectants are associated with the smell of phenol, and in fact these compounds form the basis of the greatest number of commercial disinfectants. (See Appendix B-5.4.) There are a wide variety of patented chemical derivatives, combinations and variations of formula for various purposes. It is vital to recognize that a product which gives excellent service in one application may not be adequate in another situation.

The main criteria are as follows: cost of material at an effective dilution; specificity towards particular organisms (this can vary considerably); corrosive properties towards materials to be treated; mammalian toxicity; toxicity towards other creatures (e.g. fish) if used; irritancy towards humans. Where large-scale use is intended, it is very much worth while discussing the exact details of the intended usage, and possibly monitoring to check the effectiveness. Manufacturers' information should be consulted initially.

A general feature of these materials is that their toxicity is roughly proportional to concentration. A 1% solution is about the minimum which can be expected to show a lethal effect to a wide variety of microbes. At lower concentrations the effect is one of inhibition (i.e. the organisms do not proliferate, but are not killed). At the lowest concentrations, phenols can be metabolized by some organisms, and can even act as a growth stimulant.

Phenol and cresol derivatives are largely ineffective against spores, even over long contact periods, and therefore cannot produce total sterility. The halogenated phenols are much less toxic to mammals and can be equally effective as the non-halogenated equivalents for many organisms, but they tend to be more easily deactivated by the presence of excess organic material. This is important if treating highly soiled items, in which case it is preferable to use some kind of pre-clean procedure.

A traditional cleansing and disinfecting agent is a solution of cresols in soap known as Lysol. (Lysol is a trade name in some countries. See Appendix B-5.4, note (1).) This has now been superseded by commercial formulations called 'clean soluble phenolics' which are equally effective, but less hazardous to use. Phenol itself is the most dangerous of this group, as it is a powerful blood poison which is readily absorbed through the skin. Medical treatment should be sought in all cases of even minor skin burns with phenol or related compounds.

It is difficult to monitor the activity of a bath of phenolic disinfectant, so it is important that it is frequently renewed, preferably daily.

10.2.3.4 Hypochlorite and Other Halogen Agents

Sodium hypochlorite is the active ingredient in domestic bleach, and calcium hypochlorite is the basis of bleaching powder, both of which are long-established germicides. Combinations of hypochlorite with surface active agents and with alkali are sold as proprietary sterilizing fluids for laboratory and other use. Some formulations use hypobromite, which is similarly effective. In recent years, a range of solid products have been marketed which contain sodium or calcium dichloroisocyanurate. These dissolve in water to give a hypochlorite solution. (See Appendix B-5.1.)

The strength of a hypochlorite solution is generally quoted in 'parts per million available chlorine' (or 'available halogen' for similar products) usually abbreviated to 'ppm'. The basis for this measurement need not concern us here, only its usage. Typical domestic bleach when new and full strength has about 50 000 ppm. 'Thick' bleach is made so by the addition of another chemical: it does not contain any more hypochlorite and has no advantage for laboratory use. 'Gentle' bleaches (e.g. for babies' feeding bottles) may be only 10 000 ppm. Some commercial bleaches are 100 000 ppm and certain solid products may be quoted as more than a million ppm.

Hypochlorite solutions (and to a lesser extent solids) become less active with age. More dilute solutions (and solids exposed to moisture) decay more rapidly, and must therefore be replaced even if unused. The presence of organic material removes hypochlorite by a chemical reaction, as does the presence of certain chemicals (e.g. thiosulphate).

As a rule of thumb, it is recommended that a working concentration of 2500 ppm be made up. For example, this means a 1 to 20 dilution of domestic bleach. This is sufficient for common usage. However, if heavy organic contamination is expected (e.g. blood, faeces, broths) then a concentration of 10 000 ppm should be used. Concentrations up to the neat commercial product may be used for more drastic treatment, such as coping with a release of harmful organisms.

CAUTION: concentrated hypochlorite or prolonged contact with dilute hypochlorite can be very corrosive to metals. Hypochlorite (and similar compounds) should never be used to disinfect centrifuge parts, because corrosion which is invisible to the naked eye can cause stress cracking and failure in use. This applies to any other items which may be highly stressed in use because of pressure or mechanical action.

The activity of a hypochlorite solution can be checked quite easily by dipping in a starch-iodide test paper and removing. If the paper goes blue then bleaches, there is at least 1000 ppm chlorine and the solution is active. If the paper gives a good ink blue, but does not bleach, then there is at least 200 ppm chlorine. The solution is still active, but should not be relied upon for further waste. If the test paper does not give a good ink blue within one minute, then the solution should be regarded as inactive. Any items in the solution should not be removed until further hypochlorite has been added and time allowed for it to work.

The same test can be applied to hypobromite or to solutions of the halogens in water (used industrially, but rarely in the laboratory as a disinfectant).

A recent development has been a patented two-part sodium chlorite/organic acid mixture, which effectively gives chlorine dioxide in solution. This is a more powerful agent than chlorine, giving a more rapid kill at equivalent concentrations. Its effectiveness against spores, viruses and bacteria is claimed to be comparable to activated glutaraldehyde. The principal advantage appears to be relatively low toxicity/irritancy to humans and the higher animals.

10.2.3.5 Formaldehyde (Formalin, Methanal)

Formaldehyde is the gas HCHO, also known as methanal. In the USA it is available as a gas in cylinders which can be used for fumigation. In the UK it is more usually generated by heating its polymer, paraformaldehyde, or used as the vapour displaced from its solution in water. The solution in water is commercially of 40% concentration, with about 10% methanol as a stabilizer. This solution is called formalin.

Formalin is an effective traditional medium for the preservation of tissue. It is a general sterilizing agent which kills bacterial spores. A one-tenth dilution (i.e. 4% formaldehyde) is adequate for most purposes. In an enclosed vessel (e.g. a closed Petri dish) a swab soaked in formalin will kill virtually all organisms in about 12 hours by virtue of the vapour of formaldehyde given off. Thus the action of the liquid is penetrating even into crevices which are not actually wetted. Formalin is, however, unpleasant to use: it has a highly irritating vapour and the liquid causes severe skin damage. The vapour is now believed to be a carcinogen.

Formaldehyde fumigation is useful for rooms, safety cabinets and for sterilizing filters in laminar flow cabinets prior to their removal for disposal. See section 6.4.4.

CAUTION: formaldehyde gas is extremely dangerous. Large-scale fumigation should only be carried out by people with specialist training in the technique.

10.2.3.6 Glutaraldehyde

This is used by some institutions in place of formalin. It is less irritating to use, and less penetrating, but on smooth surfaces it is actually more rapidly bactericidal. Like formalin, it is sporicidal. It can be used for fumigation, when it has the advantage of not depositing polymeric formaldehyde.

CAUTION: specialist knowledge is required. See section 10.2.3.5.

It is supplied as a 50% solution, which may be diluted down to about 2% for use in a soak bath or discard jar. It retains its effectiveness for about 12 hours in this condition, so it is essential that it be discarded daily.

There are several commercial formulations including glutaraldehyde, usually as a two-part product. This is usually a glutaraldehyde concentrate and a so-called 'activator'. There is evidence that activated glutaraldehyde may be more effective, and it is claimed that the solution can be kept for 14 or (according to one manufacturer) 28 days. However, it is not recommended by these authors to keep a solution more than one week. Discard jars should still be replaced daily, but a container of activated glutaraldehyde is a useful standby as an emergency disinfectant.

Different formulations should not be mixed, as some are acid and some are alkaline. The manufacturer's procedure should be followed in each case. Note that some activators are coloured for identification, but the presence of colour in the solution is *not* an indicator of disinfectant activity. (See Appendix B-5.3.)

10.2.3.7 Other Chemicals

Most powerful chemical rections will kill microbes, but they may be dangerous, messy or uneconomic to use. However, where compatible with a particular kind of waste, all of the following have been used.

Dichromates, permanganates and perchlorates, or even nitric acid can oxidize organic material, sometimes with explosive violence (nitric acid and ethanol mixtures are a common danger). Hydrogen peroxide, peracetic acid and ozone are powerful toxic oxidants which can also be dangerous to use, but decompose to relatively innocuous products. The halogens, chlorine and bromine and iodine, can be used as gases or solutions with an effect ranging from antiseptic to sterilizing depending on concentration.* Compounds which release them (notably hypochlorite, hypobromite and chloramines) have similar effects. Sulphur dioxide gas or sulphite and bisulphite in solution are used for food and drink, because of their action under slightly acid conditions against moulds and bacteria without harming certain yeasts. They can sometimes be used if moulds are a specific problem.

---

*Generally, chlorine is effective at 200 ppm, other halogens at 400 ppm.

10.2.3.8 Ultra-violet Light

UV radiation in the range 200 to 300 nm is mainly used to inactivate airborne microbes, e.g. in enclosed cabinets. It can be effective on clean (e.g. distilled) water to maintain sterility, but its effectiveness is greatly reduced by even a small amount of turbidity, as found in much mains water. It is effective on clean surfaces, but the shadow of a dust particle may be sufficient to protect a micro-organism. In addition, it is common for UV lamps to lose their effectiveness with age. An aged lamp may appear just as bright, but the output of biocidal wavelength is severely reduced.

It is therefore not recommended for waste disinfection.

10.2.3.9 Gamma Radiation

Gamma radiation (usually from a cobalt-60 source) is a powerful and penetrating method of disinfection which can deal with microbes in sealed containers or within a solid object. However, special facilities and techniques are required for its use, which means that it is unsuitable for most laboratories.

10.2.3.10 Laboratory Discard Jars

It is convenient to have containers of disinfectant on the bench into which certain items can be put immediately they have been used. This applies to swabs and pipettes, for example. Note that although this can be excellent practice for maintaining good laboratory conditions, such treatment should not be totally relied upon, especially where resistant or severely pathogenic organisms are involved. A clear phenolic agent is the most generally effective, except where viruses are important. Sodium hypochlorite is effective against viruses, but readily inactivated by the presence of organic material. (See sections 10.2.2, 10.2.3.3, 10.2.3.4, and Appendix B-5.)

The container should be large enough so that the object put inside is completely covered by the liquid disinfectant. Surprisingly, this appears to be a common mistake: long pipettes in short jars, for example. Similarly, items which float (e.g. an inverted test-tube or some plastic items) will have the exposed portion untreated. To prevent floating, items should be put in carefully so that air bubbles are expelled. (In fact with many disinfectants, even a submerged object will not be treated in the portion where the liquid is kept off the surface by an air bubble.) For light solids, a mesh can be used to keep items below the surface.

Glass, polypropylene and enamelled steel are suitable materials for the containers. Nowadays, polypropylene would be the first choice. The use of second-rate glassware, e.g. jam-jars or chipped and cracked beakers — as is done in some laboratories — is bad practice because it is almost certain to lead to a breakage and an injury. Commercial units are available specially for the purpose, but polypropylene beakers and jars are also satisfactory.

Each jar should be filled to three-quarters of its capacity with disinfectant at a suitable concentration for the purpose (as the manufacturer's recommendations: more concentrated for greater soiling). It is convenient to mark the jar with this working level for routine make-up. Material addition to the jar should be stopped when the liquid level rises to 90% of the useful volume. Items should not be removed from a discard jar unless the whole jar has been undisturbed for at least 30 minutes (since recent additions may have infected the contents). It is recommended that items intended for re-use (e.g. pipettes) should be left in disinfectant for 18 hours (i.e. overnight) in a jar reserved for this purpose alone, then removed and hot-washed well. It is best if swabs are taken and cultured to check this protocol. (See section 10.2.3 for 'in-use' tests.)

Normally the jars are left overnight, and the contents poured through an autoclavable mesh or colander. The liquid can then be discarded to the drain, the solid bagged for incineration or removal. As an alternative, it is possible to punch holes in a plastic bag, and pour the discard jar contents into the bag (supported in a bucket, for example). The bag can then drain and be put in another container for final disposal.

It is recommended that the discard jar is hot washed (65 °C for 10 minutes) each day before re-use, and preferable if it is autoclaved each week. This prevents the build-up of a colony of disinfectant-resistant organisms. Polypropylene will go brown with repeated autoclaving, but is still serviceable. Some laboratories find it convenient to have two sets of discard jars, to allow cleaning procedures while routine work continues.

It is strongly advised that the contents of discard jars be changed every day, even if unused. (If many unused jars are involved, consider making up fewer.) The effectiveness of the disinfectant should be checked occasionally by an in-use test (i.e. taking a sample, neutralizing the disinfectant and culturing: see section 10.2.3). Failure is more often due to bad management than to inadequate disinfectant. Common faults are as follows: overloading with too much organic material; 'topping-up' instead of renewing, lack of care in measuring the concentration of disinfectant; keeping disinfectant (as concentrate or as working solution) for too long before use.

If heavy soiling is inevitable, then the use of a non-ionic or anionic detergent may improve the performance of a disinfectant solution.

10.2.3.11 Disposal and Disinfection of Liquids

It is not adequate to pour liquid waste (other than very tiny amounts) into an ordinary discard jar, because it may both dilute and overload the disinfectant. Two alternative techniques are recommended, as follows.

Firstly, where prompt disinfection is preferable, some concentrated disinfectant is put into a flask, so that when liquid is added it will be at a recommended concentration. The waste liquid is added via a funnel, slowly to avoid splashing, and the funnel rinsed with disinfectant at the working concentration. A magnetic stirrer may be used to aid mixing. The mixture is left for at least an hour before discarding to drain. Note that with many disinfectants it is advisable to have the concentrate in the flask at 50% of the commercial strength (i.e. by diluting in an equal volume of water) as this mixes more readily.

Secondly, many solutions can in fact be disposed to drain directly (preferably a sluice direct to foul sewer), providing care is taken to avoid splashing. A funnel may again be useful. The liquid is helped on its way by following disposal with gentle hosing or running of water. It is a useful precaution to add disinfectant each night so that there will be an active concentration in the trap.

With those liquids which are unsuitable for disposal to drain (e.g. oils, emulsions) but which are biologically contaminated, it may be possible to heat sterilize, depending on the boiling point and heat stability of the material concerned.

CAUTION: bottles of solvent or low-boiling liquids should never be put in an autoclave.

Hypochlorite and clear phenolics are generally not very effective against oils, though some of the black and white cresolic disinfectants mix-in better, and may prove an adequate treatment.

Failing this, the liquid should be absorbed onto a suitable solid (which depends on the liquid wetting the solid) in a polythene container and incinerated. See Chapter 7.

10.2.3.12 Large Urine Containers

Polythene containers for a 24 hours collection of urine are now so common that they deserve special mention. Normally, it is acceptable for them to be carefully emptied to a sluice, which is then gently hosed down. If there is any concern about specific pathogens, or if the laboratory does not have access to a direct connection of a foul water (i.e. if disposal has to be to a normal sink) then the contents can be disinfected before disposal. To do this, a suitable concentrate is added, such that it will be at its working concentration in the total volume of liquid. The container should be closed and agitated gently so as to mix the disinfectant, and left overnight.

The polythene containers should be incinerated or autoclaved. They should never be washed out for re-use.

10.2.4 Allergens and Toxins

Not all diseases are due to live micro-organisms, so it is possible for people to suffer adverse reactions even after waste has been totally sterilized. The poisonous products of microbes (or the destruction of microbes) are called toxins. Toxins left on containers or in water can cause illness in people or animals, especially where preparations are made up for injection. These residues are often called pyrogens, so an adequate cleaning routine may be quoted as producing 'pyrogen-free' glassware and solutions.

In addition, materials with a low infective potential can cause intense irritation, amounting to an allergy in some susceptible individuals. Many so-called 'non-hazardous' fungi give highly allergenic spores. Proteins from animal fur or urine are a common cause of allergy. In fact, development of an allergy to a particular animal is an occupational hazard for research workers with animals.

Some toxins and allergens are deactivated by heat, hypochlorite, or prolonged contact with water. If a particular problem material is likely to be present, then its properties can be found and sometimes the sterilizing/ cleaning process adapted to eliminate the substance.

More generally, it is good practice to maintain standards of hygiene to minimize exposure to biological waste, even if sterile. This especially applies to dusts or droplets being thrown into the air.

10.3 MEDICAL LABORATORIES

The principal problem for medical related laboratories is the chance of human pathogens either entering or leaving the laboratory unsuspected. In a hospital, drains and fume ducts may have to pass through other areas, and it is obvious that a leakage could be disastrous in (say) a kitchen or special-care ward. (See Chapter 5 for comments on drains, especially sections 5.4.5 and 5.5. See Chapter 6 for fume extraction, especially sections 6.2.6, 6.2.7, 6.2.8, and 6.4.3 for comments on safety cabinets, and section 6.4.4 for the fumigation of cabinets.)

The correct attitude for infected waste was summed up in 1974 by a very authoritative source:

'It is a cardinal rule that no infected material shall leave the laboratory.' Collins, C.H., Hartley, E.G. and Pilsworth, R. (1974, 1977) *The Prevention of Laboratory-acquired Infection, PHLS Monograph No. 6, HMSO.*

Prompt and effective sterilization on the spot is the ideal method of control (see section 10.2). If this is impractical, then transport elsewhere should be as secure as possible. (See sections 3.3.3, 3.4. and 10.6.)

In the UK, the standard code of practice is usually referred to as the 'Howie Report'. This is in fact the *Code of Practice for the Prevention of Infection in Clinical Laboratories and Post-mortem Rooms* (1978) by the Department of Health and Social Security (ISBN 0 11 320464 7). This handbook will not attempt to summarize the waste disposal provisions, because the entire Code of Practice should be referred to. It is strongly recommended that copies be readily available to scientific and technical staff in any laboratory where infective hazards may occur. In addition, the code includes model rules for domestic service staff, laboratory reception staff, laboratory office staff, porters and messengers, which should be supplied to these people.

The code was aimed at routine clinical work in hospitals. Other laboratories, such as specialist research centres, may find alternative procedures appropriate to their particular situation. On the other hand, it is suggested that many portions could be taken as standards of good biology practice generally (e.g. in veterinary or teaching laboratories). For this reason, the Howie Report recommendations have been followed in this handbook, where appropriate.

The bibliography for this section gives recent standards and recommendations for the control of infective hazards. Note that the 1981 revision of the (UK) Health Building Note 15 'Pathology Departments' contains important changes in line with the Howie Report. There are recent guidelines for the hazards of AIDS and Hepatitis B in laboratory work.

There have been recent changes in the organization and general practice of disposal of clinical waste from hospitals in the UK. Table 10.1 gives the new Health & Safety Commission categories (see bibliography for references). Although certain categories are tolerated for disposal to land, present moves are towards incineration (see section 7.4). Table 3.1 gives a suggested colour code for segregation of bagged waste. See also section 3.4 for some problems in hospital segregation and control of waste, and practical methods of collection.

In concentrating on infective hazards, it is possible to overlook other dangers. Sterilization will not in general control chemical or radioactive hazards, which are increasingly found in medical laboratories. Even where the laboratory does not use a substance directly, it may handle samples from patients that contain potent drugs or radioactive isotopes. The waste disposal procedure must be adequate for these materials. (See Chapter 9 for chemical waste and Chapter 11 for radioactives. See section 12.1 for problems with carcinogens, teratogens and cytotoxic agents. See sections 15.2.4 and 15.2.5 for spill control, and section 15.3.4 for problems with fire.)

10.4 FOOD LABORATORIES

Laboratories dealing with foodstuffs are generally associated with industrial production, and can make use of some of the main waste disposal facilities. However, some food process waste is taken for conversion into animal feed or fertilizer. The laboratory waste should be strictly segregated from any waste intended for further use. In particular, unused samples should never be returned to the main process. The collection and storage arrangements must prevent mistakes by, for example, the use of differently coloured and labelled bags. (See section 3.4.)

Likewise, food samples should not be eaten (except where taste is part of the test procedure) or taken away by staff. It is also bad practice to bring food and drink (e.g. sandwiches and coffee) into the laboratory, because of the possibility of contamination or errors. (See section 3.3.2.)

Table 10.1. Categories of clinical waste as defined by the UK Health &
Safety Commission

GROUP A		Acceptable for landfill
(a)	Soiled surgical dressings, swabs and all other contaminated waste from treatment areas.	YES
(b)	Material other than linen from cases of infectious disease.	NO
(c)	All human tissues (whether infected or not), animal carcasses and tissues from laboratories, and all related swabs and dressings.	NO
GROUP B		
	Discarded syringes, needles, cartridges, broken glass and any other sharp instruments.	YES
GROUP C		
	Laboratory and post-mortem room waste other than waste included in Group A.	NO
GROUP D		
	Certain pharmaceutical and chemical waste (see note).	
GROUP E		
	Used disposable bed-pan liners, urine containers, incontinence pads and stoma bags.	YES
NOTE:	Some of Group D is Hazardous Waste as defined by the Control of Pollution Act 1974. If landfill is used, the site must be licensed for the material, and the transport (Section 17) regulations must be followed.	

The great majority of food laboratory materials do not contain significant quantities of harmful micro-organisms, so they do not necessarily require sterilization. In general, all that is required is essentially good kitchen practice, i.e. the use of covered containers, prompt and frequent removal, and scrupulous attention to washing and disinfection. However, there ought to be some procedure available to neutralize materials which are found to be harmful (e.g. if a test for salmonella proves positive.)

A typical laboratory will generate small amounts of highly processed samples (resulting from digestions, extractions and chemical treatment) which are essentially chemical waste, and a much larger amount of material which has merely had a tiny portion removed, or has suffered only trivial contamination (by having an aqueous extract taken, for example).

Much of this latter waste can be disposed to drains (see Chapter 5) by means of a kitchen-type grinder waste disposal unit fitted into a sink, or a similar free-standing macerator such as hospitals use. This should not be used for the disposal of chemicals as such, but can tolerate minor contamination providing it is frequently well rinsed. The drainage connections should be carefully arranged so that there is no possibility of contamination from the foul sewer coming back into the laboratory. (See section 5.5.2.)

General solid waste should be collected in closed or covered containers, usually in a supported plastic bag in a self-closing device. The bags should be removed (not emptied) daily or when two-thirds full, and sealed with tape or by a heat-sealing unit. It is preferable for food waste to go to incineration, but it may also go to a sanitary landfill site dealing with domestic garbage. Sharp objects or harmful chemicals should be kept out of this waste and disposed of separately. (See section 10.6.)

If facilities are available, then it may be practical to cook some wastes before disposal. Care must be taken that plastics and chemicals do not get into the oven.

Some laboratories find a compaction unit useful. This reduces the volume of typical waste several times, giving a more hygienic dense bale in a sealed plastic sack which is more convenient to store and transport.

CAUTION: a compactor or baler must not be used for waste to be incinerated. See section 7.4.5.2.

Where cultures are made, it is vital that they do not contaminate incoming samples. Storage of samples and waste should be segregated from the culture area and from each other. There should be facilities for the sterilization of cultures and for the regular disinfection of areas nearby.

Although the majority of waste from a typical food laboratory can be disposed of relatively easily, it is very important that there is a specific policy for more difficult material, whether regular (e.g. sharp objects) or very rare (e.g. spilt mercury) to ensure that hazardous items are not put into ordinary waste. (See Chapters 3 and 12.)

The apparatus and procedures normally adopted in food science are intended to protect the sample from contamination by the worker. If it is known or suspected that contaminated foodstuff is to be handled, the same procedure may not necessarily protect the worker from the sample. This applies for example in the case of laminar flow cabinets (see Chapter 6).

10.5 ANIMALS

10.5.1 Legal Aspects (UK, USA, Canada)

The use of animals in laboratories may be subject to national and local laws, both to protect the animal and to protect public health. In the UK, a licence is required from the Home Office for work with live animals which involves injections, surgery, anaesthesia, or the possibility of inflicting pain (physical or mental). A demonstration of a technique is still covered by the regulations, even though it may not be an experiment.

The premises of licence-holders are subject to inspection by Home Office inspectors, who are concerned with the general keeping of the animals as well as the actual experiments.

Where animals are kept for purposes not requiring a licence (e.g. for nature observation in schools) then it is still necessary to ensure that they are kept in adequate conditions and properly fed and watered. Failure to remove excreta, dead animals, rotting food, etc. from cages may be considered cruel, and thus render the institution liable to prosecution. (Apart from the penalties imposed by the courts, there is considerable public ill-will towards those convicted of cruelty to animals.)

Inadequate disposal of animal waste or of dead animals may be an offence under the Public Health or other laws. Escaped animals (including flies) can cause breaches of environmental health legislation — for example if they escaped into kitchens.

Current information on UK law relevant to animals in laboratories is obtainable from the Research Defence Society, Grosvenor Gardens, London SW1W 0BS.

In the USA, very similar principles hold true for the public and environmental health aspects, but cruelty laws vary considerably from state to state. Up-to-date information can be obtained from the Animal Welfare Institute, P.O. Box 3650, Washington D 20007, USA.

In Canada, legal and other aspects of laboratory animal work are obtainable from the Canadian Council on Animal Care, Suite 1105, 151 Slater Street, Ottawa, ONTARIO K1P 5H3, Canada.

10.5.2 General Hygiene

To keep animals in a healthy and scientifically worthwhile state requires a consistent standard of care and hygiene, in which adequate waste disposal is a vital aspect. It is found in practice that a clean environment and careful techniques to avoid the spread of contamination are the most effective means of controlling disease. Heavy use of disinfectants or expensive air purification will not undo the damage which can be done by simple inattention to cleanliness in dealing with the animals.

The major means of transferring disease is via dust, particularly from fouled bedding, although airborne hair and feather fluff are also carriers.

Operations such as sweeping or shaking used food and bedding are almost certain to disseminate disease around a room. Mishandling of animals so that they struggle violently can spread dust around: this is especially true of birds. Although a vacuum cleaner is a good method of picking up dust, its outlet can blow material into the air. If vacuming is done in any room where animals are kept, it is preferable to have a piped system with the fan and filter elsewhere. However, the installation should be such that blockages can be removed without great disturbance, and it is preferable if the pipes are suitable for fumigation

Unused food from a cage should not be transferred to another animal, but should be disposed of as contaminated waste.

The regular use of an ampholytic detergent (or the alternating use of a cleanser and a disinfectant) on floors and walls will normally keep the room sufficiently free of micro-organisms and parasite eggs so that fumigation will only very rarely be necessary.

For the very highest standard of work, it may be a necessity to have air and water purified to an extraordinary level. However, for most work, an air system which prevents the entry of insects and dusts provides sufficient protection. As well as being less expensive, a coarse filter requires less maintenance than a very fine or absolute system, and it may be better to have a less efficient system working all the time than a super-efficient system which sometimes fails. Similarly, water suitable for human drinking purposes is generally adequate. Some supposed sterilizing systems are largely cosmetic: those that are truly effective are demanding of money and attention. Acidification of drinking water to pH 2.5 with hydrochloric acid can be used to suppress fungal, algal and bacterial action in the water bottle for rats and mice, but not other animals.

Access to rooms where animals are kept should be limited to as few people as possible, preferably those with some training in animal keeping. This is one of the most important practical methods of hygiene.

It should be remembered that animals can be very sensitive to traces of chemicals (including disinfectants and anaesthetics) in their environment. Mice have been known to die from traces of chloroform from an adjacent room. Many laboratory animals have been kept in a state of permanent poor health due to their environment (e.g. a cage floor) being regularly treated with a persistent disinfectant (e.g. a cresol). It has even been known for one laboratory worker to use insecticidal agents in the same room where another was trying to maintain a colony of flies!

Thus it is important that hygiene means a clean environment rather than one in which microbes are suppressed by a permanent high level of chemical agents.

10.5.3 Bedding

Animal litter should be selected from materials appropriate to the animal, as given in an authoritative source, such as the UFAW Handbook. The

27

28

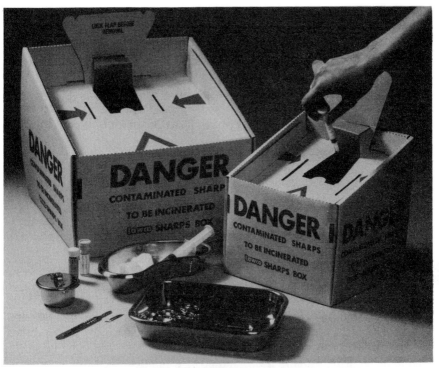

29

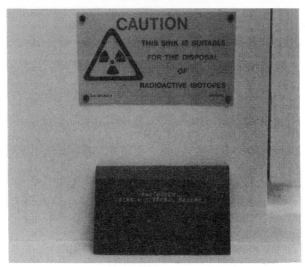

30

㉗ A bucket of disinfectant. The floating items are not actually being disinfected.   ㉘ A commercial bin for the collection of disposable 'sharps'. This one has been filled with tissues.   ㉙ Commercial boxes for the collection of medical 'sharps' to be incinerated. Photo courtesy of Lawtons Ltd, Liverpool.   ㉚ A clear sign and a record book for the amounts of radioactive substances put down this sink.

material should be that sold by a reputable supplier especially for laboratory animals. It is possible for a knowledgeable person to make up an alternative, but wood products in particular may contain irritating substances or particles. It is tedious, but possible to sterilize bedding: the mineral-based ones present fewest problems, but require adequate time (usually under vacuum) to free them of ethylene oxide if used.

Bedding needs to be changed regularly, and more often if there is any occurrence such as leaking water bottle. It has been known for staff to be reluctant to handle a particular animal (such as a very aggressive one, or one being used to breed fleas) and therefore to skip this chore. It is a management responsibility to deal with human failings of this nature.

The depth of litter can be important. In the case of fowls (which tend to peck on the floor) shallow litter will allow rapid transmission of intestinal infections and parasites. The litter should be deep and loose enough to permit faeces to fall from the surface, and should be particularly renewed in areas of high contamination such as around a feeder in a communal pen.

Bedding which is very dry is likely to give rise to dust which can cause respiratory infections. On the other hand, moist bedding (from urine or drinking water) encourages mould, permits the survival of nematodes and promotes bacterial decomposition of urine and the liberation of ammonia. Peat, as a litter material, has the advantage that is slightly acid and thus ammonia liberation is inhibited. However, the real key is to keep the humidity at a reasonable level, and change the bedding often enough to avoid saturation.

Probably the worst possible practice is still carried out in some smaller laboratories for changing bedding. This involves the use of a spare cage into which each animal is put in turn while its cage material is changed in the same room. The common contact point and the dust generation ensure that diseases are rapidly spread throughout the stock.

By contrast, the two-corridor system has proved successful with larger animal houses. Its advantages can be gained by other means on the small scale by an appropriate routine. In the two-corridor system, the animal cages are contained in a series of rooms with a door at each end leading to a 'clean' and a 'dirty' corridor. See Fig. 10.1 To change bedding, all the cages in one room are wheeled into the 'dirty' corridor, and the room cleaned as necessary. A complete set of cages with fresh bedding is wheeled in from the 'clean' corridor. The animals are then transferred one by one to their appropriate new cages. For 'shoebox' style cages, the cage lids and water bottles may sometimes be retained.

The cages are then wheeled down the corridor to a work area where the soiled bedding is removed into bags and the water bottles and feed trays are emptied. The cages are washed (by hose or machine) and passed (usually wheeled) into a double-ended autoclave. The autoclaved cages can be removed into the 'clean' corridor for the next use.

It is advisable for the person emptying bedding to wear a protective dust mask. If there are likely to be any special problems (i.e. organisms which

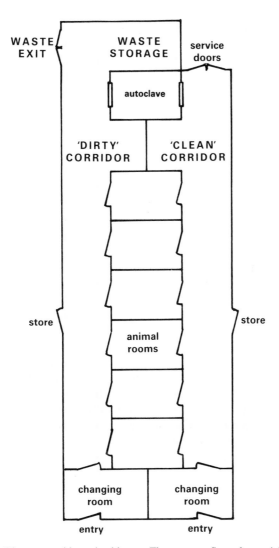

Fig. 10.1. The two-corridor animal house. The one-way flow of material from the 'clean' to the 'dirty' side means that waste does not contaminate feed, bedding, etc. Hardware is returned only via the autoclave (or a disinfectant bath in some establishments).

affect humans, chemical residues from special diets, or an individual's special allergy) then a greater degree of protection by ventilation or other means is required. Proper attention to personal hygiene after handling soiled bedding is of course essential. Gloves and clothing after this operation will definitely be 'dirty' and must not be brought into the clean area.

The bedding is collected in plastic or paper sacks of sufficient strength to be transported as necessary. The technique of double-bagging is strongly recommended if (a) the waste has to travel any distance (b) it is to be handled by people other than laboratory staff, e.g. refuse collectors (c) there are any special hazards. Primate waste should always be assumed to be especially hazardous.

Some landfill sites will accept untreated animal bedding, providing it is securely packed. This most commonly occurs in association with the disposal of hospital waste, but is relatively unusual. Some sites will accept such waste which has been sterilized, and occasionally the local authority refuse collectors will collect such waste by prior arrangement.

From a management point of view, the simplest method is burning in an incinerator on-site. This is generally reliable, but care must be taken that fumes from the incinerator (which may contain infective particles) are not drawn into the animal ventilation system. A chute or other closed system for transport to the incinerator house is preferable, especially for large institutions. However, waste must be securely bagged for any transport, even by chute, and laboratory treatment (autoclaving, double-bagging) is still advised prior to incineration of especially hazardous infected material. (See Chapter 7.)

Regular arrangements may alternatively be made for burning at a local refuse incinerator. For this purpose, bags may be stored in a closed skip or protected bin.

Very small laboratories with less hazardous animals and materials (e.g. some schools) may be able to have bagged animal waste included in the normal refuse collection. As an alternative, some can be disposed to foul sewer via a macerator or sink waste disposal unit — see section 5.3.5.

It is vital that fouled bedding is not stored (even temporarily) in an area where it might contaminate food or fresh bedding, whether directly or indirectly. Collection of refuse and delivery of goods should be carefully segregated for this purpose. Sacks should not be left unattended in an area (e.g. a corridor) where they might get damaged from traffic (e.g. the corner of a trolley).

Even if bedding is not routinely treated, it is necessary to have some arrangement (i.e. both facilities and a laid-down procedure) to disinfect particular batches, where some infection has occurred. (See section 10.2.) It may be noted that deodorizing tablets are available which greatly reduce the smell from steam autoclaving faeces and urine.

10.5.4 Live Animals

It is not in general a good idea for unwanted laboratory animals to be taken home as pets by laboratory staff. For one reason, the domestic animal is more likely to contract disease which can be carried in by the staff member to infect the stock animals. What can be even worse is for a person not trained in animal care (and often not equipped for the special needs of the animal) to take home an unusual but emotionally appealing creature, such as a monkey.

Where an unusual, special or expensive animal is in good health but no longer wanted (or can no longer be afforded) then offers may be made to other institutions. However, sick animals should be destroyed except in the most exceptional circumstances.

The person responsible for the laboratories should ensure that there is at least one person who can be relied upon to carry out the humane killing of the animals. It is in fact a good idea for all experimentalists to know an appropriate rapid technique by physical (e.g. breaking the neck) or chemical (e.g. lethal injection if the person is competent) means. This allows them to take immediate action in the emergency of an animal being in severe pain. It is essential that workers realize that the techniques are specific to the animal, and should not be applied to another species.

Special techniques of anaesthesia and killing may be necessary for particular experimental purposes. However, if the animal is merely unwanted, then the least distressing general method appears to be anaesthesia by a mixture of 70% CO_2, 30% O_2 (not air), followed by 100% CO_2. Details are available in the UFAW handbook or from one of the laboratory animal institutions.

Unless death is by decapitation, exsanguination or (for large animals) a properly applied humane killer, then final disposal of the carcass should not be commenced until rigor mortis has set in.

10.5.5 Carcasses

As for bedding waste, one of the simplest techniques is by incineration within the facility, providing the incinerator is suitable (which incinerators for other purposes are usually not).

Local authorities often operate a collection and disposal service for veterinary surgeons, which the laboratory may join. Alternatively, the laboratory may itself arrange transport to a designated local site (most probably an incinerator). Exceptionally, a disposal operation at a local hospital, abattoir, or livestock breeding centre may agree to take animal carcasses.

Carcasses or parts of animals should not be included in ordinary refuse unless specifically authorized by the refuse collection service. This will normally require the items to be both autoclaved and securely packaged.

Hospital-style macerators can be used very effectively to dispose of carcasses via the foul sewer. For very small animals (worms, fish, lizards, mice, etc.) a sink waste disposal grinder may be adequate. Modifications may be required to accept frozen animal tissue.

Carcasses containing radioactivity must be disposed in accordance with the directions of the radioactivity enforcement agency. This may include burial at a specific site. Note that radioactivity can accumulate in certain organs, and it may only be these which have to be dealt with in an exceptional fashion. (See Chapter 11.)

High temperature incineration is at present the only feasible method of disposal of animal tissue contaminated with potent carcinogens. See Chapters 7 and 12.

The method of containment and handling must be appropriate to any special hazards (human pathogens, toxic chemicals, radioactivity, carcinogens) for transport purposes.

It is often necessary to store dead animals prior to disposal. A refrigerator is adequate for overnight; a freezer is preferred for longer periods. It is preferable if the animals are placed in plastic bags prior to freezing, with the date and any necessary information. If ether or other flammable solvent is used to anaesthetize or kill the animals, then the refrigerator or freezer *must* be internally spark-proofed.

As an alternative, carcasses can be temporarily preserved in formalin, or kept in a cool place in a plastic bag containing sufficient quantity of vermiculite to absorb body fluids, and sufficient bleaching powder to inhibit decomposition. See section 11.10.

Where food animals (e.g. fish, fowl) are used for dissection or similar procedures, then the carcasses may often be disposed of as food waste (i.e. like catering waste: see section 10.4) providing laboratory chemicals and the like are excluded from the collection bags.

10.6 SHARP OBJECTS

According to one study,* infections from laboratory accidents were caused by hypodermic syringes in 25.2% of cases, and from broken glass or other sharp objects in 15.9% of cases. To guard against accidental inoculation, no sharp or fragile contaminated item should be handled if at all possible. Thus people should not move items from one container to another. The practice of salvage from mixed discarded labware is a dangerous practice which unfortunately persists, though the cost savings are usually illusory.

For bench or ward workers, there should be a suitable container within arm's reach for any repetitive work involving disposable sharps. A beaker or glass jar is better than just putting items down on the bench, but it is preferable to have a commercial purpose-designed and clearly-labelled container. These are closed and removed for disposal as often as necessary, which may be twice a day, but should not normally be less often than once a week.

Under no circumstances should sharp objects be placed in an ordinary plastic bag, though some reinforced paper or polypropylene may be strong enough for limited transport of broken glass. However, it is quite possible to place sealed containers of sharps into plastic bags for removal. One company produces reinforced cardboard envelopes for all sharps, which are shaped to fit a waste bin.

---

*Pike, R.M. (1976) Laboratory-associated Infections. Summary and Analysis of 3921 Cases. *Heath Lab. Sci.,* **13**, 105–114.

There are a variety of designs commercially available, mainly in toughened cardboard or rigid plastic. No one type can be said to be best, and a laboratory may find that units from two or more manufacturers are required. The main considerations are the size and shape of objects to be put in, and the number and sizes required for the most effective management. (See Appendix B-2 for manufacturers).

Some units are sold for the special disposal of hypodermic syringes. Many models include a device to bend the needle or break off the luer fitting. This renders the syringe unusable, in order to discourage theft for drug abuse. Where a laboratory is dealing in very hazardous materials (virulent pathogens or anti-neoplastic drugs, for example) then the syringe should be discarded whole. This is because the additional operation creates a hazard, including the possible unnoticed release of an aerosol. Instead, the most particular care should be taken that the unit is disposed of correctly. (See section 3.3.3 on safety and security.)

Sharps' containers should be closed for disposal when they are about two thirds full (some manufacturers place an indication of the working level). Some users like to place a wad of paper or cotton wool with disinfectant on top of the waste (where the design allows). If this is done, the wad should be handled with tongs and *not* pushed down with the fingers.

It is generally considered preferable for sharps' containers to be promptly incinerated. If this is done on the premises, then the incinerator must have facilities such as a special tray so that there is no hazard to the operator from sharp items in the ash. Where incineration is carried out at a large munici-pal or industrial unit, then there is invariably machinery to cope with this problem. However, steps must be taken to ensure that the material is not a hazard during transport to the incinerator. The first of these is to check that the disposal service actually goes to this destination — many services will sometimes use a domestic refuse tip, unless they are fully informed that this is unacceptable. Secondly, the various laboratory containers must be packed for safe transit: a fibre-board drum is ideal. Finally, some consideration should be given to the question of treating the materials before disposal. Autoclaving will damage some containers, and may not penetrate sufficiently. It is not usually a good idea to pour in disinfectants, which will have to be poured out, or may in any case leak. Ethylene oxide treatment is likely to be the most effective, but is impractical for most laboratories. Where special hazards exist, then special techniques are needed.

BIBLIOGRAPHY
[section 10.1]

Hartree, E, and Booth, V. (eds) (1977) *Safety in Biological Laboratories,* Biochemical Society.

Hellman, A., Oxman, M.N. and Pollack, R. (eds) (1973) *Biohazards in Biological Research,* Cold Spring Harbor Labs.

Shapton, D.A. and Board, R.G. (eds) (1972) *Safety in Microbiology,* Academic.

UK Dept of the Environment (1983) *Waste Management Paper No. 25: Clinical Waste,* HMSO.

UK Dept of Health & Social Security (1978) *Code of Practice for the Prevention of Infection in Clinical Laboratories and Post-mortem Rooms [The Howie Report],* HMSO.

UK Health & Safety Commission (1982) *The Safe Disposal of Clinical Waste.*

[section 10.2]

Borick, P.M. (ed) (1973) *Chemical Sterilization,* Dowden, Hutchinson & Ross.

Collins, C.H. (1983) *Laboratory-acquired infections,* Butterworths.

Collins, C.H., Allwood, M.C., Bloomfield, S.F. and Fox, A. (eds) (1981) *Disinfectants: Their Use and Evaluation of Effectiveness. Soc. Appl. Bacteriology Tech. Series No. 16,* Academic.

Collins, C.H., Hartley, E.G. and Pilsworth, R. (1974) *The Prevention of Laboratory Acquired Infection. Public Health Service Monograph 6,* HMSO.

Hugo, W.B. (1971) *Inhibition and Destruction of the Microbial Cell,* Academic.

Lowbury, E.J.L., Ayliffe, G.A.J., Geddes, A.M. and Williams, J.D. (1975) *Control of Hospital Infection — A Practical Handbook,* Chapman & Hall.

Sykes, G. (1967) *Disinfection and Sterilization 2nd edn.,* Spon.

Sykes, G. (1969) *Methods and Equipment for Sterilization of Laboratory Apparatus and Media,* In: Norris, J.R. and Ribbons, D.W. (eds) *Methods in Microbiology 1,* Academic.

UK Dept of Health & Social Security (1980) *Health Technical Memorandum 10: Sterilizers,* HMSO.

[section 10.3]

Advisory Committee on Dangerous Pathogens (1984) *Categorisation of Pathogens According to Hazard and Categories of Containment,* HMSO.

American National Standards Institute (1981) ANSI/NFPA 56C-1981 *Safety Standards for Laboratories in Health-related Institutions,* ANSI.

Collins, C.H.(1983) *Laboratory-acquired Infections,* Butterworths.

UK Dept of the Environment (1983) *Waste Management Paper No. 25: Clinical Waste,* HMSO.

UK Dept of Health & Social Security (1978) *Code of Practice for the Prevention of Infection in Clinical Laboratories and Post-mortem Rooms [The Howie Report],* HMSO.

UK Dept of Health & Social Security (1981) *Health Building Note No. 15: Pathology Departments,* HMSO.

UK Dept of Health & Social Security (1985) *AIDS (Acquired Immune Deficiency Syndrome): Interim Guidelines*, HMSO.
UK Health & Safety Commission (1982) *The Safe Disposal of Clinical Waste*, HMSO.
UK Health & Safety Executive (1985) *Safety in Health Service Laboratories: Hepatitis B*, HMSO.
U.S. Public Health Service (1974) *National Institutes of Health Biosafety Hazards Guide*, U.S. DHEW.
World Health Organization (1983) *Laboratory Biosafety Manual*, WHO.

[section 10.5]

Biological Council (1984) *Guidelines on the Use of Living Animals in Scientific Investigations*, Institute of Biology.
Buckland, M.D. (ed) (1981) *Guide to Laboratory Animal Technology*, Heinemann.
Inglis, J.K. (1980) *Laboratory Animal Technology*, Pergamon.
Orlans, F.B. (1977) *Animal Care from Protozoa to Small Mammals*, Addison-Wesley.
Seamer, J.H. and Wood, M. (1981) *Laboratory Animal Handbooks: 5. Safety in the Animal House*, Lab. Animals Ltd.
(UK) Universities Federation for Animal Welfare (1976) *The UFAW Handbook on the Care and Management of Laboratory Animals, 5th edn*, UFAW.
(UK) Universities Federation for Animal Welfare (1978) *Humane Killing of Animals, 3rd edn*, UFAW.
(US) Institute for Laboratory Animal Resources (1979) *Animals for Research 10th edn*, Nat. Acad. Press.
(US) Institute for Laboratory Animal Resources (1979) *Laboratory Animal Houses*, Nat. Acad. Press.

[section 10.6]

Hansford, J. (1979) Sharps and Their Disposal, *Sterile World*, **1**, No. 3, 5–8.
UK Dept of the Environment (1983) *Waste Management Paper No. 25: Clinical Waste*, HMSO.
UK Dept of Health and Social Security (1982) *Specification No. TSS/S/330.015: Specification for Containers for the Disposal of Used Needles and Sharp Instruments*, DHSS.
UK Health & Safety Commission (1982) *The Safe Disposal of Clinical Waste*, HMSO.

CHAPTER 11

Radioactive Substances

11.1 INTRODUCTION

11.1.1 General Comments

In the UK, USA and many other countries the use of the radioactive substances is rigorously controlled by law, and users have to be registered, unless exempt. In the UK, enquiries should first be addressed to the local office of the Health & Safety Executive, though registration may be required with the Department of the Environment in England, and the equivalent bodies in Scotland, Wales and Northern Ireland. In the USA, enquiries should first be addressed to the Nuclear Regulatory Commission, Washington DC, though in some states a licence is issued by the state instead of the NRC.

In order to use radioactive substances at all, many requirements have to be met in order to satisfy the appointed inspectors. An approved method of disposal of waste is one of the most crucial of these requirements.

It must therefore be emphasized that any proposed method of disposal must be discussed with the local inspector, who will normally have available the necessary information to advise on practical problems. Each case will be firstly dependent on the type and amount of radionuclide and secondly on the exact situation. This chapter is not intended to replace consultation with inspectors and local experts, but may assist in pointing out some of the matters which need to be considered.

Radioactivity is much feared by the general public, and even small incidents with negligible risk can generate embarrassing bad publicity.

206

Money and effort spent on control of waste disposal can therefore be justified for good public relations as well as for the health of employees.

In this chapter, it will be assumed that the laboratory makes minor or incidental use of radioactive sources, and that the person responsible for waste disposal is generally familiar with the nature of radiation, and the characteristics of the material being used.

11.1.2 Classification

Radionuclides are classified in terms of radiations they emit, and also in terms of their toxicity if they should enter the body. The classification of radio-toxicity recognizes four classes, with class I being the most toxic. In some texts they are referred to as 'High Toxicity' (class I), 'Medium Toxicity — Upper Sub-Group A' (class II), 'Medium Toxicity — Lower Sub-Group B' (class III), and 'Low Toxicity' (class IV).

A class I radionuclide may be a million times more toxic than a class IV radionuclide which emits radiation at the same count rate (i.e. has the same radioactivity — see section 11.1.3). This illustrates the need to pay attention to the individual characteristics of the substances in use.

As a general rule, alpha radiation is the most harmful inside the human body. Thus strong alpha-emitting substances are specifically prohibited for many waste disposal routes. The isotope strontium-90 is dangerous because of its accumulation in the bones, so it is likewise prohibited by many general and local regulations.

Beta and gamma radiation can be directly hazardous without ingestion, so for waste disposal it is important to limit the radiation level on the exterior of any container (including sink waste traps).

The half-life is important since it determines how long the material will remain active. A half-life of a month or less means that the material can be stored until the activity is reduced to a sufficiently low level. A half-life of a year or less means that (in principle at least) it can be permitted to decay buried in a landfill site (i.e. there is no permanent ecological hazard). Substantially longer half-lives mean that for human purposes the material is permanently radioactive and must be disposed of so that it does not create an environmental hazard.

Note that the physical, chemical and biological states of the material are of major importance in determining a safe waste disposal procedure.

11.1.3 Units

Until recently, the fundamental unit of *radioactivity* was the curie. This is the unit most workers recognize, and is found in all but the most recent literature and regulations. One curie (Ci) of radioactive material is the quantity which undergoes 3.7×10^{10} nuclear disintegrations per second. It is rather a large unit for most laboratories, which generally dealt in millicuries (mCi = one thousandth of a curie) or microcuries (μCi = one millionth of a curie).

Nowadays the SI unit is used. This is the becquerel (Bq) which is simply the quantity of radioactive material which undergoes 1 nuclear disintegration per second. This is a rather small unit, so laboratories generally deal with kilobecquerels (kBq = 1000 Bq) or megabecquerels (MBq = 1 000 000 Bq). It is obvious that one curie equals 3.7×10^{10} becquerels.

In general, the control agencies have simply converted from one unit to another so that a laboratory may, for example, be given a limit of 37 MBq instead of 1.0 mCi for waste storage or disposal. Suppliers may also offer sources in sizes of 37 kBq instead of μCi.

The *specific activity* has two commonly used meanings. When specified per mass (Bq kg^{-1} or Ci/gram) and for a particular isotope, it is a physical property, namely the intrinsic activity of that isotope (or a fixed mixture). (For example, ^{239}Pu has a specific activity of 2.3×10^{12} Bq kg^{-1}, which equals 62 mCi/gram.) The same term is incorrectly, but very often used for the volume concentration of radioactivity (Bq m^{-3} or Ci/litre). Other things being equal, a low concentration (a dilute waste) is less of a hazard than a more concentrated one.

The *dose rate* is used (amongst other things) to measure the intensity of radiation coming from a surface (e.g. a waste bin, a drain pipe, the wall of a store). This involves the number of radioactive particles given off, and their individual energies. For example, typical alpha particles carry 10 to 20 times the energy of beta particles, so would give 10 to 20 times the dose rate for the same count rate.

The traditional unit of dose rate was rems per hour. The SI unit is sieverts per second (Sv s^{-1}) where one sievert equals 100 rems. For typical laboratory use this unit is a million times too big, so the usual units are microsieverts per second, where 1.0 μSv s^{-1} = 0.36 rem/hr. As with activities, the control agencies have tended to simply convert units in their specifications. For example, a permissible dose rate of 0.5 rem/hr is replaced by a permissible dose rate of 1.39 μSv s^{-1}.

11.1.4 Records

A practical and reliable system of record-keeping is absolutely essential for the use and disposal of radioactive material. The enforcing authority will demand certain records to be kept in a particular format, and copies to be returned at intervals.

A practical minimum for small users is to keep a hardback book, with appropriately ruled columns, in the laboratory. Every user should record the amounts of radioactive materials brought into the laboratory, amounts returned to store (where appropriate) and estimates of the amount and route of loss or disposal. The person charged with record-keeping will have a separate record in a different room (for security in the event of fire etc.). He or she will make the necessary calculations to convert the laboratory notes into an approved form, and will update the formal record at frequent intervals. The central record may of course serve several laboratories.

A variation on this system is to have a book which records materials brought into the laboratory or removed to store elsewhere, and separate books at disposal points such as sinks or fume cupboards. Each waste disposal action should be separately noted by the person concerned. Once again, there will be a separate central record regularly updated from the laboratory books. It is usual for there to be some small discrepancy between materials brought in and disposed of, owing to the difficulties of estimating quantities.

Loose-leaf systems have some advantages. For example, a laboratory worker may be issued with a standard form when collecting radioactives from storage, and be required to return the form with details of disposal when returning unused material. However, great care must be taken that sheets are not mislaid or borrowed before the details are transferred to a more permanent record.

Laboratories with substantial use of radioactives should take special steps to secure the records of radionuclide holdings, waste disposal and human exposure. It is suggested that key records should be kept in a fireproof and secure cabinet. A duplicate copy should be made of the most important facts and kept elsewhere, preferably in another building. One method of doing this is to photograph documents every 3 or 6 months and keep the developed film in a safe place.

11.2 SEALED SOURCES

Under no circumstances should a user attempt to open, repair or modify a sealed source. Such operations should only be carried out by the manufacturer's representative or another person specifically authorized to do so, on premises designed for the purpose. If a source is faulty or damaged, it should be put into a secure store and advice sought from the supplier and the regulatory agency about its disposition.

11.2.1 Source with short half-lives

If the source has a half-life such that storage for 7 or more half-lives is feasible, then this will reduce the radioactivity by over 99%. The spent source may then often be disposed of as normal laboratory waste. However, it is expedient if it is deposited in a site or incinerator which can accept low-level radioactive waste, and appropriate people should be notified. This is to prevent any scare due to a defunct source being accidentally discovered and thought to be active.

Some short-lived sealed sources are made so that they can be safely flushed into the sewer. These are mainly items used for tracing drains, or for medical purposes which are swallowed and later excreted by the patient. There will be no doubt if a source of this kind is in use. No other sealed sources should be disposed of into the drains.

11.2.2 Sealed sources with longer half-lives

Where storage is impractical for the necessary period, the source should, if possible, be returned to the supplier (who may make a charge for the service). Otherwise it should preferably be taken by a specialist licensed waste disposal contractor or a national nuclear agency such as Harwell (Atomic Energy Research Establishment), UK. For small sources, it may be permissible for the source to be buried on a licensed site, with the approval of the regulatory agency.

Sealed sources are now to be found as part of measuring devices in many machines and instruments. Usually the sources are so small that little harm would be likely to occur if they were deposited on a refuse tip. However, when scrapping or destroying old or unwanted apparatus, advice should be obtained as to the appropriate treatment for any part containing radionuclides. It is considered good practice to damage the parts (without damaging the source) to make them less attractive as salvage.

It is not normally appropriate to include sealed active sources in waste destined for incineration.

11.3 NATURAL URANIUM, THORIUM AND POTASSIUM

These elements are radioactive in their naturally occurring isotopic mixtures, but not strongly so. It is possible to use small amounts of them accordingly. Natural potassium can be disposed of without limit from radiation regulations.

Natural thorium is sometimes referred to as 'Thorium-X', notably in certain exempting regulations. Natural uranium consists mainly of the relatively inactive isotope ^{238}U, with about 0.7% more active materials, notably ^{235}U. Compounds are now available in which the more active portion has been reduced to about 0.2%, in which case the element is referred to as 'depleted uranium'. Depleted uranium represents a negligible radiation hazard, but is highly toxic.

If the laboratory is part of a site using substantial amounts of thorium or uranium (kilograms or more) then a licence may be required from the appropriate authority. If the laboratory uses other radioactive substances, then thorium and uranium will be included in the waste disposal limits. Otherwise, amounts of less than 37 MBq (1 mCi) per day may be disposed of as chemical waste. This is equivalent to about 1 g of uranium or 10 g of thorium. There are special arrangements for thorium waste resulting from medical work. For regular use or larger inventories, it is recommended that the enforcement agency be consulted (see section 11.1).

11.4 MATERIALS WiTH VERY SHORT HALF-LIVES

Radio-isotopes with very short half-lives are produced (often to order) by atomic bombardment in nuclear installations, or sometimes as the secondary product of some other radioactive decay. They have the special property that the radioactivity diminishes rapidly in storage. Usually a period of 7 half-lives will reduce the activity to a negligible level, and 10 half-lives will virtually eliminate it. The resulting waste can then usually be disposed of as very low-level waste.

The storage conditions need to be adequate in shielding and security for the original material. Furthermore, they must take into account the chemical properties of the whole material, and any physical or biological problems. For example, the material may be flammable, corrosive, liable to putrefy, or have sharp edges. Particular care should be taken if the radioactive decay process gives rise to a gas, especially if the gas is itself radioactive (e.g. radon).

Physical or chemical treatment may be used to reduce the bulk to be stored. For example, paper etc. may be baled, empty bottles etc. may be crushed, chemicals may be concentrated by precipitation, distillation or absorption onto an ion-exchange resin. (See sections 9.3.6, 9.3.7, 14.4, 14.6.) Generally it is safer to store solids than liquids, but fine powders can be particularly hazardous: they should be well packaged and preferably treated with a suitable liquid (e.g. mineral oil or glycerol/water) so that they will not form dust clouds.

Such treatments should be carried out by knowledgeable people, using apparatus reserved for radioactive work. Care should be taken that treatment procedures do not put people at any additional risk. It may sometimes be better to simply arrange adequate storage than to perform complex procedures.

Each package put into storage must be clearly marked as radioactive, with any important details of the material and (absolutely vital) the date. A tied plastic bag is a minimum container for low activity solids. Solid chemicals and material as particles should be in separate plastic bags, plastic jars or cans within the larger sack, if used. Bottles or cans of liquids should be standing in trays to contain any possible spillage. Biological materials (e.g. animal carcasses) may need to be stored in a refrigerator or freezer. For the vast majority of radioactives the refrigerator or freezer requires no special modification other than a radiation sign and strict control as to its usage. As an alternative to refrigeration, some biological specimens may be preserved in formalin using standard techniques. Small animal carcasses can be kept from decomposition by sealing in a plastic bag with bleaching powder and an inert absorbent. It is very important that the correct technique is followed for this. Details are given in section 11.10, *Nature* (Lond.), **179,** 54 (1957) and in the HSE 'Guidance Notes for the Protection of Persons Exposed to Ionising Radiations in Research and Teaching, HMSO (London), 1976.

Exceptionally, small amounts of objectionable material (such as human faeces) are disposed promptly down the foul sewer rather than being stored. This should be agreed with the enforcement agency. For solid material compatible with refuse, the practice is sometimes adopted to take it to a landfill site and ensure that it is buried at least 5 m down. This should also be agreed.

Where regular discharges of short-lived radionuclides are to be made to the sewer, some institutions have a holding tank for the drains from the disposal sinks. This tank is usually emptied by pumping and is of such a volume that the radioactive material will on average spend a sufficiently long time in the tank before discharge to the normal sewer. Note that this is quite different from a dilution tank, and in fact the operation of a holding tank is adversely affected by running large volumes of water into it.

It is technically possible to incinerate waste containing involatile short-lived radionuclides and then to store the ash. However, this is not advised as it requires special expertise in design of the operation, and unusual care in its management to ensure safety.

11.5 CONTAMINATED CLOTHING

11.5.1 Disposables

If extensive use is made of disposable items of clothing, then separate collection bins should be used. More commonly, gloves are discarded on a single-use basis into bins which also receive paper towels and similar low-level waste. Under normal use, only low-level contamination of clothing occurs. If a higher level of contamination occurs then the item should be put into a container kept for higher activity waste. In the case of gloves it may be possible to wash the gloved hands using a decontaminating cleanser, before disposal as usual.

Where biological materials are in use, it may be necessary to arrange collection facilities so that the items can be disinfected or autoclaved before disposal as radioactive waste.

In cases of accidental contamination, it is as well to consider all clothing potentially disposable.

11.5.2 Re-usable Items

In major radioactive facilities, virtually all work clothing is kept within the vicinity of the laboratory, with a changing room and cleaning facilities. At the other extreme, where radionuclide use is only minor, ordinary laboratory coats are worn and go in the general laundry without special treatment. In the latter case it is important to have an effective monitoring procedure to ensure that radioactive contamination does not get into other areas.

There are a great many laboratories that are somewhere between these limits. The most general procedure is as follows. One set of coats or overalls is reserved for use in the designated area, and people who also work elsewhere change coats on entry. Disposable overshoes may also be supplied. The overalls may be washed in a machine within the laboratory suite itself. If so, the machine should be an automatic one with a tumble-dry finish. Its drain should go to the pipe for radioactive low-level waste, and the air vent should be treated as if from a fume hood for radioactive work. In a large institution with a centralized laundry, one machine may be set aside for items from the radioactive laboratories. It will be necessary to have a system of labelling and transport (preferably including colour-coded bags, trucks, etc.). If necessary, a special sterilizer must also be set aside for pre-treatment. The laundry must be monitored, as must the people who work in the laundry. It must be understood that the segregation is mainly to control contamination, especially when there has been a failure of normal controls. It is then important that items and areas which may have been affected are clearly recognizable. One simple expedient which can be very helpful in control is to purchase laboratory clothing in a distinctive colour for use in radioactive work areas.

Where unusual contamination of clothing occurs, e.g. by splashing, an effective procedure is to cut out the area with scissors. The piece of cloth can then be treated or disposed of more conveniently, and the item of clothing can either be repaired or be disposed of as ordinary waste.

If very short-lived isotopes are used, then a useful reduction in radioactive contamination can sometimes be achieved by storage in the laboratory facility for a few days before sending to an external laundry. This only applies to dry clothing which has not been exposed to putrescible material.

11.6 LOW ACTIVITY SOLID WASTE

This means mixed solid, no part of which has an activity greater than 37 kBq kg^{-1} (1 nCi/gram).

11.6.1 Collection

Items such as contaminated glassware will normally be washed with a suitable cleanser. This procedure is sometimes advisable for broken apparatus, if it can be carried out safely, so that the major portion of the radioactivity is disposed of down the drain. (See section 11.8.3.) Similarly, some solid chemicals may be dissolved and flushed away.

However, some contamination cannot be removed easily (notably ^{32}P from glassware) and the item must be discarded. In addition, there will be items such as filter papers, paper towels, disposable items, minor breakages, animal carcasses and powders. These must be collected in bins or other receptacles suitable for their other properties, preferably emptied on a daily basis. The collected material may be accumulated in a safe store away from the laboratory for (say) weekly disposal. See section 3.5 for collection receptacles.

Sharp objects (including broken glass) should be put into a puncture-proof container (such as a proprietary 'sharps' box, or a plastic jar or metal can: see section 10.6, Appendix B-2). Powders should be put into individual tied or sealed plastic bags. Reactive chemicals should be rendered less so, if at all possible. Biological items may be autoclaved if this is considered advisable, or they may be put into a designated refrigerator or freezer. Otherwise they should be collected in a self-closing sack holder. All containers should be unambiguously labelled for the categories of waste for which they are intended (for example: broken glass, scintillation vials, beta-emitters, hand-towels, sharps, powders, plastic items, etc.).

All waste receptacles should be lined with a plastic bag which is taken away with the waste. For beta-emitters, a plastic or cardboard bin is preferred to metal, whereas a metal bin is more effective for gamma-emitters. The waste receptacles should be located for convenience of use, but so positioned that people do not remain very close to them (particularly in the breathing zone) for any length of time.

It must be made clear which containers (if any) may be emptied by regular cleaning staff, and which must only be dealt with by registered radiation badge holders. There should be definite limits to the amount of radioactivity which can be collected in one receptacle. It is essential that there is regular monitoring of all waste receivers for radioactivity. This includes those containers in the laboratory which are not supposed to receive radioactive waste.

11.6.2 Disposal into normal refuse

It is generally permissible for small amounts of radioactive material to be included in trade or domestic refuse collected by the local authority, and for it still to be regarded as normal refuse. Local regulations may vary, and the waste disposal agency should be consulted as well as the control agency for radioactives (see section 11.1.1). However, typical guidelines are as follows.

A sack or other container with at least 0.1 m^3 of refuse may contain a total of 370 kBq (10 μCi). No single item should contain more than 37 kBq (1 μCi), unless the isotope is ^{3}H or ^{14}C only. Only very small amounts of ^{226}Ra are tolerated. Other alpha-emitters are generally excluded, as is waste containing ^{90}Sr. Thorium-X waste from medical use may be permitted in some areas up to a limit of 370 kBq (10 μCi) per week. For greater amounts, special authorization is necessary.

There are sometimes different requirements for waste which goes to an incinerator. Where a waste collecting authority uses both land sites and an incinerator, it is advisable to check that refuse containing radioactive material is always taken to the correct disposal site.

11.6.3 Disposal to landfill waste sites

Landfill sites accepting domestic trade and industrial waste will often accept larger amounts of radioactivity by prior arrangement. This will normally be brought in by the producer of the waste, his agent, a specialist contractor or rarely be specially collected by the site operator.

Where the site mainly handles domestic refuse, typical requirements are as follows. The waste should be free of alpha-emitters and of radionuclides with half-lives greater than one year. The waste should be in adequately sealed containers, each of which is limited to 37 MBq (1 mCi). The radiation level on the surface of the containers should be no greater than 55 μSv s^{-1} (20 millirem/h).

Operation of landfill sites varies, particularly between countries, so the waste should be presented in a manner as specified by the site. On some sites it is the practice to spread out such waste (i.e. to dilute it). On others it is contained as far as possible. If the latter case applies, it may be immobilized by surrounding with impervious material, or even by filling the container with a special concrete-like material to immobilize the contents. Special kits are available to immobilize low-level laboratory waste in standard metal drums for this purpose.

11.6.4 Incineration

It is unlikely that a laboratory will find it worth while having an incinerator solely for radioactive waste. However, there may well be an incinerator for other laboratory waste or for waste from elsewhere in the facility.

CAUTION: an amateur design of incinerator, or a modification to an existing incinerator, could be very dangerous. The manufacturer's advice (or that of another expert) should be sought before burning material other than that for which the incinerator was designed. See also Chapter 7.

There is no objection in principle to low-level radioactive waste of the same type as the waste generally burnt in an incinerator. However, if this is done then the incinerator chimney becomes a release point for volatile radionuclides (e.g. ^{3}H, ^{14}C) which must be licensed and have set limits. Similarly, non-volatile radionuclides (all metals and some others) will accumulate in the ash. The ash must then be disposed of as radioactive waste to a land site (or exceptionally by storage of short half-life material). Particular steps should be taken to prevent dust from the ash being inhaled. Depending on conditions, carbon and sulphur isotopes may or may not accumulate in the ash. They may be retained as carbonate and sulphate or released as carbon dioxide and sulphur dioxide. Any scrubber system must also take into account the radioactive content. For operator safety it is recommended that no more than 3.7 kBq (0.1 μCi) of class I radionuclide, or 3.7 MBq (0.1 mCi) of other classes, be included in any one charge to the incinerator.

Incineration is particularly favoured for animal carcasses and other biological material. Some facilities use it for paper and plastic waste, particularly if there are large amounts involved and the radioactivity is low. If only volatile radionuclides are involved, it may be possible to achieve disposal of all the radioactivity via the incinerator chimney, providing the release rate is permissible.

On the other hand if the material is substantially of non-volatile radionuclides, incineration has the main advantage of greatly reducing the bulk, and rendering it sterile. This is very helpful if short-lived isotopes are involved, which are to be allowed to decay on site. For disposal off the premises, ashing reduces the size and frequency of loads designated as radioactive, though it increases the concentration (specific activity) which may have to be reduced by dilution with an inert material such as kaolin. Ash is especially suitable for solidification into a concrete or polymer mix, if this is favoured by the landfill site. (See section 11.6.3.)

Counter to these advantages are the need for very rigorous control and monitoring of the incinerator operating procedure, monitoring of staff, additional training and possibly major changes to storage and transport procedures.

11.6.5 Disposal to sewers

Certain categories of solid waste can be converted to a form suitable for disposal into sewage. Note that in this case, it is only necessary that the radioactive elements be in a suitable form. Thus it may sometimes be possible to extract the radionuclide by washing with water or a solution of a chelating agent, leaving behind a solid which has little or no residual activity. This can be effective for some organics and group I or II metals in particular.

Animal carcasses and other biological materials are often put through a macerator (usually fitted in a sink waste pipe). This can, with careful thought, be extended to other materials which do not risk blocking the drains.

Generally speaking, it is practical to put into sewers material which is permanently soluble (i.e. will not precipitate out in the pipe) or is of small particle size and of specific gravity between 0.9 and 2.0. This can include small quantities of adsorbents such as clay or charcoal. The quantities must always be negligible relative to the general flow and the point of entry, and the material must not be in any way inimical to the sewage system.

This can be a very useful option, but one which must have the agreement of the drainage authority as well as the enforcement agency for radioactive substances.

11.7 MODERATE-ACTIVITY SOLID MATERIALS

11.7.1 Unused items

It occasionally happens that radioactive material is purchased, and some of it is not required. In such cases, the first approach should be to the supplier. Depending on the value and demand for the material, the supplier may purchase the goods back, but it is more likely that he will take them free, or even make a charge for the service.

If this option is not possible, disposal of the material should be as follows.

11.7.2 Other materials

If more than 3000 MBq are involved, then the only realistic option is disposal by safe burial, via one of the national agencies or a specialist contractor.

If radionuclides of reasonably short half-lives are involved, it may be possible to store the waste until it can be disposed of via the route normally used for low-activity material. However, this means that safe storage is required for some considerable time. Moreover, the amount of activity will usually be reckoned within the total permitted under the licence, without allowance for decay. There is thus some incentive for removal from the premises. Needless to say, unwanted radionuclides with long half-lives should be removed as soon as possible.

(See sections 11.5 and 11.6 for treatment prior to disposal.)

11.8 LIQUIDS

This category of waste covers three main kinds: washing-up water from sinks, other aqueous waste, and solvent-based materials. Medical establishments may also use sluices.

11.8.1 Washing-up water from sinks and handbasins

Any laboratory using radioactive materials in the form of unsealed sources must have a handwash facility with appropriate aids (cleaning agents, nailbrush, paper towels) and a means of monitoring the hands for contamination. This should be a separate basin or sink from that used for washing up laboratory ware, if at all possible. The best practice is to wash the gloved hands in the washing-up sink, remove the gloves and then wash the hands in the handbasin, monitor and, if necessary, clean the hands again. A hot-air hand dryer is not advised, since it can blow contamination into the air.

The drain line from the handbasin will be deemed to be carrying radioactive waste. It should be constructed and monitored accordingly. The same applies to the sink which is used for washing up laboratory apparatus used with radio-isotopes. Where radioactivity is an important part of an institution's work, there may be a special drainage system to which these sinks and basins will be connected.

On the other hand, if radioactive work is a very minor part of laboratory procedures, the ordinary foul sewer or trade waste drain will be used. The number of sinks designated for radioactive waste should be as few as practicable (in keeping with safe working) and the connections to the main drain should be as short and as direct as possible. This ensures that there is very little piping in which radionuclides can accumulate, and which will need special attention. The connection should preferably be to a well-used conduit which is likely to have a good flow of other water, so that the radioactive material is promptly diluted to a negligible amount.

The handbasin may be of stainless steel or glazed ceramic. Plastics and enamel are not suitable. The main disadvantage with ceramic is that small cracks may make traps which are difficult to decontaminate. A single chip may often be repaired (particularly if the use of radioactives is small) but a general cracking of the glaze (as is often found with older basins) means that the basin should be replaced.

If possible, the taps on the handbasin should be foot-operated or have elbow levers. Unfortunately, many attractive modern designs of handbasin tend to splash their contents out if the water pressure from the tap is high. This should be checked, and if necessary a flow restrictor (a very cheap device) fitted to the water supply.

Except in the smallest laboratory or in a mobile unit, there should be a separate sink for washing up. If it is truly essential that one sink be used for both washing up and hand washing, then one side must be kept clear of laboratory ware. The drainer or adjacent benching should be labelled 'handwash' on one side (the right side is most usual) and 'labware' on the other (or some similar clear wording). There should be a clear area of at least 15 cm on the handwash side (to allow for splashing) then the hand-cleaning supplies and monitor. There should be strict control to ensure that laboratory ware and gloves are not put down on the handwash side, even when washed.

In a major radiochemical facility, all the sinks will be special purpose, but many laboratories will just have one for radioactive purposes. This sink must be clearly labelled (for example: 'This sink suitable for radioactive waste') and constructed and maintained accordingly. Stainless steel is the only really satisfactory material for the sink, because it can be decontaminated by very harsh means (corrosives or abrasives) if necessary.

It is preferable if the sink has an integral drainer on both sides. If not, stainless steel-covered benching is advisable, since the sink area is the most likely to become contaminated. A single-piece stainless steel splashback is likewise preferred. It is essential that any joints around the sink are well sealed against splashes.

A very convenient arrangement is a double sink with a drainer on each side. To aid good organization, it is helpful if the left-hand drainer is labelled 'dirty', and the right one 'clean'. The left-hand sink is labelled 'wash' and the right-hand one 'rinse'.

The best materials for drain lines from radioactive sinks and basins are stainless steel or glass, since these are the easiest materials to keep clean. (Glass can adsorb ^{32}P, but this is short lived and does not usually give a long-term problem.) Glass has the special advantage that it is possible to see if there is any build-up of solid or slime. Copper and lead have the main hazard of reacting with many elements and causing accumulation of radioactivity in some cases.

Polypropylene or PVC piping can be acceptable if the radioactive work is minor and there is no likelihood of solvents being discarded down the sink. PVC is less likely to accumulate biological slime which could entrap radionuclides. Push-fit connections should not be used. Welded joints should be used on plastic piping, with the minimum number of screwed unions for ready maintenance.

The handbasin should be fitted with an s-bend, but the sink should have a small trap for small items which may fall in the drain. (See section 5.3.3.) However, there should be no open traps, grids or other points where the radioactive waste water could come into contact with the air, before the line joins the main drain. The piping should be laid out and labelled so that it can be readily inspected and maintained, and so there is little likelihood of it being confused with (or connected to) other piping. (A radiation symbol on adhesive tape may be used on traps and piping every metre or so.)

The drains in any fume cupboard designated for work with radio-isotopes in the form of unsealed sources should be counted as radioactive drains, even if their intended use is only for condenser cooling water. This allows for spillages. (See Chapter 5 for general information on drains.)

11.8.2 Sluices and latrines

Where samples of excreta or similar biological material are tested in any amount, a sluice or water closet may be used to dispose of the samples. A particular sluice (if there are several) should be designated for radioactive waste, but need not be restricted to radioactive samples only. It is preferable not to use a water closet also used by people, except when they are patients receiving radio-isotopes and there is insufficient work to justify the provision of a sluice.

The ordinary measures used to prevent contamination by infective material should generally control radioactive contamination from operation of a sluice. However, a phenolic disinfectant is preferred to chlorine or oxygen bleaches, which could conceivably cause radioactive vapours to be given off.

As with all radioactive disposal to drains and sewers, there will be a daily, monthly and yearly limit on the amount which can be disposed down a particular drain. This should be remembered when arranging for samples to be brought to the laboratory, as such materials are not pleasant to keep or transport elsewhere.

11.8.3 Other aqueous liquids

The main problem with this class of waste is the very wide range of possible options. In general, aqueous waste can be readily converted to other forms by physical or chemical treatment. However, there is no doubt that the most convenient method for the average laboratory is prompt disposal to the drains. It is usually acceptable to do this for ^{14}C, ^{3}H and isotopes with a half-life of less than 26 weeks, which are not alpha-emitters. It is recommended that effort be put into devising a safe procedure for doing this, before other options are considered. The reason is that other options are likely to pose greater risks of accidents during storage or transport.

In order to put material into the drain, it must be rendered chemically innocuous (e.g. by neutralizing acids or bases or other reactive groups) and essentially free of human pathogens (i.e. sterilized if necessary). It should not be likely to deposit substances by precipitation or reaction with the drain materials (see section 9.3.3 and Table 9.1). Above all, it should be dilute, and the amounts and concentrations must be agreed with the drainage authority and the radioactives' enforcement agency.

If very large amounts are involved, then a process will have to be fully engineered by a group of experts. This may involve on-site treatment plant or special transport, which are outside the scope of this book.

For ordinary amounts, the usual constraint is to avoid too high a level appearing in the sewers at one time. (Remember that the sewer is a workplace for some people.) Assuming the institution's effluent is joined by others in the sewer, a rule of thumb might be about 40 MBq per thousand litres of effluent, except where class I radionuclides are involved. For class I radionuclides, a further dilution factor of a million times may be specified. (However, very few laboratories are licensed for a total inventory of over 37 MBq of class I radionuclides.)

It is therefore reasonable practice to dispose of radioactive waste in small amounts as it arises during the day, and not exceeding the daily limit. There are two important exceptions to this. The first is if the laboratory operates at unusual hours, e.g. overnight or during holidays when there may not be much flow in the drains. In this case, it may be necessary to hold the liquid until a better time (the drainage authority will normally advise when this is likely to be). The second exception is if the manner of work is such that the waste can only be collected in larger amounts once a day or less often.

For the latter situation, it would be bad practice to simply pour the daily limit into the drains as a single charge, and much worse if a weekly or monthly amount was involved. To deal with the waste may require transfer and storage in small containers for a responsible person to pour into the drains at appropriate intervals. If the procedure is regular, then a holding tank will be required with a small dosing pump. Where any substantial inventory is involved, it is important that the discharge to the drains must be positive, i.e. by a pump rather than gravity, or by a person removing a bottle from a leakproof bin. Thus a failure of the system will not permit uncontrolled entry to the drains.

Where materials with very short half-lives are involved, there may be some useful reduction in activity by storage prior to disposal to sewers. Effective storage may be in the laboratory or in a special holding tank with a suitable residence time. (See section 11.4.) However, it can be argued that radionuclides with short half-lives are not a long-term environmental hazard, and the authorities may consider that the time taken in drains and sewage treatment is adequate protection, and better than the risk of accident while in storage at the laboratory.

If the aqueous waste is unsuitable for the drains for any reason, one possibility is to use the chemical properties of the radionuclide to separate the radioactive material from the other waste. The two wastes can then be treated separately. Typical techniques are the use of ion-exchange resins, electroplating, precipitation (of hydrated oxide, sulphate or fluoride), and displacement (e.g. copper from solution by iron). It may occasionally be worth concentrating the solution by distillation. The separated or concentrated material may then be transported for disposal to land. Such techniques pose the hazard of working with higher concentrations of radionuclide, and give the possibility of an accident during the additional operations. Most particular care should be taken that radioactive gases, vapours or mists cannot be emitted during these processes. (See sections 9.3.6, 9.3.7, 14.3, 14.6 for some techniques.)

Some aqueous waste may be directly acceptable at certain landfill disposal sites. However, it is also possible to convert aqueous solutions to solids by adding setting agents such as plaster of paris or cement. At least one company has a kit available for this purpose.

It is not usual to dispose of aqueous waste by incineration. However, small amounts can sometimes be added to a solids' incinerator licensed for radioactives as follows. Sawdust is put into plastic bags in quantities of about 500 g. To each bag is added up to 100 ml of liquid, and the bag is sealed. The closed bags are added to the waste to be fed into the incinerator. It will be necessary to calculate the amounts involved to comply with the limits for the incinerator and for the people involved. There should preferably be no more than 3.7 MBq (100 μCi) present in the incinerator at one time (reduced to 3.7 kBq for class I radionuclides).

11.8.4 Solvent-based waste

In the past, much of this waste was legitimately deposited on landfill sites. However, in recent years the site operators have become reluctant to accept such material, and other methods are normally required.

Small amounts of water-miscible solvents may be permitted into the drains. If the radioactivity is from the solvent itself (i.e. a labelled compound) rather than a solute or suspension, then small amounts can be allowed to evaporate under controlled conditions (see section 11.9). Otherwise the main options are incineration, chemical treatment, land disposal or storage (if the half-life is short enough).

Storage of radioactive waste solvents should be minimized. Not only is there a question of safety from different hazards, but the storage practice has to comply with several different pieces of legislation (which are not always conveniently compatible). Generally the storage should be: *secure* — to limit the access to proper persons; *shielded* — as necessary for the type of radiation; *contained* — so that spills are controlled; *fire-resistant*; *ventilated*;

monitored for radioactivity; and *recorded* so that activity can be accounted for. Clearly, a glass bottle in a fume cupboard does not entirely meet these requirements, and must be regarded as a short-term expedient. If used, the bottle should be in a tray (or better, an outer container) of a material unaffected by the solvent. Both the bottle and the outer container must be labelled adequately.

If solvent waste is to be burnt, note that certain elements (notably C, H, O, N) are likely to end up in the gaseous emissions, whereas the metals and phosphorus generally end up in the ash. Sulphur and the halogens may occur in both gases (as SO_2 and HCl, etc.) or in the ash (as sulphates or halides), depending on the exact conditions. A scrubber, if fitted, will of course absorb many elements into the water phase.

In any case, any unit used for burning radioactive material on site will need to be licensed, though this does not necessarily mean it must be complex. There is at least one unit in the form of a perforated metal drum filled with stones which is used to burn very low-level waste on open ground near the laboratories.

Some boilers and incinerators will accept solvents being sprayed into the burning zone. However, a proper knowledge of burner technology is required to do this safely. An amateur attempt could be extremely dangerous, both from the risk of explosion and from the risk of radioactive contamination.

Most usually, the solvents are burnt in a unit used for incinerating solid waste. Before feeding in, the solvent is absorbed on about 3 to 5 times its own weight of vermiculite or a similar non-combustible absorbent. Sawdust is not recommended, as the combination of solvent and sawdust can aggravate the fire risk, and will in any case burn too fiercely in the combustion chamber. (See, also, Chapter 7.)

Solvents are often suitable for recovery by distillation. (See section 14.6.) Where a particular kind is in frequent routine use, then it is worth considering a dedicated still. A typical example is some scintillation 'cocktails'. These may be separated to give a very clean toluene and liquid waste which is of much smaller volume but of relatively low activity. The reduction in volume may well be as valuable as the saving of solvent.

CAUTION: solvents containing dioxan or other ethers may form explosive peroxides, which can cause accidents during distillation. See section 14.6.7, and Tables 14.1 and 14.4.

To minimize handling of large numbers of vials, ampoules or bottles, small crushing units are commercially available. If used, the crusher should be clearly labelled as being for radioactive materials, and operated so that the liquids are collected. The broken glass may be rinsed (e.g. with recovered solvent) before disposal as solid low-level waste.

Depending on the substance involved, chemical techniques may be used to extract the radioactives from the solvent. This can be into water (from non-miscible solvents) using suitable reactive or complexing agents, or onto solids such as carbon, alumina, fuller's earth, or certain porous polymers (which are sold for chromatography and like purposes). If the radioactive material is in the form of a suspension or emulsion, this can be broken by chemicals or perhaps heat, and the suspended material settled out. (See section 9.3.6.)

11.8.5 Liquid Scintillation Wastes

These are generally glass or plastic vials containing a few cubic centimetres of a 'cocktail' of solvents, scintillants, emulsifiers and samples of ^3H or ^{14}C isotopically enriched substances. They deserve special mention because they are so widespread, and appear to be causing increasing difficulties for laboratories. Landfill sites for radioactives are becoming reluctant to accept them, because of their fire and toxic dangers. Toxic waste sites are refusing to accept them because of their radioactive content. In the UK the concept of 'nuclear free zones' has closed certain sites to this waste, for reasons based on a total misunderstanding of the applications of radioactivity, rather than any good technical or even moral reason.

There are three basic methods for the individual laboratory. (1) The vials and their contents can be disposed of unopened. (2) The vials can be emptied and disposed as solid waste, while the liquid is recovered or disposed separately. (3) The vials are emptied and cleaned for re-use, the liquid again being treated separately.

Clearly, the first option provides minimum handling within the laboratory. If such waste can be incinerated nearby, then it is a fairly safe option. The incinerator will of course have to be technically suitable and licensed. Plastic vials are commonly chosen for this route, but it should be noted that they are significantly permeable to their contents. The work procedure and workplace should be checked for any vapour hazard. It is an appropriate method for some techniques which use very small quantities of liquid, because of the difficulty of removal of the cocktail from the vial. Laboratories using this method for land disposal should consider the possible closure of the facility, and plan for an alternative disposal route.

Virtually all other methods (unless highly automated) involve some risk to people handling the waste. Work procedures are required to minimize human exposure to the liquid and vapour, allowing for the chemical hazards (fire, acute toxicity, possible carcinogens) as well as the radioactive ones. The main radioactive danger comes from volatile substances being inhaled.

For a smaller laboratory with several workers using modest amounts, a sensible procedure is for each worker to be responsible for emptying his or her own vials. It may be necessary to segregate the liquid waste according to categories for disposal or recovery. For larger laboratories routinely using many hundreds or thousands of samples, it may be necessary to make vial disposal a significant portion of one person's job.

In planning large-scale or routine liquid scintillation work, the waste disposal procedure should be one of the most important considerations. For example, a cocktail can be chosen which is water immiscible but easy to distill, in order to recover valuable solvent and reduce the bulk of the waste. Conversely, a mixture of water-miscible solvents and emulsifiers will enable liquid to be flushed to drain (local regulations permitting). Plastic vials have many advantages, particularly if the final disposal is by incineration. On the other hand, glass vials are easier to clean, whether for re-use or for disposal as non-radioactive waste.

Managers should beware of cost arguments which are not based on the proper costing of manpower, and they should seriously consider the health, safety and environmental aspects. The variety of liquid scintillation work means that the correct choice of waste disposal will vary.

An alternative to emptying vials is to load them in a crushing machine. Such machines can be quite small hand-operated devices, which are commercially available. For any home-made versions, the greatest care should be taken that aerosols and vapours are not released. As would be expected, they work best with glass, but can in fact treat some plastics. A typical technique is for the liquid to be filtered off and partially distilled. Some of the distillate is used to rinse the crushed solid. The solid is then rinsed with a solution of sufactant which is flushed to drain. The solid is usually then (after monitoring) acceptable as non-radioactive waste.

Some degree of automation is very desirable for the washing of vials, whether for disposal or for re-use. McElroy, Sauerbrunn and Eckelmann have described a comprehensive washing system for a medical centre, with costings and tests of efficiency. Note that alkaline cleansers are preferred to acid ones, because they give less risk of volatile compounds of radionuclides being formed.

Recovery of solvents is normally by batch distillation. It is usually a wise precaution to wash the liquid with 10% aqueous NaOH. This stabilizes many ^{3}H compounds from conversion to volatile forms by exchange, e.g. with water. An alternative or supplemental technique is to treat the liquid with activated charcoal, which will absorb many volatile radioactive compounds. The process will give relatively pure (although wet) toluene and xylene. With a good-enough column (e.g. a spinning band equivalent to 20 or more theoretical plates) these solvents can be good enough to re-use. The majority of the radioactivity remains in the still pot with the emulsifiers and other chemicals. This residue is generally water-miscible and non-flammable, if the process is completed efficiently.

CAUTION: where dioxan or other ethers are present, it is, in principle, possible for explosive peroxides to form. No portion of the solvent should be evaporated to near dryness. If in doubt, test with a commercial test-paper or other colorimetric test for peroxide. It has been known for scintillation mixtures containing dioxan to explode when stored for a long time. See section 14.6.7.

Even where distilled solvent is not re-used, there is some advantage in separating the flammable hazard from the radioactive hazard, and reducing the bulk of the latter. However, financial savings in solvent purchase is the most likely justification for the capital equipment investment in an automatic still, which is essential.

See, also, sections 9.3.6 and 14.6.

11.9 GASES AND VAPOURS

If substantial amounts are involved, or there are special circumstances, some gases can be absorbed into liquids (e.g. CO_2 into alkali) and some vapours can be concentrated by adsorption onto solid (e.g. organics onto charcoal). These must be disposed of, or perhaps stored.

However, in the majority of cases it is perfectly permissible to arrange controlled discharge to the atmosphere. Many radionuclides produce radioactive gases as the products of primary or secondary decay processes, so the disposal of radioactive gases is implicit in their storage and use. If a store is isolated and the gas production small, then natural ventilation may be adequate. Otherwise a fan system will be required.

In all cases, the general aim is that discharges should be away from people (and from air intakes for machines or ventilation) and that in any case the concentration in the discharged air should approach that which can safely be breathed. (See also Chapter 6.) A discharge point may be a pipe specially for that purpose, a duct from a fume cupboard or a chimney of a boiler or incinerator. In all cases it will be approved by the enforcement agency, who will specify limits as to the daily amount and the concentration of radioactivity which can be emitted.

Where permanent gases are involved, the flow rate should be metered (at least approximately) using, for example, a cheap commercial tube flow meter. For low flows, a soap bubble flow meter can be made quite easily (see Fig. 9.2). If the gas is not under pressure, it can be displaced from its container at a controlled rate by filling the container with water (or other suitable fluid).

Organic liquids which boil near room temperature should be chilled before placing in an evaporating basin in a fume cupboard. For materials of higher boiling point it is preferable to aid evaporation by placing them in a good draught, *not* by heating.

A gas dissolved in a liquid may be displaced by chemical action or by purging (i.e. bubbling another gas through) with air or nitrogen.

As a general rule the amount placed into a fume cupboard for disposal should be no more than half the amount of radioactivity permitted for a daily discharge. No other work should be carried out in a fume cupboard in use for this purpose.

Except in emergency, it is best if a venting method for radioactive substances can be tested first with the equivalent non-radioactive material The chemical and physical properties must always be taken into account. See section 9.3.7 for more practical details of venting.

11.10 BIOLOGICAL MATERIALS

These mainly give rise to the problem of putrefaction, but there can be other hazards from aerosol or dust formation, clogging of drains, and escape of animals. See, also, Chapter 10.

Where pathogens or troublesome micro-organisms are used, then facilities should be available for immediate chemical sterilization — if possible within the glovebox or experimental area. For other materials, steam sterilization can be used, providing the autoclave drain is treated as a radioactive discharge point, and venting is to outside the building.

An on-site incinerator is a common satisfactory method of dealing with a variety of waste. The material is collected in a plastic bag which is sealed and then put inside another bag for transport to the incinerator. This technique is particularly useful for fouled animal bedding. If the collection procedure is carefully controlled, sterilization may not be required. (See section 11.6.4 and Chapter 7.)

Autoclaved low-level waste such as Petri dishes and paper towels can be disposed of as in section 11.6. Some tissue samples and fluids can be disposed of as in section 11.6.5.

Where radionuclides of short half-life are involved (which is frequently the case) the material must be stored safely for a period of time before disposal as very low-level or non-radioactive waste. Some items can be sterilized and kept in sealed containers for the necessary time, but the most common practice is to store in a refrigerator or freezer. Note that it is not always necessary to store the whole item. For example, many elements will accumulate in particular organs in an animal, and it is only these organs which need to be kept. (The rest of the carcass should, of course, be monitored before discarding.)

It is vitally important that each item in a freezer or refrigerator is labelled with the date, type and approximate amount of radionuclide. Items should not be loose, but in sealed plastic bags. (Excising animal claws is useful to prevent punctures of the bags.)

The freezer must be labelled and its use rigorously controlled. It should preferably be in a room which is not a general workplace, and which is not accessible except via the radiation laboratories.

If freezing is not practicable, or waste has to be kept unexpectedly, then tissues may be temporarily preserved in formalin, or in a plastic bag with bleaching powder and in an inert absorbent. Animal carcasses should have the claws cut off and the limbs tied. The abdominal and chest cavities should be slit open and the carcass eviscerated. The major organs (lungs, heart, intestines, stomach and liver) should be pierced or slit. (Most important on larger animals.)

For storage in a plastic bag, about one fifth of the weight is required of bleaching powder, and a similar amount of mineral absorbent (e.g. vermiculite). As the bleaching powder can generate heat by its reaction, the carcass should be tumbled in half the vermiculite first in the bag, then the remaining vermiculite mixed with the bleaching powder added. Avoid blowing dust about.

For further details of this technique, see particularly the paper by Boursnell and Gleeson-White in *Nature* (Lond.), **179**, 54 (1957).

11.11 SPILLAGES AND DECONTAMINATION

A major release of radioactive material is of course an emergency, requiring immediate evacuation of the area, and possibly extreme measures to cope with the situation. (See Chapter 15.) The following notes are intended to help with much smaller spills and losses which are inevitable in even the best-run laboratories.

11.11.1 Spill control

Good work practice can do much to reduce the problem of drips and splashes. Proper handling of vessels, avoidance of sudden movements or vigorous agitation, and working over a suitable surface will prevent many losses of material. A tissue or paper towel wrapped round a beaker before pouring will catch any drips which happen to run down the outside. Standing the receiving vessel on a filter paper will help to retain any drops which miss in dispensing. For any work involving many transfers, it is a good idea if it can be carried out over a sheet of plastic-backed absorbent paper, or a tray of absorbent mineral.

When these measures fail, there may be spills of liquid or solid to pick up. Paper towels, tissues or filter papers are universally used for very small drops of liquid (excepting mercury and violently-active substances such as strong nitric acid). For a discussion of absorbents, see Chapter 15. Note that dry sand can be used to form a barrier round almost any liquid or solid spill, and activated charcoal will inhibit vapour emission from volatile materials. A large spill will require a large amount of absorbent, or an arrangement for transferring the liquid to the drain, if appropriate.

It is important that dealing with a spillage does not expose the worker to contamination in the breathing zone. A cartridge respirator can give some protection if correctly selected and used. Sweeping of powders with a brush should be avoided as it can generate breathable dust. A piece of stiff card or plastic should be used instead. A rubber squeegee (as sold for wiping motor-car windows) is excellent. If powder is spilt onto paper (whether a notebook or a filter paper) it is best if it is not swept off, but simply folded up into the paper and discarded in the packet.

Dealing with a spillage generates a certain amount of waste. Paper towels and the like may sometimes be permitted to dry off in a fume cupboard. In very rare cases it will be thought worth while recovering valuable substances from the collected material. More generally, goods such as paper, powders, absorbents, gloves, etc. will be collected in plastic bags, then dealt with as solid waste. Broken jars or other sharp objects will require a more substantial container, such as a plastic jar. Aqueous liquids are most commonly rinsed down the sink, and paper towels can also be rinsed before being discarded. Absorbed solvents may be sent for incineration, permitted to desorb in a fume cupboard, or washed out with other non-radioactive solvents (possibly using laboratory technology such as a Soxhlet extractor).

It is important that an estimate be made of the amount of radioactivity involved in a spillage, in order to assess the seriousness of the situation and also to record (if more than a truly trivial amount). It may well be necessary to dispose of collected material over several days in order to remain within discharge limits.

11.11.2 Decontamination

Decontamination of people is a matter of either good practice in personal hygiene, which should be taught, or a medical emergency which is outside the scope of this book.

Contamination may occur from a particular known incident, or be discovered by routine monitoring. It may apply to an object such as a tool, beaker or door handle.It may be on a surface such as a bench or wall. It may be on an impervious surface (such as dried spillages or deposits in sinks) or may have penetrated to a greater or lesser extent.

Two important rules apply. Firstly, if the contamination is serious the best technique is to remove and dispose of the item as a whole. This is often the only way where contamination has penetrated fabric or furniture. Secondly, the decontamination process should not make matters worse — either by spreading it over a greater area, or by making it more available, e.g. as dust in the air.

If the nature of the material is precisely known, this may strongly suggest a means of removal such as a solvent. Otherwise, the solvent which is tried first is always water. This is followed by an aqueous solution of a commercial cleaning agent recommended for radioactive decontamination (there are several good ones readily available). Only if these fail is attention given to solvents or corrosives.

Small items can be immersed (or even boiled) but fixed items or large surfaces should be swabbed, that is they are rubbed with pieces of absorbent cotton wool soaked with the cleaning agent. The procedure is to work from the perimeter of the contamination in towards the centre, so as to avoid spreading material.

Abrasives such as wire wool or scouring powder can be used on some surfaces, but they must always be used wet, so as to avoid dust formation. Note that stainless steel is one of the few materials which retains its surface properties after scouring: many other materials become more likely to retain contamination.

Strippable waxes are available for surfaces such as linoleum or wood. The provide ordinary protection to dirt, but the whole wax layer (and the contamination if it has not penetrated further) can be removed by a proprietary cleanser when required. Peelable paints are used for a similar reason on vertical surfaces.

A technique, which is occasionally surprisingly effective, is to wet the surface with water and detergent, then work in fuller's earth to make a paste, and leave for a few hours. The absorbent (which has ion-exchange capacity) draws material up from the surface and can then be removed.

REFERENCE
[section 11.8.5]

McElroy, N.L., Sauerbrunn, B.J.L. and Eckelman, W.C. (1982) The Feasibility of Recycling Glass Liquid Scintillation Vials, *Health Physics*, **42**, no. 2, 236–238.

BIBLIOGRAPHY

Duncan, W.P. and Susan, A.B. (eds) (1982) *Synthesis and Applications of Isotopically Labelled Compounds*, Elsevier.

Faires, R.A. and Boswell, G.G.J. (1981) *Radioisotope Laboratory Techniques 4th edn*, Butterworths.

Hughes, D. (ed) (1970) *Design of Laboratories for Work with Radioactive Materials*, Brit. Radiol. Protect. Ass.

Hughes, D. and Collingworth, R. (1971) *The Design of Laboratories for Radioactive and Other Toxic Substances*, Koch-Light.

International Atomic Energy Agency (1965) *Safety Series No. 12. The Management of Radioactive Wastes Produced by Radioisotope Users*, I.A.E.A. (Vienna).

International Atomic Energy Agency (1966) *Safety Series No. 19. The Management of Radioactive Wastes Produced by Radioisotope Users: Technical Addendum 1966*, I.A.E.A. (Vienna).

International Commission on Radiological Protection (1976) *ICRP Publication No. 25. The Handling, Storage, Use and Disposal of Unsealed Nuclides in Hospitals and Medical Research Establishments*, Pergamon.

Martin, E.B.M. (1982) *Health Physics Aspects of the Use of Tritium. Occupational Hygiene Monograph No. 6*, Science Reviews Ltd.

Pearce, K.W. (1971) The Treatment and Disposal of Radioactive and Objectionable Solid Wastes, *Chem. Ind.*, **22**, 590-592.

Stewart, D.C. (1981) *Handling Radioactivity*, Wiley.

UK Dept of Health & Social Security (1972) *Code of Practice for the Protection of Persons against Ionizing Radiations arising from Medical and Dental Use 3rd edn*, HMSO.

UK Health & Safety Executive (1976 Code of Practice) *Guidance Notes for the Protection of Persons Exposed to Ionising Radiations in Research and Teaching*, HMSO.

US National Bureau of Standards (1951) *Handbook No. 48. Control and Removal of Radioactive Contamination in Laboratories*, US Govt.

US National Bureau of Standards (1964) *Handbook No. 92. Safe Handling of Radioactive Materials,* US Govt.

US National Bureau of Standards (1969) *Handbook No. 69. Maximum Permissible Body Burdens and Maximum Permissible Concentrations of Radionuclides in Air and Water for Occupational Exposure,* US Govt.

CHAPTER 12

Special Problems

12.1 CARCINOGENS, TERATOGENS AND CYTOTOXIC AGENTS

12.1.1 Introduction

The risk of cancer or birth defects due to chemical exposure has caused a lot of concern in recent years, and created a certain amount of misinformation. This handbook is not the place for a treatise on the causes of cancer, but materials which are known carcinogens (tumour causing) or teratogens (causing malformed foetus) or cytotoxic agents (damaging dividing cells) can create waste disposal problems unless people are adequately informed.

It must first of all be understood that like all toxic properties, there is an enormous variation in the potency of materials. So-called 'known carcinogens' range from those which are virtually certain to cause a tumour after a single application to those which only give a statistically increased possibility after prolonged heavy exposure (e.g. cigarette smoke). In fact for the majority of carcinogens the risk is probably less than for smoking, even for occupational exposure. As waste disposal is usually of much shorter duration than routine use, it ought (if carried out with proper care) to give less exposure and thus less risk than using the compound.

Laboratories which carry out work on cancer, or on agents which cause cancer (chemicals, viruses, radiation) or chemicals related to potent carcinogens should have secure waste disposal methods built in to their procedures, and should not commence work with a material until the disposal method and emergency procedure have been established. Some

231

sources of information on methods are given in the references to this section.

For other laboratories, the problem is one of carcinogens being incidental to the work. There may be old chemicals to be removed, which are not used today because they have been proven potent carcinogens. It is quite possible that some chemical stores contain old stock of material which has been prohibited under legislation for many years. In other cases, materials may be in use despite their known carcinogenic properties because there is no adequate substitute. Finally, carcinogens may arise as by-products of chemical reactions, quite unknown to the laboratory worker. The chemist in charge of synthesis should be aware that there is some possibility of this occurring, and arrange containment and waste disposal accordingly.

Cancer may arise many years after exposure, with no obvious ill-effects before. With relatively weak carcinogens such as benzene, exposure to a relatively high level over some years seems necessary. Good work procedures are therefore required, and the use of such materials is now restricted as far as is practicable. However, a single exposure during waste disposal would be unlikely to give any measurable cancer risk (though the risk of fire or acute poisoning may exist). The main danger is that the fear of cancer may cause inappropriate measures to be taken, or the material just to be put to one side or disposed of illegally.

A list is given in Appendix C-3 of some of the most potent carcinogens. If it is found that a laboratory has one of these compounds in use or in store, it is strongly recommended that specialist advice be taken (e.g. from one of the cancer research establishments) and the disposal procedure handled by an expert with full facilities.

The evidence linking chemicals with birth defects is less complete than that for cancers except in a few cases, namely anaesthetic gases and a range of pharmaceuticals. As damage can occur before the woman concerned knows she is pregnant, it is a wise precaution to take extra steps to minimize chemical exposure for women who may be intending to have children. A woman in this position may prefer not to take part in disposal procedures, and when pregnant should certainly keep clear of ethers, halogenated solvents, and pharmaceutical agents.

The term 'cytotoxic' has recently been adopted as a hazard warning for certain chemical substances. They are poisons which particularly attack cells in the process of dividing. Human exposure can lead to blood disorders, infertility, poor recovery from illness, and of course harm to the unborn child.

Cytotoxic materials are used commercially as pesticides and in the treatment of disease, especially cancer (see section 12.1.3). Research laboratories may evaluate cytotoxic compounds for their commercial potential in these areas, or may use them for unrelated purposes. Medical laboratories may handle samples from patients given high doses of cytotoxic drugs, which are therefore present in body tissue and fluids.

Control measures for cytotoxic substances are still under development. However, the general advice is to handle and dispose of them as highly toxic materials, where the harm they cause may not be apparent until some considerable time after exposure. In this respect, the precautions used for potent carcinogens are appropriate, with additional care for women *and men* of reproductive age, and anyone with other health problems (such as recovery from injury or chronic disease).

12.1.2 Disposal Techniques

In general, substances legally controlled as carcinogens may not be discharged to drains or to land in any known quantity. Those not specifically included are certainly 'Hazardous Waste' in the USA, and 'Special Waste' in the UK, and thus subject to the relevant legal controls (see Chapter 2).

Incineration is an effective method for most organic substances, but special facilities and a high degree of control are required. Open air burning or the use of ordinary incinerators can be very dangerous. (See Chapter 7.) Where there is a low concentration of a low or moderately potent agent on a combustible substrate (e.g. animal bedding) then in-house incineration is sometimes practicable. New technologies are currently under development, such as molten salt combustion.

Chemical treatment is the most general technique, though it must be pointed out that this typically converts the substance to another chemical which is still toxic, if less carcinogenic. Sometimes it is possible to ensure that the substance is completely converted by the chemical procedure for which it is used, or by a similar reaction on the residue.

There is little information on spillage treatment, other than recommendations for highly enclosed systems to limit the spread. Activated charcoal is probably the most effective adsorbent, as it prevents vapour release (see Chapter 15).

The International Agency for Research on Cancer publishes regular monographs on carcinogens, which include handling recommendations and specific destruction techniques for individual substances, and would welcome further information of this nature. A review is given in IARC monograph no. 33.

Several reviews of destruction techniques and new technologies are given in a recent book edited by Walters.

12.1.3 Anti-neoplastic Drugs

These are very potent pharmaceutical agents with long-term risks, which are used to control cancerous conditions. The long-term risks may be acceptable to the patient, but not to staff who may handle the drugs or contaminated materials. All such staff should be aware of the cytotoxic effects (see section 12.1.1) and the fact that most of these drugs may be cancer-causing themselves.

Disposal of unused materials, emptied containers, medical samples and particularly of used hypodermics should be stringently controlled. It is recommended that only technically qualified personnel handle containers (e.g. 'sharps' boxes of syringes) of this waste. The greatest care must be taken with animal bedding or other waste liable to generate dust. See sections 3.3.3 (safety and security), 10.3 (medical laboratories), 10.5.3 (animal bedding) and 10.6 (sharps).

12.1.4 Polychlorinated Biphenyls (PCBs)

These are chemically inert fluids which can be highly hazardous to man, including the risk of cancer. They may be found in substantial amounts in older electric and electronic apparatus in transformers and condensers, and may well be present as old stock in some physics, electrical or chemical laboratories.

They are particularly difficult to burn, requiring a temperature of 1200 °C for at least 2 seconds, which is achieved by few incinerators. Ordinary attempts at incineration are liable to simply release large amounts to the surrounding atmosphere. They should be removed by a competent contractor for destruction in an approved incinerator.

Large-scale use (see Table 12.1) and disposal are legally controlled in the USA and Europe.

Table 12.1. Uses of PCBs

Though usage is now diminishing, and is in many cases prohibited, laboratories may find samples or goods containing polychlorinated biphenyls in stock from the following products, particularly fire-resistant grades.

Dielectric fluids (transformers, capacitors)
Heat transfer fluids
Hydraulic fluids
Plasticizers
Microscopy fluids
Carbonless copy paper
Pigments
Flame retardants

12.1.5 Asbestos

Laboratories which routinely handle asbestos products will have access to commercial disposal methods, and will be required to observe the airborne control limits for dust (see Table 6.1 and section 6.1). For other laboratories, there are three typical problems: (1) disposal of laboratory goods such as asbestos rope; (2) action over constructional asbestos (benches, ducting, fume cupboards, etc.); (3) disposal of samples (from mineralogical or occupational health analyses etc.) found to contain asbestos.

Present occupational health practice is to regard all forms of asbestos as equally dangerous (i.e. to be avoided) and equally likely to give cancer. However, it is widely thought that blue asbestos is especially dangerous in this respect.

CAUTION: the mechanism of lung damage is complex. Harmful particles are not generally visible, and may remain in the air after other dust has settled. It is possible to inhale harmful amounts of asbestos without any immediate discomfort.

The only method of determining asbestos levels in the laboratory air is to have measurements made by experts. Government agencies, commercial consultancies and universities offer services of identification and measurement. This is strongly advised if it is thought that asbestos fibres may be being released from (for example) deteriorated constructional material.

It is technically possible to destroy asbestos by chemical or thermal means, but few laboratories will have the specialist knowledge or desire to do this. Instead, the local waste disposal authority should be contacted for details of the nearest disposal site accepting asbestos. In principle, this should be done for any amount, no matter how small. In practice, some asbestos-containing products (such as vehicle brake linings) are commonly disposed of as normal trade refuse. The laboratory should check local practice with the enforcement agency.

The most hazardous items likely to be found in a laboratory are asbestos rope, asbestos paper, soft asbestos sheet, asbestos pulp (filter aid). However, some filter cartridges (for water etc.) manufactured before 1974 included substantial amounts of blue asbestos. All these items must be removed.

A simple procedure is to wet the item and its surroundings with a water spray, then place the item into a plastic bag. If possible this bag should be heat-sealed and labelled 'asbestos'. At least one company has standard labels commercially available. These individual bags are placed inside a strong coloured plastic bag, clearly labelled 'asbestos waste'. (In the UK it should be red: in the USA it should be yellow.) If substantial amounts are involved then the bags should be placed in a metal drum, of the same colour and also labelled.

The member of laboratory staff collecting up such items should (at the very least) wear a disposable apron, gloves and an efficient filter respirator. At the end of the collection process, the apron, gloves and filter cartridge are sealed into plastic bags and added to the asbestos waste.

If there are substantial quantities, then a licensed contractor *must* be appointed, to collect up the goods, decontaminate the surroundings and safely transport the waste away. Any other arrangement is likely to be illegal.

Hard asbestos sheeting is less of a hazard because the fibres are held in place by cement or other components. However, for peace of mind, any loose items may be removed. Note that 'mineral board' composites have been in use for some time. A piece of commercial apparatus intended for high temperature is quite likely to have an asbestos substitute rather than asbestos board, though the two may seem similar.

CAUTION: constructional asbestos poses the greatest hazard when it is being removed. No-one other than a licensed contractor should perform this job, and it should not be taken in hand too hurriedly. However, items with obvious surface damage (due to heat, frost, chemicals) may shed a significant amount of harmful fibres, so early attention is advised.

Where asbestos-based ducting, benching or hard insulation is present, it may be treated to reduce the chance of fibre release. An epoxy-based paint is suitable for ducts and surfaces in good condition, where very high temperatures are not encountered. Commercial products are available (used on chemical plant) for treating ducts and hard insulation. On benches and fume cupboard linings, a mixture of waterglass (liquid sodium silicate) and water may be applied. Three applications of a 1:2 mixture of waterglass and water should be used, with each coating being allowed to dry out overnight.

Where asbestos is covered by paint etc., it should be clearly labelled 'asbestos' so that no-one in the future can drill or cut into it without realizing its nature.

CAUTION: in preparation for treatment or in any proposed decontamination, dust removal should be by a wet technique. Ordinary vacuum cleaners may greatly increase the health hazard by leaking fibres from joints and from the filters. Special machines are available which do retain dust efficiently, and which are used by specialist contractors.

Any laboratory intending to handle asbestos samples should ensure that it has adequate containment (e.g. glovebox for sample preparation) and waste disposal available. Techniques such as encapsulation in clear plastic have been developed for safe examination by microscopy.

CAUTION: the use of vacuum cleaners, safety cabinets or extract ducts with filters can produce a dangerous concentration of asbestos on the filter. The most rigorous precautions should be taken in changing filters and in disposing of them.

This applies both to laboratories which have been deliberately handling asbestos, and to laboratories which discover some unintentional source, such as deteriorating constructional material. See section 6.4.7.

In the UK, asbestos is a listed substance under the 'Special Waste (Section 17) Regulations'. (See section 2.6. and Appendix C-1.) This means that prior notification is required before disposal of waste containing (a) any detectable amount of blue asbestos or (b) 1% or more by weight of free asbestos fibres or dust. In addition, operations such as the removal of structural asbestos must be carried out by a licensed contractor under the Asbestos (Licensing) Regulations 1983.

In the USA, asbestos is *not* listed as a 'Hazardous Waste' under the RCRA regulations. However, the Environmental Protection Agency has laid down requirements (under 40 CFR 61.25) for waste disposal sites accepting asbestos. Other sections of 40 CFR 61 list the requirements for notification and working procedures when stripping out constructional asbestos. Packaging and labelling of waste asbestos are regulated by the Occupational Safety and Health Administration under 29 CFR 1910.1001.

12.2 EXPLOSIVES

Laboratories in the specialist explosives industries (fireworks, munitions, blasting products, military) will necessarily have facilities for disposal of items up to whatever size they handle. The facilities will be controlled by special government inspectors, and a licence will be required.

The detailed requirements are outside the scope of this book. However, other laboratories may find themselves dealing with explosives as an incidental matter: possibly as samples to be analysed, possibly inadvertent mixtures, or most often chemicals which have explosive properties but which are used for their other reactive properties.

A select list of some laboratory compounds which are explosive in nature is given in Appendix C-4.

Where commercial explosive formulations are treated, then precise information is essential. Some can be safely ignited and will burn quietly — others will detonate under the same conditions. Some in fact require an initiating charge of another explosive to ensure their complete destruction. If an occasional use or testing of these materials is envisaged, full technical advice should be sought from the manufacturer.

CAUTION: whether a commercial explosive or a laboratory reagent is involved, it is essential that no attempt is made to cause an explosion, except by a person genuinely knowledgeable and experienced in the correct handling of explosives. Amateurs with limited knowledge (e.g. from military service) may prove a great hazard.

If at all possible, a known explosive should not be handled in amounts greater than 5 g. Many substances are rendered less sensitive by being kept wet, and in this condition may be chemically treated. Some explosives which are actually intimate mixtures of an oxidant and a fuel (e.g. gunpowder) can have the oxidant removed by prolonged washing with water. The residue can then be burnt in small amounts with little danger of explosion.

CAUTION: if a material normally kept wet — such as benzoyl peroxide or picric acid — has been allowed to dry out, then even the handling required to re-wet it may cause explosions.

The effects of even 10 g of material detonating can be disastrous. If the laboratory has any doubt about its ability to handle a particular situation, then it may not be an extreme reaction to enlist the help of the military or police bomb disposal experts.

12.3 UNIDENTIFIED MATERIAL

At some time or other, virtually every laboratory will discover a container whose contents, ownership and age are doubtful. The vast majority of such finds are in fact relatively innocuous, but this cannot be relied upon. The following checklist is intended as a guide only, to help responsible people. It must be used in the light of any particular local situation.

(1) Do not move the item unless absolutely necessary. A photograph taken at this time may assist in the identification or in planning disposal or in an inquiry afterwards, should something go wrong.

(2) Any movement should be carried out by volunteers wearing suitable protective clothing. It should be as gentle as possible. Take particular care that the container is not faulty. Cardboard boxes, metal cans and even glass bottles have been known to leave the base behind when lifted.

(3) Beware of any obvious assumptions. Two examples are a 'chart recorder' which discharged several kilograms of mercury when tipped onto its side, and a 'distilled water' wash bottle which injured a technician because it contained hydrofluoric acid.

(4) Make every effort to identify the person who was responsible for the container. If that person is no longer available, his or her supervisor should discuss the possible and probable nature of the material. For the sake of safety, it is quite reasonable to contact someone at his or her new place of employment or at home if he or she has moved elsewhere or retired. This is the most important and valuable step.

(5) If the owner is still available within the organization, it should be made his or her responsibility. This may entail some inconvenience if the person concerned has moved to another area or department, which is the price paid for not having dealt with it before.

(6) If the above has not given absolute identification, then there should be a discussion with any staff who may be able to contribute. They should decide (a) what of known hazards are at all possible, i.e. if substances have ever been used which are infective, explosive, etc. (b) if there are any possibilities which can be ruled out, e.g. there has never been work with micro-organisms.

(7) If the item can obviously be treated by some routine procedure such as autoclaving or incineration as it is (e.g. if it is a plastic culture bottle containing some broth or gel) then this can be done, using the most severe conditions to ensure full destruction.

(8) If there is any possibility that the material is an explosive, or is a

chemical which has degraded to an explosive condition, then the container should not be opened and not handled. It should be inspected for any signs such as deposits around the lid or stopper, crystals within liquid, or dried-out solid in a jar. Advice should be taken from experts in the chemicals concerned and in explosives. See also section 12.2 and Appendix C-6.

(9) On the balance of the evidence, it should be decided whether the container is safe to open. It may be advisable to open the container under water, in a bath of hypochlorite, or in an inert atmosphere. It is best if any lid or screw-top is held by a mechanical device (e.g. a retort clamp or strap wrench) rather than by hand, in case of any explosion or violent reaction. For a bottle with a rubber or cork stopper, a greased hypodermic syringe needle should be inserted down the side to relieve any pressure inside.

At the very minimum, the person opening a suspect container should wear a heavy apron, full face shield and gloves, and work in a fume cupboard over a tray.

CAUTION: if the container does not open easily, it should not be held against the body to exert greater force. A case is known where a small bottle of isopropyl ether caused lethal injuries when opened in this fashion.

Sealed glass ampoules should be cooled in a mixture of ice and water prior to opening by the correct standard technique. Cooling is also useful for other containers which may have developed internal pressure or may contain a very volatile substance. (See Appendix C-6.)

(10) If the container cannot be opened then it should either be treated as in (11) or be broken open. Plastic closures can be sawn or scribed and then cracked. Metal closures or seized glass stoppers will normally require the container to be broken. This can be achieved by scribing round the neck with a diamond pencil, then applying a small blob of molten glass to the scratch. This usually causes a clean break. In case the contents are a flammable liquid, the flame for heating the glass should not be in the fume cupboard, the container should be cooled down before applying the technique, and the molten glass should be immediately removed from the vicinity.

(11) Containers which are unsafe to open may be treated thus: (a) disposed to an incinerator, notified as the most dangerous compound of the possible ones; (b) broken outside (bullet, explosive charge or remote impact, all with appropriate safeguards). This should be in a situation where immediate treatment can be carried out, e.g. above an open container of a suitable neutralizing agent, or on a pile of absorbent for burning or collection.

(12) A container which has been opened should be observed for any immediate reaction with the air, e.g. fumes with moisture, or starting

to glow as air is admitted. This will indicate a general class of compound and the major hazard (respectively corrosive with a violent water reaction, and pyrophoric).

(13) The storage arrangement will give some clues. For example, alkali metals and phosphorus are stored under liquids, as are certain explosive compounds. If there are obviously several physically separate components, it may be necessary to sample each one. A disposable plastic bulb pipette is suitable for liquids; a plastic spatula should be used for solids. (A metal spatula may cause certain unstable compounds to detonate; a wooden one may ignite with some oxidizing agents).

(14) A tiny amount should be heated rather strongly on a metal spoon. If there is an audible crack or obvious sudden explosion, then the material should be treated as an unstable explosive. Otherwise it should be noted if it is or is not flammable.

(15) A tiny amount should be dropped into a large excess of water to see if there is any violent reaction. Liquids which float are generally suitable for incineration. Solids which do not dissolve may be acceptable to land disposal. Substances which dissolve may be diluted to drain, providing they are not too toxic.

(16) On the basis of these simple tests, the majority of samples will be identified by the laboratory as a recognized substance sufficient for waste disposal. Failing this, it will be necessary to use the services of an analytical chemist (in-house or consultant) to identify the material well enough for legal disposal.

Some waste disposal companies will provide a certified analysis of unknowns as part of the disposal service. A commercial kit (Hazkit) has also been used to identify unknowns, particularly in urgent circumstances.

12.4 CLEARING OUT A 'DEAD' STORE

The following technique is typical of that which might be used by experts to make safe a disused store of chemicals. It may be adapted to suit other requirements, such as annual clear-outs of cupboards, or the aftermath of an incident.

12.4.1 Before Entry

1. Ascertain from people and records just what materials are likely to be present, probable amounts, containers and methods of storage. Make a list.
2. Ask what materials should not be present, but could just possibly have been left there. Make a second list.
3. From the above lists, note materials which are especially hazardous and

those which can deteriorate to a dangerous condition on storage. See Appendices C-3, C-4, C-5, C-6.

4. If a plan of the room is available, make copies.
5. Find out details of the room's construction, exits, shelves, sprinklers, machinery or any other information which may be important in an emergency. Get confirmation if the lighting is spark-proof.
6. Clear roads and aisles in the vicinity to provide safe access and for emergency procedures.
7. Agree a general course of action with all people who may be involved, including security, maintenance, etc.
8. Make ready and check all apparatus and necessary items, including room keys. Protective clothing for the expected hazard and people trained in its use will often be required. Additional fire-fighting arrangements are always advisable. A flammable gas detector is routine. Monitors for other vapours may be necessary. If there is no safety shower nearby, there should be another supply of water immediately available. Portable safe lights may be required.

12.4.2 The Inspection

1. One person should carefully open the door and test the atmosphere, while the others stand back.
2. If the test is satisfactory, he should advance one pace into the room and repeat the tests.
3. If there is any doubt about the lighting, or if there are flammable fumes present, then only certified safe inspection lights should be used.
4. If fumes are detected, consideration should be given to the use of breathing apparatus and/or forced ventilation. In most cases where highly toxic materials are not free, accumulated vapours can be allowed to disperse by normal ventilation. For the purposes of inspection, people may make short visits into the room, providing the fumes are not above the Maximum Allowable Concentration. A cartridge-type respirator, of the correct type and used by a trained person, is sometimes an appropriate precaution. In any case, some time should be allowed for natural or forced ventilation before the full inspection takes place.
5. If there is more than one exit, at least two doors should be open before anyone enters. This also assists ventilation.
6. The first inspection should last no more than one minute, and should be by a person who remains within clear view of the door. He should then come well out of the room.
7. Following this visit, there should be a brief discussion to check if the original information was correct. This is often a useful time to identify problems which had been forgotten (e.g. additional drums or a blocked exit).

8. A sketch map should be made (or a room plan marked) showing bays, shelves, racks, etc. If there are no signs or inadequate signs, a complete numbering system should be devised (e.g. rack A, shelf 1, etc.) so that it is possible to refer unambiguously to any area to be cleared. This should be marked on sketch maps. If possible, labels should be attached to the shelves. It is sometimes useful to mark off the floor in an imaginary or actual grid, on the plan and even on the floor, so that a precise sequence can be followed.

9. For the main part of the inspection, there should be only one or two people in the room at a time, and there should always be an observer just outside.

10. The main inspection should produce a list of all movable containers, noting any that are in bad condition or where identification is doubtful. Items should not be moved at this point. If some items are obscured, this should just be noted.

12.4.3 The Removal Process

1. The general strategy is to remove easy items first, and those materials which are precisely understood, then proceed with greater care for more difficult items, having cleared the area. The logistics of the removal should be discussed, and disposal methods arranged for known materials.

2. A place or places should be made ready to receive items to be removed in suitable categories. It will often be necessary to remove some items first and leave others until special arrangements can be made.

3. Any necessary handling equipment should be got ready. This may include drum carriers, trolleys, sack trucks, bottle carriers, plastic bins and extra containers for insecure or hazardous items to prevent spillage.

4. All people involved should be adequately protected and should fully understand what is required of them. A written work programme is best.

5. A final check should be made that the general vicinity is clear of people, vehicles or obstructions which could interfere with the work or emergency action.

6. General clutter and harmless items are removed first particularly those which aid access (e.g. goods in passageways, around entrances or in front of shelves), providing this does not entail moving chemicals or unknowns.

7. A competent person should then inspect goods revealed by this first clearing up, and make any notes or other arrangements necessary.

8. Items which are unknown or particularly hazardous should be specially marked with a bright coloured adhesive label, on or by the item. (An orange disc will do.) Unlabelled or doubtful items should also be marked to be left, with the same or a different colour.

9. Goods may now be moved out by category and by sections. (For

example: category 'acids', systematically go through sections 1, 2, 3, etc. on the plan.) Appropriate measures for handling and protection will be taken for each category. Removal should be done without moving other items. For this reason, it will often be necessary to repeat a 'category' procedure. For example, if drums of solvent are to be removed and a container of acid is found among them, it will be necessary to wait until the acid has been removed by an appropriate technique before moving the solvent drums from behind it. This is most important. Only one category should be in hand at one time. Items adjacent to containers marked as hazardous should usually be left in place.

10. This will perhaps leave a largely empty room with a few items which (a) are unknown (b) are in unsafe containers (c) are extremely toxic (d) may be shock-sensitive explosives. There will also be some items which have been left next to the problem containers. These adjacent items should now be moved, taking the greatest care not to disturb the problem items.

11. It is usually found that by closer inspection and discussion of available information, most of the unknowns can be identified well enough for disposal, though they may well be counted explosives or severe poisons.

12. The extremely toxic materials should be removed by a skilled person. The safest general technique is to place the container into a larger plastic container, padded with an inert mineral (sand or vermiculite).

13. The possible explosives should now be considered. It is not extreme to request the assistance of the military, the police bomb squad or other specialist group. If such an item is to be moved, without special equipment, then the safest general technique is as follows. A large, open container is one-third filled with dry sand, and moved on a trolley to as close to the suspect item as possible. The item should then be moved gently to stand on the sand. Further sand is poured slowly around the item until it is mostly, but not completely, covered. The open container may be covered with a plastic sheet, but not with any solid lid. A 2-gallon bucket will serve for jars of chemicals up to 100 g, a 5-gallon bucket for jars up to 500 g, and an open-topped drum or plastic dustbin for larger jars.

14. Materials in insecure containers can now be dealt with. Drums of liquid or cracked bottles should be siphoned out. It may be possible to strengthen cracked jars with self-adhesive tape sufficiently to enable them to be eased into a bag or other container. The spillages should be dealt with immediately.

12.4.4 Afterwards

1. The room should be inspected, preferably by two people, to see if there have been any items overlooked, and any evidence of spillages or corrosion.

2. Spillages and dust should be treated with great care and removed with all the precautions appropriate to the chemical concerned. With certain kinds of contamination (notably asbestos, mercury, allergens, neurotoxins, and fire-promoting agents) it is advisable to have a specialist company arrange decontamination of the premises.
3. Wood which has been exposed to chemicals may well be unsafe. If oxidizing agents (e.g. nitrates, perchlorates, permanganates, peroxides, etc.) have soaked in, then the wood may be easily ignited and can burn almost explosively. Metal shelving may have been weakened by exposure to fumes. On occasion, the services or the structure of the building may have been affected. In any of these cases, serious consideration should be given to the stripping out and removal of affected items for disposal.
4. A check should be made that materials removed have in fact been properly dealt with, and not simply moved to another dead store.

12.5 SMALL LABORATORIES AND MOBILE UNITS

The following comments are necessarily general. They must be taken in conjunction with the type of laboratory and its exact circumstances.

12.5.1 Small Laboratories

A typical small laboratory is here envisaged as a service to some non-scientific facility, with limited facilities and only a couple of employees (or the equivalent on a part-time basis). Observations on similar operations have yielded the following tips.

1. Return all the unused portion of samples to process waste, or at least to the originator.
2. Return used samples and any other material to process waste if at all possible. The small amounts involved often cause no measurable impact, but this must be agreed with the process management.
3. See if there is a boiler or incinerator available on site (or nearby, if you have friends) which can tolerate small amounts of waste.
4. As far as possible neutralize and convert chemicals so that they may be flushed down the drain.
5. Find out what materials would be acceptable if well dispersed into trade waste. See if there is any regular waste which would be especially suitable to accept laboratory waste (e.g. fly ash, gypsum, clay, etc.).
6. Choose laboratory techniques which generate less waste, even if they are slightly more expensive. This may sometimes mean buying packaged commercial products: in other cases it may mean *not* buying labour-saving disposables.
7. Avoid materials (e.g. radioactives) which are the subject of complex legal controls — unless the procedure is absolutely vital, *or* the usage is so small that it is exempt.

8. If fume extraction is poor or non-existent, consider the use of a glove bag, glovebox or self-contained portable fume cupboard.
9. Do not be persuaded to collect solvents in a large can, unless the can is well away from the laboratory, and someone else's responsibility.
10. Do not attempt to recycle solvents unless there is a good fume cupboard available, and it will not seriously interfere with your need for work under fume extraction. However, with highly repetitive work, solvent recycling is often quite possible.
11. Do not use mercury if it can be avoided. There are many other fluids which can be used in manometers and many excellent commercial devices for pressure measurement. Alcohol thermometers, metal thermometers, thermocouples and electric resistance thermometers are often superior to traditional mercury-in-glass thermometers.
12. Be sure there is adequate absorbent and other facilities for spill control. (See Chapter 15.)
13. Take care to label waste and waste receptacles. This is often thought unnecessary where only one or two people are present, but has led to accidents.

12.5.2 Mobile Laboratories

A typical mobile laboratory is here envisaged as a converted caravan or vehicle, without access to main services, and somewhat cramped. However, it is normally associated with a well-equipped facility elsewhere. It is relatively uncommon, and generally made up specially. The following suggestions are offered to designers and users.

1. The airflow in such a confined space means that it is unlikely that a standard fume cupboard or microbiology safety cabinet class I would operate correctly (i.e. to give the specified protection factor) under all weather conditions. It is possible that a class IIA would be suitable and adequate for some purposes, but a class III cabinet would be necessary for any work which might involve pathogens. See Chapter 6.
2. For work with chemicals with hazardous vapours or unpleasant smells, an adequate system might be made consisting of a chamber exhausted to the outside by a low-speed fan, and front access via a clear panel with two holes. That is, a negative pressure glovebox, without the gloves being fixed in place. The entry ports should have covers for when one or other is not in use.
3. Wall-mounted roll dispensers of plastic bags are obtainable, along with quick-seal devices. (They are often used in supermarkets.) They may be used to seal up individual items of waste as they arise.
4. Rectangular plastic boxes with lids which seal are available for food use. They can be colour coded and labelled, and fitted into drawers for waste collection. If the laboratory stays in one location for some time, these boxes can be regularly removed (to the central institution) and replaced. One can be used for broken glass.

5. If sharp items (scalpels, needles) are in use, then a cylindrical sharps' box can be located in a hole in the bench. This will improve bench access, and prevent items being knocked over.
6. A similar arrangement can be made for liquid waste receptacles. It is essential that these are plastic or metal.
7. In general, there should be several small recceivers for each waste category (say 3) rather than one big one, so that one can be removed for emptying without disturbing the work of the laboratory.
8. It is essential that appropriate and sufficient materials for dealing with spillages are immediately available. Commercial kits are most likely to be suitable. See Chapter 15.

See, also items 6, 7, 11 and 13 in section 12.5.1.

REFERENCES
[section 12.1]

International Agency for Research on Cancer (Lyon, France) *Publication 33 (1979). Handling Chemical Carcinogens in the Laboratory. Problems of Safety.*
Walters, D.B. (ed.) (1980) *Safe Handling of Chemical Carcinogens, Mutagens, Teratogens and Highly Toxic Substances. Vol. 2,* Ann Arbor.

BIBLIOGRAPHY
[section 12.1]

Barbeito, M.S. (1979) Laboratory Design and Operation Procedures for Chemical Carcinogen Use, In: Scott, R.A. (ed.) *Toxic Chemical and Explosives Facilities,* ACS Symp 96, Amer. Chem. Soc., 191–214.
Craig, N.R., Jr. (1976) Disposal of Carcinogens and Highly Toxic Chemicals, *Safety Newsletter (National Research Council, USA),* May 1976.
Fairchild, E.J. (1978) *Suspected Carcinogens,* Castle House.
Fisbein, L., Flamm, W.G and Falk, H.L. (1970) *Chemical Mutagens,* Academic.
Howe, J.R. (1975) A method of Recognizing Carcinogens in the Laboratory, *Lab. Pract.,* **24,** 457–467.
International Agency for Research on Cancer (Lyon, France), *Publication 33* (1979) Handling Chemical Carcinogens in the Laboratory. Problems of Safety.
International Agency for Research on Cancer (Lyon, France), *Publication 37* (1980) Laboratory Decontamination and Destruction of Aflatoxins B_1, B_2, G_1, G_2 in Laboratory Waste.
International Agency for Research on Cancer (Lyon, France), *Publication 43* (1982) Laboratory Decontamination and Destruction of Carcinogens in Laboratory Wastes: Some N-Nitrosamines.

International Agency for Research on Cancer (Lyon, France), *Publication 49* (1983) Laboratory Decontamination and Destruction of Carcinogens in Laboratory Wastes: Some Polycyclic Aromatic Hydrocarbons.

Medical Research Council (UK, London) (1981) *Guidelines for Work with Chemical Carcinogens in Medical Research Council Establishments,* MRC.

Sax, N. (ed.) (1980) *Cancer Causing Chemicals,* Reinhold.

Searle, C.E. (ed.) (1976) *Chemical Carcinogens, ACS Monograph,* **173,** Amer. Chem. Soc.

US Department of Health Education and Welfare (1979) *Guidelines for the Laboratory Use of Chemical Substances Posing a Potential Occupational Carcinogenic Risk,* US Govt.

Walters, D.B. (ed.) (1980) *Safe Handling of Chemical Carcinogens, Mutagens, Teratogens and Highly Toxic Substances. Vol. 2,* Ann Arbor.

[section 12.1.4]

UK Department of the Environment (1976) Waste Management Paper 6: Polychlorinated Biphenyl (PCB) Wastes, HMSO.

US Federal Regulations (1979) 44 FR 31514 (E.P.A. Regulations for Use and Disposal of PCBs).

[section 12.1.5]

Asbestos Information Centre (1984) *Review of UK Asbestos Legislation,* A.I.C.

Michaels, L. and Chissick, S.S. (eds) (1979, 1983) *Asbestos: Properties, Applications and Hazards Vols. 1 & 2,* Wiley.

UK Department of the Environment (1984) *Asbestos in Buildings (Rev.),* HMSO.

UK Department of the Environment (1984) *Waste Management Paper 18: Asbestos Waste (Rev.),* HMSO.

UK Health and Safety Executive (1983) *Work with Asbestos Insulation and Asbestos Coating: Approved Code of Practice and Guidance Note COP 3,* HMSO.

UK Health and Safety Executive (1984) *Asbestos and You,* HMSO.

UK Health and Safety Executive (1984) *Asbestos — Control Limits and Measurements of Airborne Dust Concentrations. Guidance Note EH 10,* HMSO.

UK Health and Safety Executive (1984) *Guidance on the Requirements of the Asbestos (Licensing Regulations 1983 Health & Safety Series HS (R) 19,* HMSO.

US Environmental Protection Agency (1980) *Asbestos Containing Materials in School Buildings: A Guidance Document,* US Govt.

[section 12.2)

Federoff, B.T. (ed.) (1960–) *Encyclopaedia of Explosives and Related Compounds,* Dover.
Forsten, I. (1973) Pollution Abatement in a Munitions Plant, *Environ. Sci. Technol,* **7,** 9, 806–810.
Gould, R.F. (1966) *Advanced Propellant Chemistry (ACS Monograph 54),* Amer. Chem. Soc.
Pollock, B.D., Fisco, W.J., Kramer, H. and Forsyth, A.C. (1977) Handling, Storability and Destruction of Azides, In: Fair, H.D. and Walker, R.F. (eds) *Energetic Materials Vol. 2,* Plenum, 73–109.
Schumacher, J.C. (1960) *Perchlorates: Their Properties, Manufacture and Uses,* Reinhold.
Swern, D. (ed.) (1970–) *Organic Peroxides,* Wiley.
Urbanski, T. (1964–7) *Chemistry and Technology of Explosives (3 vols),* Macmillan.

CHAPTER 13

Educational Institutions

13.1 INTRODUCTION

Educational laboratories of all disciplines have two important special characteristics. Firstly, students are by definition inexperienced, and the laboratory operation (including provision for waste disposal) must take this into account. Secondly, the experiments may in principle be altered or discontinued if they pose special problems. An exception is where an examination has to be carried out according to the specification of an outside body. Apart from this, it is usually possible to cope with a difficult waste disposal by suspending the experiment creating the waste.

Some places integrate the waste disposal operation into the educational process, and make it an important feature of the experiment. Regrettably there are others who take virtually no account of it in planning experiments or operating laboratories. Some schools in particular find that they have very little resources in the way of manpower, facilities or even basic information. On the other hand, there are universities who have mistakenly assumed that the presence of expensive modern apparatus is all that is required to teach good scientific technique.

13.2 TEACHING WASTE DISPOSAL

Part of the aim of education must be to produce socially responsible citizens who are aware of the world around them. Waste disposal is a major feature of technological society and should be recognized as such. Moreover, its effectiveness is vital for all societies, though developed and other countries may have different needs. Perhaps most important of all, if large-scale waste disposal is mismanaged it could create major damage to the environment. This will not be prevented by ignorance, only by understanding.

Pollution control has become fashionable as a topic of education in recent years, and may be used to illustrate the teaching of more traditional subjects. Students in a variety of disciplines and of all levels will often appreciate a visit to a sewage works or a refuse disposal site. Viewing special industrial waste control equipment may be more difficult to arrange, but will often be worth the effort.

Clearly, there is some ethical imperative for the educational institution to demonstrate its concern by setting a good example in waste disposal. Certainly, the effect of teaching by example is so great that it should be carried through to the behaviour of individuals. This is difficult to achieve, but is well worth while, as a thoughtless action by a teacher or other staff member can create the impression that the rules are not really so important.

13.3 CLASSWORK

13.3.1 General Requirements

Waste disposal must of course be appropriate to the materials involved. In the classroom it must also be adjusted to the teaching methods employed and the abilities of the students. For example, in a school or very junior class it may often be appropriate for the students to leave certain disposal procedures to the technical staff, whereas advanced college students may be expected to carry them out personally. The amount and types of material arising will depend on the overall course structure. That is, if the whole class does the same experiment on the same day, there will be a large amount of similar waste at one time, whereas the same amount will be collected over many weeks where groups of students take it in turns to carry out specific experiments.

In timetabling laboratory classes, some attention should be paid to the waste disposal requirements.

Laboratory work is commonly scheduled for afternoons, often including Friday afternoon. This means that technical staff have the morning to prepare, but conversely there may be no time after the class to safely deal with waste within the normal hours of work. Practical experience suggests that it is unwise to rely on the sense of duty of teachers, graduate demonstrators or technical staff, particularly if this is called upon too frequently.

Responsible people should critically examine set experiments on a regular basis, for their value in education and for practical reasons. If an experiment is too long to be carried out at a reasonable (student's) pace, then there is likely to be a rush at the end, with a possibly dangerous inattention to waste disposal in the hurry to clear up (often under pressure from staff members who wish to finish on time).

Assuming that the experiment is valuable enough to retain (some remain on schedules for essentially historical reasons though they have long since ceased to relate to the course) then there are several methods of improvement. A critical path analysis (or a related management technique) is useful for educationalists who know the method. This should look for technical ways to shorten the longest steps — for example a mechanical aid such as a high-speed mixer may eliminate a long manual process. Another alternative is to break the experiment into two parts — either by stopping it at a certain point and continuing the next time, or by creating two new experiments which achieve the aims (such as use of specific laboratory procedures or demonstrating certain features of the subject) of the original. Failing this, some consideration should be given to the start and end of the experiment. Is it possible that it could be partially prepared by the laboratory staff? If not, is it worth making a special exception to the usual waste disposal practice, and leaving this to the staff? In the latter case, everyone should understand the reason.

13.3.2 Practical Considerations

Waste disposal should be considered at the design stage of experiments and should be specifically included in the instructions. For example, 'the washings should be flushed down the sink', 'this portion of the extract should be discarded to solvent can A (halogenated solvents)', 'all tissues should be left on the working tray for a technician to remove'. A vague request for materials to be disposed of carefully is not adequate.

Waste disposal should be promoted as part of good laboratory practice according to the discipline. In a chemical class, the possible ways of neutralization can be discussed. In a biological class the importance of orderly work and attention to hygiene can be stressed. In an engineering laboratory, the importance of good housekeeping can be pointed out as a major aid to safety.

The carrot and stick approach may be used to give some incentive to students. For example, a certain amount of the marks may be allocated to satisfactory cleaning up, on the judgement of the technician responsible for the laboratory. (See, also, section 13.3.3.) On the more positive side, students can be asked (either on a general basis or as a specific exercise) to critically examine the laboratory waste disposal procedures and suggest improvements. They usually enjoy criticizing the institution, but are then personally obliged to demonstrate more care in their own actions.

Technical staff will normally check the waste disposal facilities as part of the normal preparation for a class. However, the supervising teaching staff (or graduate students where appropriate) should personally inspect the waste disposal facilities at the start of each class. That is, they should check that the solvent receptacles are not full, that bins are in place and empty, that local receptacles (e.g. sharps' boxes on the benches) are ready, that fume cupboards and safety cabinets are working, and that any 'one-off' special arrangements for the particular session are in order. This check should be an obligatory start to any laboratory class.

The types of receptacle and the organization of them are discussed in Chapter 3. However, decisions will have to be made for each laboratory as to the general strategy and its particular implementation. In principle, there can be a set of waste receptacles for each student (or student pair, trio, etc.) or the students may use common receptacles. The latter is simpler, but will almost inevitably result in wastes being wrongly mixed. An intermediate strategy is possible — for example having one flammable solvent can per bench or set of students. This limits the damage and inconvenience. Where all students are carrying out the same experiment then it may be possible to have a set of containers specially for that class with very clear labels (which may be changed for the next class). For example a set of plastic bags in appropriate holders may be used.

Where students rotate around a fixed series of experiments, then the safest procedure is to have waste receptacles for each experiment as part of the experimental apparatus. These will be emptied or treated by the technical staff as part of the clear-up process. The container may be a disposable item itself (e.g. for waste needing to be autoclaved) or may be large enough to accumulate a term's worth of material (e.g. precious metal precipitates). As a permanent fixture, there is no excuse for the receptacle not being of suitable design and properly labelled.

Students issued with a personal set of apparatus and locker may put these items away at the end of the session, since they will personally suffer if cleaning is inadequate. However, where they use common apparatus it is preferable that they leave all items (clean, but not necessarily dry) on the work-bench. It may be a chore for the technical staff to dry and put the items away, but it is a opportunity to check for breakages, dirty items, missing parts, etc. Similarly, any empty bottle of chemical should be left on the bench so that staff are aware it needs replenishing, and can return the bottle to the supplier if appropriate.

13.3.3 Low-waste Techniques

The problems of disposal of certain kinds of waste (and recent concerns about safety) have led some educationalists to try to develop laboratory courses which produce much less waste or less difficult waste than those which they replace.

One option is to replace practical experiments with video recordings, computer simulations or demonstrations. These have particular value for experiments which are really outside the safe capabilities of the laboratory. They may also be used as a preparation for practical work, demonstrating correct techniques. However, there is a real value in actual 'hands-on' work in the teaching of science and technology. It is our view that students in many courses would benefit from the experience of handling hazardous materials in a controlled, safe manner. Thus these techniques should not completely displace practical classes.

Quite small changes in operation of experiments can often make a major improvement to the waste disposal. Many experiments call for chemicals (e.g. solvents) now considered hazardous and/or difficult to dispose of. It is often not difficult to replace them with more acceptable materials, or those which can be more readily recycled. (See Chapter 14.)

A useful saving can often be made by adjusting the experiments to fit the pack size of available materials — for example, adjusting the recipe so that a term's work will (with allowance for spillages etc.) largely use up one jar of chemical. This may mean an adjustment up or down in the instructions, but saves the waste of material which cannot be easily kept from one year to the next. A reserve may perhaps be kept in a much smaller bottle. With very difficult materials it may be worth while buying small packs and altering the experiment so that a fresh pack is opened each time. If the manufacturer does not supply a small enough size then packs can be prepared in advance by a skilled person using a dry-box, glovebox or other appropriate device. It is a fact of life that students can be expected to leave lids insecure, which can lead to considerable losses of air- or moisture-sensitive materials.

Animal and plant tissue also comes in standard packs — namely the whole organism. If an animal is to be sacrificed for one purpose, then with some co-ordination it may in fact provide organs or tissue for other laboratories (in other years, courses or even departments — it would be feasible for pharmacy, biochemistry and biology students to carry out different practicals on the same day using different parts from the same animal). For some purposes, tissue and organs can of course be frozen or otherwise preserved.

For biological work in particular, the current trend is towards disposable items for treatment in a large autoclave followed by incineration. A small and relatively isolated laboratory such as a school may perhaps find the occasional disposal of small quantities of such waste awkward. Thus it may in fact be more effective to use laboratory ware which is intended for re-use, such as glass Petri dishes rather than plastic ones. A very small autoclave may be sufficient to render items harmless for hand-washing, so that gels and broths can be disposed via the sink. If a sink grinder unit is fitted, then most tissue can also be disposed to drains, and all other waste should be capable of processing to a condition acceptable as normal refuse. (See sections 3.9.3, 5.3.5, 10.2, 10.5.4.)

CAUTION: the practice of trying to clean disposable plastic items for re-use is very bad. It can be dangerous and makes very small savings in return for a lowering of standards in the laboratory.

In chemical work in particular, but to some extent in other areas, it may be possible to reduce the waste disposal problem by reducing the scale of the experiments. There has been a general move from traditional scale to so-called 'semi-micro' working for synthesis, analysis and physical chemistry. This has been extended in some quarters to courses which are mainly carried out on the 'micro' scale — that is using quantities of less than a gram of chemicals in a few cm^3 of solvent.

The sensitivity of modern instruments means that quantities of a tenth of a gram or less can be readily characterized. In fact, some facility in working with very small amounts of material is essential for skilled use of most analytical instruments. However, it is not possible to simply scale down, whether from macro or semi-macro, or semi-micro to micro. Each scale has its own peculiarities of technique, and the experiments must take them into account.

For laboratories already working on the semi-micro scale, there may be little savings in reagent costs, since the minimum purchase quantity may be several years' supply already. However, there is a notable drop in solvent requirements (and hence solvent waste). This may very well produce such small quantities of solvent waste that evaporation is an adequate technique, largely eliminating the disposal problem. Other chemicals are likewise easier to get rid of in very small amounts.

Perhaps the most spectacular improvement given by the micro scale of experiment is the reduction in fume emission and thus ventilation requirement. This is a valid alternative to upgrading some extract systems to give an acceptable working air quality. (See section 6.4.5 and Chapter 6 generally.)

Some institutions have successfully integrated re-use and recycling into the experimental programme. For example, a student may perform a synthesis and use its products for another experiment (possibly another synthesis, possibly physical chemistry) following what may be a long sequence up to final disposal.

In the 'Zero Effluent Laboratory' project at the University of California — Riverside, each student was issued a few hundred cm^3 of the principal solvents needed in the course. He or she was expected to recover and retain each solvent by distillation or other procedures for further use in the course. Finally there were marks awarded for the quantity and quality of solvent returned to the institution. It was found that the final quality was in fact good enough to be bulked for use with the next year's class, and there were considerable savings in solvent purchase, as well as educational benefits.

Note that if recycling is well organized, it may be possible to use a more expensive solvent (for safety or for desirable technical properties) since the savings more than justify the initial extra cost. (See sections 14.5 and 14.6.)

The low-waste concept may initially be formulated to cope with practical difficulties of waste disposal in the laboratory. However, the idea can well be taken into the students' careers, since recycling is very important in many industries, and is likely to become increasingly so.

13.3.4 Examinations

Practical examinations can be expected to generate waste in two ways. There should be a general clear-up before any exam, in which old reagents are discarded, waste receptacles emptied, and facilities such as fume cupboards cleaned and tested. Following this, the examination will itself produce waste from the experiments.

As far as possible, it is recommended that each student has in effect his or her own waste disposal facility. That is, bench containers (jars, plastic bags, bottles, etc.) should be provided at each workplace along with the necessary apparatus and materials. This has several advantages for the student, and is not too onerous for the technical staff.

An examination is a very stressful time for the students, who are bound to be nervous. The great fear is that one person's experiment may be spoilt by another's accident or mistake. A collision around the waste bin or solvent can is, unfortunately, likely in the hurry to complete the experiment, and will lead at the least to frayed tempers. Individual facilities greatly reduce this possibility and ensure that one student's mistake is likely to affect his or her work alone. (Supervising staff may then attempt to retrieve the situation without having to act as arbiters in disputes.)

If full provision is considered impracticable, then at least there should be sufficient receptacles for local use — for example one can/bin etc. per bench.

Examinations set by outside bodies sometimes cause difficulties for institutions (commonly schools) who find that the resulting waste products are outside their normal arrangements. If this is the case, then the institution should press its own controlling body as well as the body setting the examination for assistance. It is not unreasonable to ask for the waste to be collected, or a competent person to come from another institution to supervise disposal.

13.3.5 Accidents

Educationalists should take a positive attitude to accidents. They are bound to occur and often teach the students something about reality. If proper precautions are observed (such as the wearing of eye protection and the use of shields and fume cupboards) then injury can be avoided, leaving only the problem of clearing up. (See Chapter 15 for emergency procedures.)

In planning the operation of the laboratory, some kind of plan should be made (and materials provided) to cope with a small fire, a substantial flood of water, and the escape of the working quantity of any material (including living organisms) in use or in store in the room. If no plan can be imagined for a particular material, then there should be a serious question over its inclusion in the laboratory at all. For example, it may not be a good idea for a teaching class to be carried out in a research room containing exceptionally hazardous material, which could be released by an accident nearby.

It is most important that all members of staff (and supervising students) are fully aware of the procedures for dealing with spillages. Apart from the obvious safety problem, contradictory orders give a very bad impression. By demonstrating their ability to deal effectively with an unexpected situation, the staff will command the respect of the class, and show that hazardous materials need not be dangerous if there is sufficient skill and foresight.

With every breakage or spill, the staff members should make a quick but sensible appraisal of the situation. It is not necessary to evacuate the classroom when a beaker breaks, but on the other hand it may well be dangerous to continue other experiments in close proximity to a spill or minor explosion of chemicals. The senior staff member should decide how much of the classwork needs to be stopped, and who should do the clearing up. If possible, the students themselves should clean up — not as punishment, but as an exercise in responsibility and in manual laboratory skills. However, it will often be safer or quicker for the procedure to be carried out by a staff member. In any case, the students should discuss the clean-up procedure as it is carried out, including any hygienic or chemical processes involved.

Attention should be paid to the extent of the contamination of people and apparatus. A piece of broken glass or a droplet of mercury going into a machine may well cause great damage when it is switched on. Biological material may become lodged in places not normally disinfected to provide a reservoir of infection. In some cases it may be advisable to remove some clothing and autoclave it on the spot. Harmful chemicals on clothing may not be noticed immediately. Sharp objects may lodge in cracks, with a small amount projecting to cause injury to the hand. If not removed, these hazards may affect other classes or laboratory staff.

13.4 PROJECT WORK

Laboratory projects are those extended studies carried out by one or two students, ranging from a week to several years (e.g. a Ph.D. project). In all cases it is the responsibility of the member of teaching staff concerned to ensure that there is an adequate method of waste disposal *before* the project commences. This is especially vital where large amounts or very hazardous materials are involved. In the latter case, there should be regular checks (as part of the normal supervision process) that waste disposal is being correctly carried out. Waste disposal materials and procedures should be included in the budget for the project.

It is advisable for the chief technician or other person responsible for laboratory operation to ensure that he or she is notified of the termination of projects. He or she can then check that all waste has been disposed correctly. If necessary, any problems such as unlabelled containers can be dealt with while the students are still available. In some institutions it is the practice for project reports to be accepted for marking only if they have a signature from the appropriate technician to say that the work has been properly tidied up.

It is very often worth while for project students to have waste disposal facilities separate from those provided for regular classes, e.g. their own solvent can.

BIBLIOGRAPHY

American Chemical Society (1984) *Forum on Hazardous Waste Management at Academic Institutions*, American Chem. Soc.

Association for Science Education (1981) *Safeguards in the School Laboratory 8th edn*, Assoc. Sci. Educ.

Morton, T.H. (1983) *Zero Effluent Laboratory*, Dept of Chemistry, University of California, Riverside.

Orlans, F.B. (1977) *Animal Care from Protozoa to Small Mammals*, Addison-Wesley.

Scottish Schools Science Equipment Centre (1979) *Hazardous Chemicals — a Manual for Schools and Colleges*.

Notes

1. The *Journal of Education in Chemistry* has published many articles on chemical waste disposal relevant to educational institutions in the last few years.
2. In the UK there is a special School Science Service operated by CLEAPSE — the Consortium of Local Education Authorities for the Provision of Science Equipment. As well as providing information and publishing safety notes, the service can arrange measurement of fume cupboard air speeds and monitor mercury vapour levels. Details from CLEAPSE, Brunel University, Uxbridge UB8 3PH.

CHAPTER 14

Materials Recovery

14.1 GENERAL COMMENTS

The salvaging of worthwhile material from waste has much to recommend it, providing it does not interfere with the safe and efficient operation of the laboratory. Management should realistically cost the time and effort involved, compared with the costs of disposal and purchase. For example, a complex distillation process may not be justified to recover a cheap solvent in modest quantities. On the other hand, a chemically and physically identical process may be well worth while to recover a deuterated or ^{14}C labelled version of the same solvent.

The hoarding of items which are unlikely to be used is not uncommon, generally on the grounds that 'it is a pity to throw it away'. Such practices are not economically justified, and can detract from the safety and efficiency of the laboratory by cluttering the place with broken apparatus, unwanted chemicals and often a large amount of combustible plastic and cardboard.

It is suggested that the following items are *not* kept, unless an actual financial saving can be demonstrated, and there is no hazard in retention.

(1) Non-functioning electronic apparatus where replacement parts are no longer available.
(2) Electric devices which cannot easily be made to conform to current legal safety requirements.
(3) Gas cylinders near to (or past) their regular pressure testing date. (See section 9.4.)
(4) Part-filled gas cylinders which are unlikely to be used for any specific purpose in the next 6 months.

258

(5) Boxes, bottles and used containers beyond the obvious uses in the next two weeks.
(6) Used disposable plastic items such as syringes and single-use pipettes. It is impractical and unsafe to try to clean them and defeats the advantage of disposables.
(7) Rubber and plastic items (tubing, gloves, etc.) which are visibly perishing with age or usage.
(8) Chemicals and assay kits or other limited life items which are well past their stated expiry date.
(9) Obviously degraded chemicals.
(10) Bottles or jars of chemicals which have been more than three-quarters emptied and where the residue is worth less than £1 sterling or $1 US.
(11) Damaged glassware which cannot be repaired within the institution (or where experience has shown that it will not be) in a reasonable time to a safe standard.
(12) Exceptionally hazardous substances beyond the immediate need for them.

14.2 GLASSWARE

Many technicians and scientists have the skill to make small repairs to glassware using a simple gas torch. However, it is important that individuals do not attempt repairs beyond their skill or facilities. Work with ground glass joints or with more complex items normally requires inspection by a professional glassblower, who is able to judge whether the item can be safely repaired.

CAUTION: items for use under vacuum must be properly annealed before use. It is preferable if they are also inspected with a polarized light viewer for strains. Annealing is also recommended for repaired apparatus where a breakage would be costly or dangerous.

It is usually easy and worth while repairing small chips in the top of measuring receivers such as beakers and measuring cylinders. With a diamond saw it is possible to cut down a damaged (say) 100 cm^3 measuring cylinder to give a squat 50 cm^3 one. However, it is generally impractical to repair calibrated items such as volumetric flasks, pipettes or flow measuring tubes.
Items with ground glass joints are usually worth repairing owing to the value of the joints themselves. Even if the apparatus is too badly broken, a glassblower may be able to cut off joints for use in making other items.

CAUTION: it is the responsibility of the laboratory to ensure that items presented to a glassblower are clean, dry and free of harmful chemicals or organisms.

14.3 SCRAP METAL

Many laboratories find themselves with substantial quantities of metal goods which may have a scrap value. Even where the value is low (as is the case with most steel) there is some satisfaction in receiving a small payment instead of giving it. When many kilograms of copper, brass, tin or some alloys are involved, then there may be a useful sum involved. This most commonly applies to electric motors, resistance boxes and the like. Some electronic apparatus has significant quantities of precious metals, which specialist scap dealers are prepared to extract. An unwanted stock of photographic film, prints or X-ray plates has some scrap value for its silver content. Lead from acid batteries can be sold, and the same dealers are usually interested in relatively small amounts of the elements antimony, arsenic and bismuth in various forms.

It is rarely worth while a laboratory trying to separate out the metal from items such as those mentioned above. The process can be hazardous and are better left to the scrap dealer. As a rule of thumb, no laboratory manager can expect to get the better of a scrap dealer in a commercial transaction.

14.4 PRECIOUS METALS AND MERCURY

14.4.1 General Comments

Mercury, silver, platinum and palladium are frequently used in laboratories. Gold, osmium, iridium, rhodium and ruthenium have specialized usage. These metals are all regularly traded by precious-metal dealers. The lanthanides (also called 'rare earths') are expensive but mainly of interest to specialist chemical suppliers.

Recovery procedures are of two kinds: (1) where the laboratory is saving a metal for further use itself; and (2) where the material is to be sold. In the latter case, some suppliers will accept waste of certain specifications and make an allowance against future orders. This is usually easier than trying to sell material on the open market.

In the first case, where a laboratory intends to re-use the element, it is essential that the laboratory has sufficient knowledge of the chemistry to be able to prepare the desired compounds in the necessary quality. An example is the preparation of certain kinds of metal catalyst.

In the second case it is mainly required to concentrate the metal and present it in a form acceptable to the purchaser. Note that for metals of very high value, some companies will buy solutions, and most will buy precipitates containing a few per cent of precious metal. It is by no means necessary to prepare the metal, though it does of course command a higher price.

14.4.2 Recovery from Solutions

A simple method which is often overlooked is to evaporate down solutions of valuable metals and ignite the residue. If necessary the residue can be taken up in further chemical solution. This is not always practical, and should not be carried out where the likely solids are either explosive (as is the case with some nitrogen compounds including cyanides) or volatile (as is often the case with mercury, osmium and some ruthenium compounds, and many metal-organic compounds). However, concentration by evaporation is frequently a useful preliminary to chemical treatment.

Sodium borohydride is a particularly useful reagent since it will reduce aqueous solutions of most valuable metals direct to the metals. This can be applied to a wide range of solutions, often by simply stirring in an appropriate quantity of borohydride solution (sometimes heating helps, and for strongly basic solutions some acidification is required). Metals recoverable in this way are platinum, palladium, rhodium, gold, silver, mercury. Less valuable metals which may be simultaneously produced are lead, copper and cadmium. Note that other species present may also be reduced — for example nitrate NO_3^- is reduced to ammonium NH_4^+ — which will require some equivalents of the borohydride. Some transition metals may precipitate out as their borides.

Instead of production as a powder, the metal can be deposited onto a solid (so-called electroless plating) such as a piece of the metal itself or copper sheet or mesh. The solid should be clean, best of all etched with acid, and usually gives best results if it is treated with (i.e. immersed in, or brushed with) an aqueous solution of tin(II) chloride or palladium(II) chloride immediately prior to the borohydride process.

CAUTION: there may be a fire hazard from hydrogen production and some metal powders may be pyrophoric (spontaneously combustible in air). The presence of arsenic, antimony, bismuth, germanium, selenium, tellurium, tin or very strong acid may in some cases result in toxic gases (the hydrides) being given off.

Some solutions are only stable under acid conditions so the metal or its oxide can be produced by careful treatment with alkali. Alternatively, an insoluble compound can be precipitated out by the addition of the appropriate ion — for example, sulphide will precipitate mercury sulphide; chloride will precipitate silver chloride. Such processes are most effective within a limited pH range and may be hindered by the presence of complexes of the metal, such as cyanide complexes. Thioacetamide is more convenient to use than hydrogen sulphide, though it is slower and is a weak carcinogen. Zinc dithiol usually gives a better quality precipitate than sulphide, with only marginal contamination by the zinc.

For a relatively noble metal under neutral-to-acid conditions, it is often possible to displace it from solution with a cheaper and less noble metal. For example, iron is used to displace silver from solution in commercial recovery from photographic and other waste.

Electroplating can be carried out with commercial apparatus where there is regular need for recovery. However, the basic requirements are very simple, and many laboratories can use existing d.c. apparatus to recover small quantities, where electric efficiency and speed are not important. For example, a piece of silver or platinum foil may be built up (perhaps over a period of years) to a saleable weight by the regular recovery of quite tiny amounts. A range of 2 to 4 volts is usually sufficient. A milliammeter is useful to indicate the rate of deposition: when this begins to fall off the solution is depleted, and the plating should be stopped. Best results are obtained from slow deposition. Graphite is suitable as the other electrode because it does not add metal to the system, though it can degrade in strong sulphuric acid. Lead is suitable in sulphuric acid and in some other solutions. The receiving electrode does not have to be identical to the metal to be plated: both copper and stainless steel have been used successfully. The process is usually most efficient within a particular pH range.

It is possible that liquid–liquid extraction might be appropriate for some solutions. It should be considered if there is sufficient value in the metal to warrant the extra trouble, and where other methods are not suitable. Usually an organic complexing agent is used to form a compound with the element, the compound being soluble in an organic solvent into which it passes. (See section 9.3.6.4.) The appropriate reagents are generally expensive, and the chemistry involved is somewhat specialized, but it has some possibility of giving a very clean separation from a complex or difficult solution.

A promising alternative is certain plastic resins with a selective affinity for some metals. The solution can be passed through a bed of resin, or beads can simply be stirred into a container and filtered off. The metal is recovered by ashing the resin. At the time of writing, the authors know of no resins readily available in laboratory quantities. However, there is very active commercial development for industrial recovery and pollution control, and products are available in tonnage amounts. The resins are generally referred to as 'macroporous' or 'macroreticular' and should not be confused with ion-exchange resins. Unlike the latter they will absorb valuable metals from solutions of as little as a few parts per million in the presence of large amounts of many other ions. They typically contain active aromatic groups such as vinyl pyridine linkages.

14.4.3 Recovery from Solids

Sludges, precipitates, collections from drains and floor sweepings from spills (or vacuum cleaner contents) may well require some separation before they can be sold or used. The exact process will require some knowledge of the mixture, the chemistry of the metals involved, and a good dose of commonsense.

The following example illustrates some of the techniques which might be used. Suppose that the platinum precipitated in a recovery process has been spilt and swept up with the aid of some sand.

The solid is stirred with warm water and a little detergent. This dissolves some dirt and chemicals, and suspends dust and similar material, which can be decanted off after a few minutes settling. This can be repeated several times, allowing the heavy metal particles time to settle to the bottom. The process is then repeated with warm dilute nitric acid to dissolve out any other metals such as iron. The remaining solid is filtered with ashless paper and washed with water. It is then dried and heated in a furnace to 450 °C to burn off any organic residues. The solid is allowed to cool and dissolved in aqua regia. The resulting solution is filtered through glass-fibre paper to remove any sand. The relatively pure solution can now be recovered by electroplating, precipitation or ashing.

14.4.4 Silver

Because of its extensive use and considerable value, silver is the element most commonly worth recovery. There are many commercial and personal methods, of which the following is only a selection of the more common.

For recovery from Chemical Oxygen Demand (COD) test solutions, the following method has been recommended. The amount for one test is small, but as it is routine in many laboratories it can accumulate useful quantities. It can be applied to many other solutions of silver salts.

The solution is brought to about pH 2 (use test paper) by the addition of 6 mol dm^{-3} HNO_3 or dilution as appropriate. (Neutralization with base may cause some silver to come out as colloidal oxide.) A near saturated solution of sodium chloride is stirred in, to precipitate out silver chloride. A 10% excess is adequate. The precipitate is filtered off and washed with dilute H_2SO_4, then twice with de-ionized water, and dried. The resulting solid is saved until there is sufficient for recovery, when it is ground into a powder. The powder may be sold as such, or it may be dissolved in photographic fixer and added to photographic solutions for recovery. Otherwise the following method can be used.

100 g of the powdered silver chloride is mixed with 50 g of granulated zinc (not zinc dust — the reaction may be too violent). This is then stirred into 500 cm^3 of warm dilute (2 mol dm^{-3}) H_2SO_4 in a fume cupboard. The zinc dissolves with vigorous evolution of hydrogen. The liquid is decanted, and the solid treated with a further 50 g of zinc and 500 cm^3 of H_2SO_4 as before. When the second batch of zinc has dissolved, then about 5 cm^3 of concentrated H_2SO_4 is added and the mixture heated with stirring to 90 °C. The mixture is allowed to cool and the silver metal filtered off, and well washed with de-ionized water until no sulphate is detected by the $BaCl_2$ test.

Any residual silver chloride can be tested by dissolving a sample in concentrated nitric acid, which should give a clear solution. Turbidity indicates silver chloride, which can be removed by a further treatment of the solid with zinc and sulphuric acid.

The principal usage of silver is in photography. Any laboratory making large-scale use of photographic solutions should really have a commercial electroplating unit. This can be rented and a service arranged whereby payment is made each time electrodes are replaced (on a time or weight schedule). For smaller or intermittent users there are several systems possible.

The simplest system is a replaceable cartridge which fits into the drain line. It contains steel wool which exchanges with the silver, iron going into solution and a sludge of silver falling to the bottom of the unit where it is collected. This is inexpensive and little trouble. It requires an effluent between pH 4 and pH 6.5, which is usually the case. (Below pH 4 the steel dissolves away; above pH 6.5 the process is not very efficient.) The initial efficiency is high, but becomes lower as the cartridge nears exhaustion.

CAUTION: if a steel wool cartridge is removed before it is exhausted it should not be allowed to dry out. The rusting reaction on its extended surface can generate sufficient heat to become a fire hazard.

Another commercial device uses a different electrochemical principle. Zinc and stainless steel are placed in electrical contact in the solution. The zinc dissolves and silver is plated onto the stainless steel. The device is usually placed in a fixing bath. As well as recovering silver it has the benefit of extending the useful life of the thiosulphate fixing solution.

Silver can be recovered on a batch basis from photographic solutions by precipitation. Sodium chloride will produce a mixed chloride and dichromate precipitate from used 'bleach-fix' solutions. This can be sold to a refiner direct, or dissolved in fixer for recovery by one of the above devices. Sodium sulphide will precipitate-out silver sulphide from alkaline fixers.

CAUTION: on no account should sulphide and acid come into contact, as they liberate the extremely toxic gas, hydrogen sulphide.

It is better not to add an excess of sodium sulphide as this could give rise to smell and other problems with the effluent. It may take some days for the silver sulphide to settle properly. This can be improved by stirring in some silver sulphide (which was made previously) before adding the sodium sulphide.

It is possible to convert the silver sulphide to metal, but it is more usual to sell it to a refiner.

The silver in fixing baths may also be reduced directly to the metal by the use of either sodium dithionite or sodium borohydride. Note that the former gives off toxic sulphur dioxide gas. Both in practice can give problems, with the silver coming out as too fine particles which tend to stick to the walls of the vessels.

CAUTION: solutions of silver containing ammonia (e.g. mirror silvering solutions, Tollen's Reagent) should not be stored for recovery. They should

be decomposed on the same day on which they are made, or explosive compounds may form.

14.4.5 Mercury

Mercury can be a very troublesome material. Its compounds are highly toxic and thus unwelcome in the environment. The elemental form is extremely difficult to clean up if spilled (see also section 15.2). Note that tiny drops of mercury metal, or even the vapour from mercury spillages can have a devastating effect on delicate electric and electronic apparatus. It is therefore wise to take particular care to avoid (or at least to contain) spillages and to ensure that as much as is practicable is collected for recycling.

Many university chemistry departments (and some physics departments) operate mercury recovery procedures. They may be willing to accept small quantities (say 10 g or more) from local schools or other laboratories. A quantity of a kilogram or more may even have some value for a local scrap merchant (see section 14.3).

CAUTION: the preparation of pure mercury metal from dirty mercury or its compounds can be extremely hazardous. It should only be carried out by a knowledgeable person with special facilities in a controlled environment, such as a commercial refiner or the laboratory of a professional chemist.

This handbook does not set out to instruct such experts, and therefore details of final refining are not given here.

Where an organization has many separate laboratories using mercury metal (e.g. an academic institution, a group of schools, a research institute, or group of industrial laboratories) it is strongly recommended that a centralized mercury service be set up. If the organization has the necessary facilities and expertise the mercury may be recovered in-house, otherwise there should be an arrangement with a local refiner.

The mercury service has obvious economic and safety benefits, but can also improve the scientific effectiveness of some work. In particular, it can ensure that the liquid mercury in certain apparatus is in good condition and therefore the apparatus works correctly. Old mercury usually has oxide film (and often other contamination) which alters its electrical conductivity, its flow properties, meniscus shape, etc.

The mercury service should have the following features.

(1) A central record of mercury purchase, waste removal and distribution. This should be examined to see if there are any losses which may be due to faulty equipment or practice. (The authors know of one institution which discovered more than 100 kg in floor drains due to cracked dashpots which were periodically topped up.)

(2) In issuing mercury, steps can be taken to ensure that users are properly

equipped (with trays, spill control aids, etc.,) and aware of the proper
procedures for safety and for returning used material.

(3) In addition, the amounts issued can be just as much as are required. It
will not then be necessary to have part used commercial containers kept
in the general laboratories.

(4) There should be a probable date when the mercury is to be returned for
each issue. If this is more than a year, the mercury should (if possible) be
inspected annually.

(5) Apparatus containing mercury metal should, where practical, be
inspected annually by a competent person. If the mercury is dirty it
should be replaced. If the apparatus is unlikely to be used in the
foreseeable future, consideration should be given to return of the
mercury to the secure central service store. (Possibly with the
apparatus.)

(6) The majority of users should not attempt to clean mercury themselves.
The best technique is for the dirty mercury to be exchanged for a similar
quantity of clean material. There is great merit in deciding that only
certain persons should actually handle mercury. In particular, it may be
possible to avoid spillages by people who are not experienced in its
unusual liquid properties. For example, it may be the procedure that a
student reports that mercury appears dirty, and for a technician to
actually drain and refill the apparatus.

Mercury metal waste is best collected in strong polyethylene or
polypropylene bottles or jars, which have been made without a seam
(because the seam may burst). Glass jars should only be used for items such
as broken thermometers, where the pieces and droplets are collected under
water for proper treatment later.

The isolated user can sometimes clean up mercury sufficient for his needs
without fully refining it, by the following techniques. Particles of glass from
broken apparatus can be strained out with a plastic mesh (e.g. a tea
strainer). As mercury is heavier than virtually all the common items of dust,
debris and its own oxide, much contamination can be removed by simple
gravity sedimentation. The mercury is placed in a suitable device, such as a
glass separating funnel, and the tap opened to allow the mercury to drain
away slowly, leaving oxide and dirt in the top layer.

**CAUTION: the funnel must be strong, well supported and have a secure tap
which cannot come loose.**

A modification of this technique is to place a filter paper in a funnel, load
with mercury then carefully pierce the point of the cone with a stainless steel
pin. The mercury drains through the pin-hole, and oxide and dust are
retained on the rough surface of the paper.

Where contamination by other metals such as zinc or copper is known to
have occurred, some users like to convert these to oxide by gently bubbling

air through the mercury for a day or two. The resulting oxide crust is then removed as above.

To remove any remaining traces of oxide, it is possible to treat the mercury with dilute nitric acid. Some device is required to expose the maximum surface area. This is usually a fine dropper leading into a tall column (0.5 to 1.0 m) of the nitric acid solution.

Note that the solutions and oxides collected in the above procedures must be disposed of as toxic mercury waste by a technique appropriate to the amount and local situation.

Regular amounts of solution, or a single large batch, can have mercury removed by the methods given in section 14.4.2. Electroplating and reduction by borohydride are quite practical, but the most common technique is to treat the solution with hydrogen sulphide, sodium sulphide or thioacetamide under acid conditions, so that the sulphide is precipitated. A solution of 13% thioacetamide is commercially available especially for this purpose.

CAUTION: hydrogen sulphide is highly poisonous. Sulphide and thioacetamide may release hydrogen sulphide if the acid is very strong. Thioacetamide is a cancer suspect agent. Proper protection must be used against these dangers.

The sulphide precipitate should be sold or given to a competent refiner. Note that it will contain some pure mercury metal if the solution contained the mercury(I) ion. Excess sulphide reagent (particularly sodium sulphide) may cause some of the precipitate to redissolve.

As an alternative, solutions of mercury(II) and chloride can be reduced with sodium bisulphite to give insoluble mercury(I) chloride precipitate. Solutions of mercury(I) compounds give the same precipitate if treated with dilute hydrochloric acid, providing they are not strongly complexed. The chloride precipitates are less acceptable to refiners.

Where mercury metal is used in any great quantity, it is likely there will be special local exhaust ventilation. If the exhaust air is passed through a suitable absorber then mercury will accumulate to a level which can be worth salvaging. Ordinary activated charcoal is surprisingly effective. Iodized charcoal may give more efficient capture, but it is less easy to salvage. Sulphurized charcoal gives at least as good efficiency of capture, and it is easy for the refiner to recover the mercury from it. Sulphurized charcoal cartridges are commercially available for this purpose. They can be fitted to extract ducts or the outlets of vacuum pumps or other mercury apparatus.

14.4.6 Gold and Platinum Group Metals

The very high scrap value of these metals means that even tiny scraps are usually worth keeping. It should not be forgotten that they are used in various instruments, and it may be worth salvaging broken electrodes, heating coils, ionization heads, etc. There is little the amateur can do with scrap items other than collecting them in separate metal categories, and offering gram or larger amounts to a precious-metal dealer. Unused chemicals are normally most saleable if presented in their original containers. This applies to alloys, compounds and preparations such as precious-metal coated catalysts.

Regrettably, many textbooks on practical chemistry with precious-metal catalysts do not give any advice at all about recovery of the used catalyst, and it is likely that a significant amount is discarded. The properties of catalysts vary, and the chemistry of the elements more so, so a single recipe cannot be given. However, as a rule of thumb, solid catalysts should be kept wet after use, should be well rinsed to remove any of the reagents or by-products and should be kept under water until a recovery procedure has been found.

Vogel's is probably the most widely available textbook which gives detailed advice on the preparation and recovery of different platinum catalysts, though there are others. For recovery of metals from compounds it is necessary to consult with books or people having specialist knowledge of the chemistry of these elements.

14.4.7 Lanthanides

This group of elements provides a number of specialist reagents, notably due to their electronic and spectroscopic properties. A laboratory handling significant quantities of a particular element for some particular work is advised to look into the necessary chemistry for recovery.

Generally they can be precipitated from aqueous ammonia, although the presence of complexing agents such as citrate may prevent this. It is convenient to prepare the carbonates by precipitation with sodium hydrogen carbonate, as the product can be readily dissolved in acid and is more easily filtered than the hydroxide. Note that sodium carbonate should not be used, as the precipitate tends to redissolve in the reagent.

The cost of individual elements varies considerably, but there is not a ready market for waste. The facilities of the individual laboratory and the use of the element will determine if it is practical to re-use it.

14.5 GENERAL CHEMICALS

It often happens that stocks of chemicals become slightly dirty, or suffer surface degradation. It is sometimes worth while purifying the chemical. This will of course cost money, effort and the time of skilled staff. On the other hand it will save the cost of repurchase and the cost of disposal. More importantly, it will reduce the quantity of harmful waste and the attendant difficulties of legal disposal. There may be some incidental benefit in (say) training of junior staff, or in finding interesting projects for certain categories of student.

As an example, take sodium metal. It is quite cheap to buy, but rather difficult to dispose of (see section 9.3.8.1). Stock bottles can often become unusable owing to the formation of oxide, hydroxide and carbonate crusts on the sticks or pellets of metal. Clean sodium can be got out of the crust in the form of 'shot' or 'eggs', which is in fact preferred by many chemists.

The pieces are cut in half under paraffin, the excess paraffin is removed with filter paper, and the pieces are placed in a flask of dry xylene fitted with a stirrer, nitrogen purge and reflux condenser. The flask is then heated on an electric mantle with stirring. Above 97 ° the sodium melts and forms globules. The speed of the stirrer determines the size: when this has been adjusted, the flask is allowed to cool down.

A similar procedure can be carried out for potassium (melting point 64 °C), but care should be taken not to cut up samples showing an orange or beige coating of peroxide, as they may explode. It is possible, but not usually practical, to carry out an equivalent procedure for lithium, owing to its higher melting point and lower density.

For sodium, it is better to prepare the shot in amounts needed as the experiments demand. For potassium there is an argument for occasional preparation of clean metal so that oxide formation is not allowed to progress to the dangerous superoxide stage.

General methods of purification of laboratory chemicals are given in Fieser and Fieser's classic volumes. A specific book is that of Perrin, Armarego and Perrin. Many practical organic chemistry books give purification methods, but most inorganic texts do not. An excellent example of one that does is Brauer's.

14.6 SOLVENTS

14.6.1 General Comments

Organic solvents are the most obvious chemicals for recovery. The technology is not difficult, and has many advantages in reducing the cost and trouble of waste disposal. Many laboratory workers feel some satisfaction in conserving resources and reducing the volume of chemical waste they contribute to the environment. Educational institutions often feel that student involvement gives both technical and moral teaching which is worth while even though the cost savings may be trivial.

It cannot be stressed too strongly that a successful programme of solvent recovery requires careful organization. Solvents must be kept segregated prior to the recovery process, and care must be taken that unnecessary dirt is not added through poor labelling, poor practice or a careless attitude. For example, a laboratory accumulating a single solvent, very slightly contaminated from a routine test, could have the whole recovery process ruined by a single beaker of mixed solvents being added to the can.

There is a widespread belief among non-scientific workers (and some less experienced scientific ones) that it is possible to separate a mixture of many solvents into its pure components by simple distillation. In fact, it is virtually impossible. The ideal candidate is a single solvent containing a non-volatile (and non-degradable) dirt, for example toluene containing a wax. A typical process would produce relatively pure toluene, and a residue which is not pure wax, but a mixture of wax and toluene, probably 10 to 40%, depending on the viscosity.

As a rule of thumb, it is usually possible to separate a mixture of two solvents (or two solvents and a non-volatile dirt) providing their boiling points differ by at least 10 °C. Using the same criterion, it is usually possible to separate out one pure component, often two and sometimes three from a mixture of three solvents. Mixtures of 4 or more solvents are increasingly difficult, and will generally require considerable knowledge in planning the separation, and quite a lot of skilled attention during the recovery process.

Mixtures of solvents with similar boiling points (e.g. n-heptane 98.4 °C and iso-octane 99.2 °C) cannot be separated by ordinary distillation. In addition, some materials form constant boiling mixtures, or azeotropes, of which the best-known example is ethanol and water. Thus pure ethanol cannot be distilled from a mixture with water, even with the most efficient apparatus: the best that can be achieved is 95.6%. Azeotropes are quite common. Another example is cyclohexane (boiling point 81.4 °C) and iso-butanol (boiling point 108.3 °C). These would be easy to separate were it not for the formation of azeotrope containing 86% cyclohexane, which distills off at 78 °C. A random mixture of 4 or more solvents has quite a good chance of an azeotrope combination, which will greatly affect the feasibility of distillation.

Many laboratories, even with experienced chemists, forget that there are many processes other than distillation which may be used. These are sometimes simpler, can often make distillation more effective if used as a pre-treatment, and may even make distillation unnecessary.

In organizing a solvent recovery programme, the person responsible should consider the use of the solvent and the recovery techniques possible. A first consideration should be the changing of use or solvent to aid recovery. For example, if a slightly more expensive solvent would do the job just as well, but be easier to recover, then it may be a better choice. Similarly, when there is a routine procedure for recovery of a particular solvent, then other laboratory work may often be adapted to make use of that solvent instead of the traditional one.

A primary consideration is safety. Storage of solvents for recovery should be in suitable containers, correctly labelled, and located so that they are convenient, safe and legal. See sections 2.4 and 3.5.

In addition the recovery apparatus must be used with regard to the available facilities (e.g. fume extract, drains, flame-free areas) and to the technical skills of the people concerned. It is best if any new solvent recovery procedure is checked in advance by a person with sufficient knowledge of the chemistry and practical experience to spot any hazards and deal with them. As an aid, a check list of some common hazards is given in Table 14.1.

14.6.2 Strategies for Recycling

There are many ways in which a solvent can be used several times. Some examples are given below.

(1) Hexane was used to extract a substance from a solid product, which was then measured by UV spectroscopy. The waste solvent from this process contained a small amount of relatively inert solute, but was still of high purity, exceeding the specification for reagent grade chemical. The hexane could be used without treatment in a preparative chemistry laboratory.

(2) A high boiling petroleum fraction became contaminated with several per cent of (low toxicity) plasticizer. It was used to clean paint brushes during laboratory modification.

(3) Spectroscopic quality carbon tetrachloride was used to extract a dye for measurement by visible light absorbance. The dye was removed by treatment with activated charcoal and the solvent could be used again.

(4) At the beginning of a chemistry course, each student was given a limited amount of the necessary solvents. The student had to recover solvent after use for his needs in further experiments, which encouraged careful work and was an exercise in itself.

(5) A regular experiment used 50% aqueous acetone to wash the product. Instead of using pure acetone, it was found that the acetone–water azeotrope (88.5%) could be diluted to the working concentration. This azeotrope was easily recovered from the waste washings, supplemented by acetone waste from another experiment.

(6) Scintillation cocktails were bulked by collection in a suitable can, then treated with activated charcoal to remove volatile radioactives. This gave 3 fractions by batch distillation. The first fraction was toluene (and water which settles out) which was free of radioactivity (i.e. not significantly above background) and pure enough for immediate use. The second fraction was xylene (and water) with a little toluene contamination and radioactivity a little above background. This was added to boiler fuel for disposal. The still residue was water, emulsifiers and the majority of the radioactivity. As this residue was non-flammable, it was acceptable for land disposal. (In some areas it may have been acceptable for sewer disposal.)

(7) A regular student experiment required material to be processed in a large bulk of solvent which was then evaporated down, losing the solvent. The instructions were changed so that the major concentration was carried out on a rotary evaporator. Each student saved the solvent collected, which was periodically bulked and distilled by a technician. Overall recovery was 60%.

Table 14.1. Hazards in solvent recovery

The following is a reminder of potentially dangerous characteristics of some solvents and mixtures, as they affect recovery by distillation, adsorption and some other treatments. Examples are given of typical chemical groups, in which it should be assumed that all compounds including that group have the specified hazard. In addition, there is a partial list of common solvents where the usual name does not include the group name (e.g. tetrahydrofuran is actually an ether) or where the hazard is unusual.

(a) Solvent groups

Chemical group	Examples	Hazards
Acrylic	Acrylic acid, acrylonitrile, methyl acrylate	1, 3
Aldehyde	Acetaldehyde, crotonaldehyde	3
Allyl	Allyl chloride, diallyl ether	3
Bromo-	Bromoform, carbon tetrachloride,	
Chloro-	trichloroethane,	1, 4
Fluoro-	fluorotrichloroethane	
Diene	Cyclohexadiene	2, 3
Ether	Diethyl ether	1, 2, 5
Ethoxy-	Dimethoxyethane,	2, 5
Methoxy-	ethoxyethyl acetate	
Glyme	Diglyme, monoglyme	2, 5
Isopropyl	Di-isopropyl ether, isopropyl alcohol	2, 5
Nitro-	Nitromethane, nitrotoluene	6
Vinyl	Vinyl acetate, divinyl benzene	1, 3

(b) Some individual solvents

Common name	Other names	Hazards
Acetonitrile	Methyl cyanide	3
Acrolein	Acrylaldehyde, propenal	1, 3
Cellosolve	2-ethoxyethanol, ethylene glycol monoethyl ether	2, 3
Cumene	Isopropyl benzene	2, 5
Dekalin	Decahydronaphthalene	2, 5
Propan-2-ol	Isopropyl alcohol	2
Pyridine		5
Styrene	Vinyl benzene	1, 2, 3
Tetrahydrofuran		1, 3, 5
Xylene		5

(c) Hazardous mixtures

Substance 1	Substance 2	Hazard
Any solvent liable to peroxides, i.e. those with note 2. above	Any ketone, e.g. acetone and methyl ethyl ketone	The ketone aids the formation of peroxides. Substances such as cumene or propan-2-ol which are usually only slowly oxidized can become explosive in a short time
Bromo-, chloro-, or fluoro- solvents, especially chloroform	Ketones, especially acetone	Violent reaction in the presence of base (e.g. KOH) or some other catalysts
	Alcohols + NaOH or KOH	Violent reaction with base, aided by the alcohol as co-solvent
Nitro-solvent, especially nitromethane	Amines, e.g. methylamine, aniline and diaminoethane	Mixtures may detonate on impact or friction
	Bases, e.g. KOH and NH_4OH	
	Acids, e.g. H_2SO_4 and HCOOH	
	Chloroform or bromoform	

Hazards
1. Normally supplied with an inhibitor. Recycled material may have lost the inhibitor, and therefore be especially likely to undergo the following hazardous reactions.
2. Forms peroxides on storage. Distillation may lead to an explosion. Air contact during processing, e.g. an air bleed to a distillation flask, may lead to high peroxide concentration in the product. Materials stored without an inhibitor may become shock-sensitive explosives.
3. Liable to polymerize. This can cause degradation of product, or may even be explosive, particularly on heating or in the presence of certain chemicals.
4. Violent reaction with aluminium (and its alloys), titanium or zinc when subject to heat, friction or impact.
5. Laboratory residues often contain sodium wire, or solid KOH as drying agents.
6. These compounds should not be distilled, as they are liable to explode. Can form shock-sensitive explosives with amines or alkalis.

(8) A quality assurance laboratory used chloroform to extract wax from a commercial product for infra-red analysis. The solutions were accumulated until 2 litres were available, then batch distilled (with no column). A little ethanol was added to the wax-rich residue before placing in the discard can. An occasional check was made that the recycled chloroform contained sufficient ethanol (1 to 2%) as stabilizer. This normally distills over in the first part of the recovery process.

(9) A group of Research and Development laboratories agreed to standardize as far as possible on a few solvents. Each scientist used his knowledge to judge if a particular waste was recoverable. If so, he carried out any preliminary removal of solids, neutralized any acids or bases, etc. It was common to carry out a rough separation into 'clean' and 'dirty' solvent by partial distillation or rotary evaporation. These solvents were then collected in separate 'recoverable solvents' cans, distinct from the discard cans. When near full, a particular solvent was distilled in a standard preparative still by an experienced technician. The product was checked (by gas chromatography) and labelled as to its purity. A stabilizer was added to one solvent. Some batches were of excellent quality; others were suitable for less critical work.

(10) A major hospital histology laboratory collected its xylene waste, and distilled it on a commercial automatic still. Care was taken that the waste did not become contaminated with other solvents or materials from other laboratories or procedures. This produced a 60% saving in solvent purchase, and a similar reduction in disposal costs. The payback time for the capital cost of the still was less than 6 months.

(11) Experiments on droplet coalescence were found to have considerable variation, even if a fresh bottle of solvent was used each time. This was probably due to very tiny quantities of surface active and unstable compounds. However, it was found that it the experiments were repeated with the same solvents recycled by distillation, the results became consistent.

(12) A routine procedure involved 3 washes of a material with solvent. The third wash liquid was saved and used for the first wash on the next procedure.

14.6.3 Preliminary Treatments

It is generally a good idea to consider other physical and chemical methods prior to distillation. Section 9.3.6 gives some suggestions. In particular, where contamination is very small, say less than 1%, then other procedures may be more economic, safer and less trouble. A chemical reaction may sometimes be found which converts contamination to an insoluble form, or conversely to a water-soluble form which can be extracted. Adsorption into alumina or charcoal is effective for many contaminants, but only really economic for low concentrations. Adsorption is, however, a very valuable technique for removal of peroxides or other dangerous substances.

Many uses of solvent produce waste containing acidic components (e.g. acetic acid, acyl halides), where it is generally a good idea for these to be neutralized before recycling. In fact, most chlorinated solvents tend to become acid with time, owing to breakdown of the solvent, and this acidity should be neutralized with solid sodium carbonate or an aqueous solution of alkali. Note that some solvents are hydrolysed on heating with alkali, so it is equally important to ensure that solid or liquid or dissolved base (NaOH, KOH, Na_2CO_3) is removed before distillation.

14.6.4 Distillation

The following are notes on practical aspects of distillation for waste recovery processes only, including some very rough and ready short-cut methods of calculation. It is assumed the reader is familiar with basic laboratory practice, but may not be used to the design and operational characteristics of distillation apparatus. Some gross oversimplifications will be obvious to anyone with a chemical engineering background, but the simple approach is generally adequate for this purpose.

A typical still consists of a heated flask in which the mixture is boiled, a vertical column in which the actual separation occurs, an adaptor or 'head' of various designs on the top of the column which takes some vapour to a side arm, and a water-cooled condenser to convert that vapour to liquid. In batch distillation, the flask is filled to its working level with liquid, and vapour boiled off through the column until (a) sufficient top product has been collected or (b) the quality of the top product is no longer good enough, or (c) the level in the flask has fallen to its minimum safe working level.

Continuous distillation is more efficient in several ways but is more complicated to set up. As its name implies, liquid is continually added to the system, while an equal total quantity is removed from the top of the column and from the flask liquid. To achieve this, the column may be made in two parts, with a central section for addition of feed liquid. It is best if this liquid is pre-heated to near the column temperature. The main practical problem is to control the rates of liquids going in and out, but this is quite possible on an automatic unit, which may be set up to process a large volume through quite a small apparatus.

An approximation to continuous distillation can be made where a low-boiling solvent is to be removed from a small amount of higher-boiling material, e.g. if 99.5% hexane is to be removed from its oil contamination. For this purpose, there is a continuous or frequent addition to the flask to keep it at its working level. The low-boiling material comes off the top of the column, while the high-boiling material accumulates in the flask. This can be achieved by the use of a small dosing pump, preferably passing the liquid through some kind of heating coil (e.g. a coil tube in a water bath) so that it is fairly close to the still temperature.

An important practical point is that all columns work better if heat loss is reduced. Commercial stills may be fitted with a vacuum jacket or an electrically-heated shield. For a unit made from standard laboratory items, then a wrapping of aluminium foil is better than nothing. A convenient method of insulation is the pre-formed sections of lagging used for industrial and domestic pipes. The plastic foam ones are suitable for lower temperature distillation, but the fibre-glass ones are suitable for all temperatures. They are much more convenient than wrapping loose fibre-glass around the column.

The effectiveness of a distillation column is measured in terms of units called 'theoretical plates' which is based on the idea of liquid and vapour coming to perfect equilibrium on an imaginary industrial-type plate column. Real plates may be equivalent to 0.4-0.8 theoretical plates. A flask with no column, just a still head and condenser, will have an efficiency of just under one theoretical plate. The number of theoretical plates in the column will vary with design, flow rate, and the properties of the mixture to be separated. It can even vary significantly during a batch distillation.

Specialist manufacturers will usually be able to quote the theoretical plate capacity of a column for certain standard mixtures such as benzene and toluene. As a rough guide, a typical interchangeable column (24 mm joint) of 200 mm working length has just over half a theoretical plate when empty, 3 plates with a coarse packing, up to 5 with a fine or efficient packing. A 300 mm Vigreux column of the same diameter gives about 3 to 4 theoretical plates. Table 14.2 gives some values for common laboratory columns.

Other things being equal, the number of theoretical plates is proportional to the height of the column. Narrow columns have more theoretical plates but lower capacity. One of the most efficient designs is the spinning band column, which is often used in commercial units. A 500 mm height of 6 mm diameter column of this type is likely to be the equivalent of 20 theoretical plates.

A commercial unit or one built by a person with specialist knowledge will have some means of controlling the reflux ratio. Simply defined, this is the ratio of liquid which is passing down the column at any time, to the amount of liquid which is being taken off as top product. A larger unit will typically control this by condensing the vapour at the top of the column, and then splitting it using a so-called 'dividing head'. As a rule of thumb, the reflux ratio should be similar to the number of theoretical plates. For very critical separations it will be necessary to control the reflux ratio precisely, as it is a major factor in column operation. For less critical work it is suggested that the above guide be used initially, then experiments are carried out to see how much more material can be taken off without lowering the product quality too much.

Table 14.3 shows how the column efficiency can vary with reflux ratio. Note that efficiency is maximum at total reflux (i.e. no product) and that efficiency falls off drastically for reflux ratios much below the maximum number of theoretical plates.

Table 14.2. Characteristics of distillation columns

The following are typical of values that might be found for commercial glass columns used for laboratory distillation. Actual values can vary considerably, particularly with reflux ratio. All are of 25 mm diameter or thereabouts.

Column	Throughput cm^3/min	Plate efficiency: theoretical plates per actual plate
Plate columns		
Bubble	10	0.8
Oldershaw	15	0.8
Other columns		Column height (mm) to give one theoretical plate
Vigreux	5	100
Packed: 6 mm Raschig rings	20	100
Packed: 3 mm Fenske helices	30	40

Table 14.3. Effect of reflux ratio on column efficiency

Data from a 1 inch × 48 inch Oldershaw column, separating n-heptane and methylcyclohexane.*

Reflux ratio	Theoretical plates
(Total)	28
120	25
80	24
40	21
20	16
10	10
4	5

*Goldsbarry, A.W. and Askevold, R.J. (1947) Evaluation of Laboratory Batch Fractionating Columns, *Proc. Am. Petroleum Inst.*, **26** (III), 18–22.

To decide on the column needed, or how an existing still can be operated, it is necessary to know something about the mixture to be distilled, and also the realistic requirements of the product. For example, there is no need to use a very high efficiency still with low capacity to produce extremely pure product, when a 90% purity would be adequate.

Properly, it is necessary to have full information about the vapour–liquid

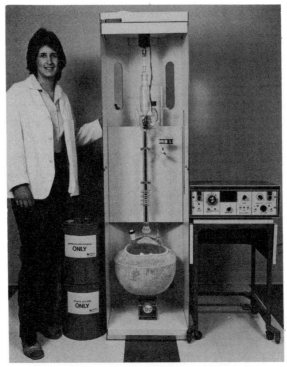

31

32

equilibrium behaviour of the components, and also to have someone such as a chemical engineer or a specialist chemist who is capable of interpreting the data and carrying out some sophisticated calculations. However, the following short-cut method is often adequate and illustrates some important features of batch distillation.

Take the example of a mixture of 2 solvents. A full calculation would require step-by-step analysis of data on the vapour pressure of mixtures. Instead, it is assumed that the materials have a constant relationship called the relative volatility α. For example, at its boiling point, a solvent A has a vapour pressure of 760 mm Hg. At the same temperature, solvent B has a vapour pressure of 380 mm Hg. Thus the relative volatility α_{AB} is 760 ÷ 380 = 2.0.

Obviously, in this case solvent A would be removed at the top of the column. Roughly speaking, every theoretical plate will enrich the product by the ratio α_{AB}. For example, if the flask has a mixture of 50% A, then the liquid has a ratio of 1:1. After one theoretical plate it will be 2:1 or 67%. After 5 theoretical plates it will be 32:1 or 97% pure. Note that these are actually mole ratios and percentages.

This can be expressed in the following formula:

$$N = \frac{\log((x_a/1 - x_a)(y_a/1 - y_a))}{\log \alpha_{AB}} \qquad (14.1)$$

where
N = number of theoretical plates
a_{AB} = relative volatility of A to B
x_a = mole fraction of A in the flask
y_a = mole fraction of A in the top product

In the example, the expression is:

$$5 = \frac{\log((0.5/0.5)(0.97/0.03))}{\log 2.0}$$

As a flask with no column gives nearly one theoretical plate, 5 theoretical plates would be given by a still with a Vigreux column or well-packed column of 4 theoretical plates. However, if the reflux ratio is not well adjusted there can be a loss of up to one theoretical plate, so for a conservative calculation it is wise to consider only the column, and we could therefore only rely on a top ratio of 16:1 or 94% pure.

Note that the composition in the flask has a very important effect. For example, when the majority of component A has been distilled over, its concentration in the flask drops to (say) 10 mole % or $y_a = 0.1$, and a ratio of

---

(31) An automatic still for the recovery of xylene from a hospital histological laboratory. Photo courtesy of B/R Instrument Corporation, Pasadena, Maryland, USA.   (32) Items kept specially for spillages in a convenient corner. Absorbent pillows, neutralizing agents, buckets, mops, brooms, dustpans and plastic bags.

1:9. A column of 4 theoretical plates would enrich this to 16:9 or 64%
pure. A column of 5 theoretical plates would enrich it to 32:9 or 78%
pure.

Thus during a batch distillation, the purity of the top product tends to
fall, even though the column performance may remain the same. It is
therefore necessary to stop the process or cease collection when the purity
drops to a certain value. A useful alternative is to collect the first portion
with an acceptable purity, for example 90%, and collect a second portion
of lower purity. The first portion can be redistilled to a higher purity. Its
residues and the second portion of distillate are saved and added to the
next batch of waste for recycling. In this way, both a good yield and a good
purity are achieved.

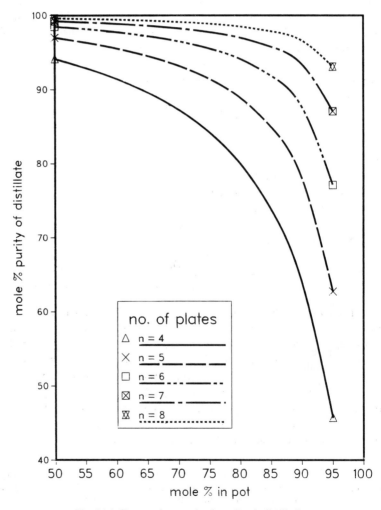

Fig. 14.1 Top product purity from batch distillation.

Fig. 14.1 shows the drop in purity of product as the distillation proceeds, for columns of 4 to 7 theoretical plates, and for a mixture of relative volatility 2. The product composition for any other relative volatility, number of plates, and composition of liquid can be predicted by rearranging equation (14.1) thus:

$$\frac{y_a}{1 - y_a} = \left(\frac{1 - x_a}{x_a}\right) \alpha_{AB}^n \qquad (14.2)$$

This can be solved on most scientific calculators (or microcomputers) in two stages, for example:

LET Z = (A ↑ N) * (1 − X)/X
LET Y = Z/(1 + Z)

Vapour pressure data is available in a number of compilations, including the Chemical Rubber Co.'s (CRC's) *Handbook of Chemistry and Physics*, and Perry's *Chemical Engineers' Handbook*.

It is possible to carry out these calculations for mixtures of many components, providing the vapour pressure of each substance is known at the same temperature. One of these is arbitrarily set to one, and the ratios of the others taken as before. Then if x_i is the mole fraction of the ith component and α_i is its relative volatility, a still of n theoretical plates will give a top product with components in the ratio

$A : B : C : D : E \ldots$

as

$$x_a a_A^n : x_b a_B^n : x_c a_C^n : x_d a_D^n \ldots \text{etc.}$$

It is unlikely that a calculation for many components will be exactly correct (though it may be a useful guide). This is partly due to some approximations in the method, but also on acount of the interactions between the components which may occur. A very important and extreme interaction will be briefly discussed in the next section.

14.6.5 Azeotropes in Distillation

A good description of the theory of azeotropes will be found in many physical chemistry and chemical engineering textbooks. For waste recycling, it is mainly necessary to know if an azeotrope can be formed or not, given the composition of the solvent mixture. The CRC's *Handbook of Chemistry and Physics* gives a very full list of 685 combinations of 2 solvents and 119 combinations of 3 solvents and the properties of the azeotropes which can form. This should be checked before planning a solvent recovery routine, as the presence of an azeotrope has a major effect on the distillation.

As was previously mentioned, an azeotrope is a mixture of two or more solvents which cannot be separated by distillation (with one proviso — see later) and which comes over as if it was a compound. To predict the course of a distillation where an azeotrope is involved, it is easiest to imagine that the azeotrope is effectively an additional component.

As an example, take one litre of 95% acetone, 5% water to be batch distilled with an efficient column. According to the CRC's Handbook, acetone has a boiling point of 56.2 °C, and water has a boiling point of 100 °C. However, there is an azeotrope which is 88.5% acetone, 11.5% water which has a boiling point of 56.1 °C. what will happen?

If the azeotrope did not exist, then acetone (being the more volatile component) would be expected to distill off as the top product at about 56 °C. In fact, what distills over first is 88.5% acetone, that is, a *lower* purity than that in the flask! Moreover, since its boiling point is virtually identical to that of acetone, the inexperienced laboratory worker might easily assume that the distillate was pure solvent. In this case the azeotrope will distill off until there is no more water left. Thus, contrary to commonsense, the water content in the still flask steadily decreases. A calculation will show that if half the sample were distilled off, the residue would be very pure acetone.

Therefore it is clear that acetone with a small amount of water can be dried by distilling off the azeotrope. This is a well-known method for primary drying of solvents and applies to many common ones, including alcohols, ketones, aromatics and chlorinated solvents. If the waste is near to the azeotrope composition, then distillation will not perform any useful separation (except from involatile compounds and other solvents not involved in the azeotrope). It may sometimes be useful to separate an azeotrope from a mixture of lower concentration. For example, a water waste containing 20% acetone could have the majority of this removed as the azeotrope. The aqueous waste might then be easier to dispose of, or there may be some use for the azeotrope. (See section 14.6.2(5).)

As the CRC's handbook shows, a few azeotropes have the useful property of separating out on cooling. Particularly where the azeotrope is with water, it may often be an advantage to make such an azeotrope. For example, toluene (boiling point 110.6 °C) forms an azeotrope with water which boils at 85 °C and contains 80% toluene, 20% water. When this is distilled off from a mixture allowed to cool, it separates into two layers. The bottom layer is water containing a trace (0.06%) of toluene. The top layer is toluene containing a trace (0.05%) of water. If necessary this trace of water can be removed from the recycled toluene by chemical drying with calcium sulphate.

For difficult, expensive or important waste it may be worth while changing the pressure or chemical conditions. For example, distillation of an ethanol-water mixture at ordinary pressure gives an azeotrope which is 95.6% ethanol. However, distillation at a reduced pressure of 95 mm Hg gives an azeotrope which is 99.5% ethanol. If an azeotrope with water is collected and re-distilled in the presence of water-soluble salts, then some enrichment may occur. Potassium citrate seems particularly effective.

Alternatively, it is sometimes possible to find a third solvent which can form a ternary azeotrope. This is done commercially where benzene is used to remove water from 95.6% ethanol as the benzene–ethanol–water azeotrope.

14.6.6 Operation of a Batch Distillation

The following is a description of how a batch distillation for waste solvent recovery might be carried out, making some compromises between speed, purity and convenience.

The flask will depend on available apparatus. It should be filled to about 2/3 full and the distillation should be stopped when the flask still has a reasonable amount of liquid in it — say 10% of its volume.

The column will depend on available apparatus, the degree of separation required and the liquid throughput. See Table 14.2. Note that when the distillation is over, the column will still have some liquid in it. A Vigreux column retains very little: a packed column quite a lot. The Oldershaw plate column is self-draining — some plate designs are not.

It is not necessary to have a large enough still to take the whole waste. Sometimes it is more convenient to use a smaller still and process several batches. This should be particularly considered where the solvent is highly expensive or possibly very dangerous, as it limits the consequences of an accident.

It is a very good idea to add some boiling aids (i.e granules or chips of inert solid with a rough surface) to the cold liquid. They should never be added to hot liquid, as sudden explosive boiling may occur. It is very preferable to use fresh boiling aids each time, as the surface tends to become inactivated with use. However, some waste solvents boil very smoothly without any aids—the behaviour varies and is not easy to predict.

It is preferable if the flask is fitted with a thermometer. For all but the easiest separations, there must be a thermometer in the distillation head. It is best if a head is used which allows the take-off rate to be controlled, including zero take-off (i.e. total reflux).

If advisable, the apparatus is purged with nitrogen. The flask heater is switched on at a high setting. If the waste is very viscous, then it should be heated more slowly, or there is a risk of 'bumping' (local explosive boiling) or cracking the flask. The thermometer in the flask will give an indication of the progress of this warm-up. When the liquid is nearing its boiling point, the heat input should be reduced so that it comes to the boil gently.

When the liquid starts to boil, condensing liquid will start to appear in the column. This will gradually rise up the column. At this time, the still head and condenser should be checked to ensure that they will give total reflux. If necessary, the heat is adjusted until vapour reaches the top of the column and refluxes on the condenser.

The heat input can now be adjusted to give the best column conditions. A packed column should be fully wetted but without being flooded, i.e. there should not be sections full of liquid. A plate column should have visible liquid on all plates but without splashing. For maximum throughput, the column can be brought to the overloaded condition, then the heat reduced to give a satisfactory condition a little below this.

The still head is adjusted to give a very small amount of take-off. Even with a supposed clean solvent it is usual to collect the very first portion separately. This is known as the 'fore-runnings' and usually contains volatile impurities, plus some dirt from the distillation apparatus. After a few minutes the still-head thermometer should have stabilized to a reading similar (not necesarily exactly equal) to the boiling point of the first solvent or azeotrope. This can now be collected.

The rate of take-off can be increased to a suitable value. The column condition should not appear to change, and the still-head thermometer should not alter its reading by more than half a degree. With a regular distillation it will be found by experience what rate can be tolerated with a satisfactory product purity.

It is sensible to collect the product in a series of receivers. Thus if the distillation overshoots then only one receiver is contaminated.

When the take-off temperature changes significantly, then a new receiver is put in place and material collected until the temperature stabilizes on a new value. This is the next product (or azeotrope). A new receiver is fitted and the intermediate liquid discarded.

In principle, though rarely in practice, a whole series of solvents can be distilled off and collected substantially free of contamination. The fore-runnings should always be disposed of, but the intermediates may be saved and added to the next batch. Where the initial waste was relatively pure then the pot residue may be retained for the next batch. Where it is desirable to recover as much as possible from a batch of expensive solvent, then it may be possible to add a cheaper, less volatile solvent (e.g. paraffin) to give the flask a reasonable liquid content when most of the solvent has been removed.

CAUTION: it is very dangerous to allow a still to run nearly dry. The flask may break and some solvents may explode owing to concentration of unstable impurities.

At the end of the distillation, the apparatus is allowed to cool down, then emptied and cleaned as necessary. With tarry residues it may be necessary to remove them while hot, or reflux with a little low-viscosity solvent (which may itself be recovered waste, e.g. intermediates from a distillation).

The column may be left wet if it is to be used for the same job later. However, if the solvent is prone to form peroxides (see Table 14.1) the column should be rinsed with a compatible solvent which does not form peroxides, e.g. methanol, hexane. It is likewise advisable to remove polymerizable liquids such as styrene from a column.

Table 14.4. Solvents which should not be recovered by distillation

Reported accidents suggest that the following solvents and mixtures are especially likely to be unstable owing to peroxide formation or other reactions. It is suggested that they should not normally be recovered from waste, as the economic benefit does not justify the hazard of explosion. Any distillation of these materials in the course of use should be carried out by a skilled person who is knowledgeable about the hazards and the precautions necessary.

Individual substances:
    Di-isopropyl ether (isopropyl ether)
    Nitromethane
    Tetrahydrofuran
    Vinylidene chloride (1,1-dichloroethylene)
Mixtures:
    Chloroform + acetone
    Any ether + any ketone
    Isopropyl alcohol + any ketone
    Any nitro compound + any amine

14.6.7 Peroxides, Polymers and Other Problems

The principle hazard in distillation of flammable solvents is obviously fire. A vapour leak from an insecure joint or a cracked piece of apparatus is very likely to ignite. It is therefore obvious that solvent recovery stills should have explosion screens and be in a position of forced ventilation, i.e. under a hood or in a fume cupboard. However, even in the absence of a leak, explosions are possible owing to the presence of peroxides. These are unstable compounds which can explode or detonate on heating. It is possible that a peroxide solution (from a chemical experiment or a plastic adhesive) may be included in solvent for recovery, but it is more common that peroxides are formed by the action of air on the solvent.

Table 14.1 gives groups of solvents which are prone to peroxide formation. In addition, Table 14.4 gives a list of some solvents where the hazard is especially great. In fact it is recommended that none of the solvents listed in Table 14.4 should be distilled for waste recovery. It is possible for them to be distilled under certain circumstances by an expert who has good reason, but otherwise the savings do not justify the considerable danger.

CAUTION: if peroxides have been deliberately added to a solvent mixture they should be chemically destroyed before the solvent is discarded. Under no circumstances should such mixtures be distilled until made free of peroxide.

In order to remove peroxides from stock bottles of ether and similar solvents, the best general method is to treatment with activated alumina (activity I, basic). The amount needed varies: 100 g alumina will typically treat 1 litre of diethyl ether or 100 cm^3 of dioxan. However, this is not usually efficient for waste mixtures because the alumina may be clogged or deactivated by other components.

For solvents which do not mix with water, the solvent can be stood in contact with a fresh 5% solution of ferrous sulphate in water, acidified to 1 or 2 mol dm^{-3} with sulphuric acid. About 10% of the solvent volume is usually adequate. The two-phase mixture is gently agitated and left overnight. The solvent layer is tested for peroxides (see below): if present, further agitation or more reagent may be used. It is usually advisable to neutralize any remaining acidity with calcium hydroxide (solid) or sodium carbonate (aqueous) before further treatment of the solvent. (N.B. A stronger solution of ferrous sulphate can be used — see section 9.3.7.5 — but the weaker solution for a longer time is safer for waste which may have reactive components, perhaps due to mistakes or poor practice.)

Many other reducing agents can be used. Note that lithium aluminium hydride has been used, and has led to a number of fires, so it is not generally recommended. Tin(II) chloride is effective both as solution (as for FeSO$_4$ but acidified with HCl) and as a solid. The solid is most effective as a fine powder, and requires a trace of water (which is usually present). This chemical also leaves a trace of acidity which can be removed by treatment with solid calcium hydroxide.

Freshly precipitated cerium(III) hydroxide has been reported to be effective. It is made by the addition of a slight excess of NaOH to a solution of a cerium(III) salt. The precipitate is washed with water and used wet. The active nature of the fresh surface means that the reaction is fairly fast. For a stirred mixture, 15 minutes has been found to be sufficient.

Peroxides are also converted by sodium bisulphite or by extended contact with a strong alkali, if the solvent mixture will tolerate it. Other methods involving heating are not recommended for waste recycling.

A simple test for peroxides is moist starch-iodide paper. A positive response means that distillation will probably be safe, providing the residual liquid is not concentrated too greatly — either by distilling to a small volume or by keeping residues in the still and possibly accumulating peroxides.

A better and equally convenient test strip is commercially available (Merck, Darmstadt). This gives a semi-quantitative indication, even at lower levels where the starch-iodide paper does not respond. The best methods of testing for peroxides are by liquid colour change. Two well-known ones are as follows: (1) take 10 cm^3 of solvent and shake with 1 cm^3 of slightly acid KI solution containing a little starch indicator; (2) take 10 cm^3 of solvent and shake with 1 cm^3 0.1% Na$_2$Cr$_2$O$_7$ and a drop of dilute sulphuric acid. In both cases a blue colour is a positive reaction. In the first case it is usually in the water layer; in the second it is usually in the solvent layer, if immiscible.

Solvents which tend to form peroxides, or otherwise degrade should have a stabilizer added, if possible. The manufacturer's specification (label or catalogue) will usually give the chemical and concentration to use. This should be put into the recovered solvent as soon as possible. Otherwise, containers should have the air in them displaced by nitrogen, and be kept in the dark in a cool place. The presence of (freshly acid-cleaned) copper gauze can inhibit both peroxide formation and polymerization to some extent, and is easy to remove, unlike some other agents. However, no agent can be relied upon to be completely effective, so recovered solvents with stability problems are best controlled by recycling every 3 months even if they have not been used.

The processes of polymerization and depolymerization can be the cause of some practical difficulties. In the first case, any substance which can polymerize may do so in the flask, column, still head, condenser or receiving vessel. (See group 3 in Table 14.1.) This can be dangerous, even explosive, but more usually creates a sticky mess and a ruined product. It is generally a wise precaution to add an appropriate inhibitor (quinol often works) to the flask even if the waste is thought to be inhibited. It is advisable to have some inhibitor in the receiving vessel. If the uninhibited product is required, this can be prepared from the recovered liquid immediately before it is wanted, using a small distillation flask with a splash head or a very small (e.g. Vigreux) column.

When the distillation of a polymerizable liquid is finished, the column should be kept on total reflux. A little compatible solvent containing some inhibitor can then be run down through the reflux condenser. It is best if it is of slightly lower boiling point, as it will then displace the polymerizable substance in the column as the apparatus cools down. A metallic column packing (stainless steel or copper) tends to inhibit polymer formation in the column. Metal mesh is widely available, as is the traditional stainless steel pan scourer, which can be unravelled and packed into a plain column. However, metals will tend to be corroded by acidic solvents, and copper in particular can catalyse the breakdown of halogenated solvents (e.g. chloroform) to hydrochloric acid, phosgene or other harmful materials.

Depolymerization can occur when a waste solvent is being recovered from a solution of a plastic or resin. If the resulting monomer is of lower or similar boiling point to the solvent, then it will appear in the product (where it will probably polymerize again). This can be avoided by carrying out an initial separation of solvent from solute in a rotary evaporator. The concentrated residue is discarded, and the condensed liquid distilled. Alternatively, the whole product can be distilled at reduced pressure so that the flask does not reach the temperature for depolymerization to occur. This temperature varies, but 80 °C is generally safe.

It is sometimes worth while including an inhibitor for other kinds of unwanted reactions. For example, some easily oxidized substances such as aniline are best recovered with a few zinc granules in the flask to maintain a slightly reducing environment.

14.6.8 Final Purification
Recycled solvent may be brought to the most stringent purity requirements

by the use of chemical and physical procedures specific to the individual solvent and its intended use. Details are given in many of the better books on preparative organic chemistry or on certain techniques such as liquid chromatography.

A selection of books which give explicit instructions for the purification of a large number of solvents is given in the bibliography.

14.7 OILS

There is a substantial industry involved in the recovery and refining of waste oils, but it is rare for a laboratory by itself to produce sufficient to be of interest, unless it is already associated with the industry. Generally for relatively clean oil, a quantity of 50 litres may be taken away free; a quantity of 500 litres may be bought. It is sometimes possible for a laboratory to lower its disposal costs by carefully segregating oils from solvents for the waste collector.

REFERENCES
[section 14.6.4]

Perry, R.H. and Green, D.W. (eds) (1984) *Chemical Engineers' Handbook 6th edn,* McGraw-Hill.
Weast, R.C. (ed.) (1984) *Handbook of Chemistry and Physics 65th edn,* Chemical Rubber Co.

BIBLIOGRAPHY

Brauer, G. (1963 and 1965) *Handbook of Preparative Inorganic Chemistry, Vols 1 and 2, 2nd edn,* Academic.
Fieser, L.F. and Fieser, M.F. (1967–1984) *Reagents for Organic Synthesis, Vols 1–11,* Wiley.
Fieser, L.F. and Williamson, K.L. (1979) *Organic Experiments 4th edn,* Wiley.
Gordon, A.J. and Ford, R.A. (1972) *The Chemist's Companion,* Wiley.
Janz, G.J. and Tomkins, R.P.T. (1972) *Non-aqueous Electrolyte Handbook, Vol. 1,* Academic.
Perrin, D.D., Armarego, W.L.F. and Perrin, D.R. (1966) *Purification of Laboratory Chemicals,* Pergamon.
Vogel, A.I. (1978) *A Textbook of Practical Organic Chemistry 4th edn,* Longman.
Weissberger, A. (ed.) (1970) *Techniques of Chemistry Vol II:* Riddick, J.A. and Bunger, W.B. (eds) *Organic Solvents — Physical Properties and Methods of Purification 3rd edn,* Wiley-Interscience.

CHAPTER 15

Emergency Procedures

15.1 INTRODUCTION

It is vital to distinguish between true emergencies and those occurrences which are merely unfortunate but require reasonably prompt attention.

In the event of a real emergency (such as an explosion, fire or release of harmful material) the very first priority for any individual must be to protect his or her own life. The second is to inform others as quickly and as accurately as possible of the situation. These actions should take priority over rescue attempts, first-aid or attempting to deal with the danger oneself.

Prompt raising of the alarm can save valuable time and hence lives in evacuating areas. Passing on accurate information can mean that management and emergency services know the nature and extent of the problem, and can therefore deal with it most effectively.

Whatever one's instinct to try to help, the best action is often to leave the area. After all, it is no good if the only person who knows what chemical was spilt is rendered unconscious trying to rescue a friend or to deal with the spill himself.

Where any accident occurs involving fire, explosion, release of harmful (or unknown) materials, or damage to major or important equipment (e.g. a crack in a large vessel, or a broken controller) the situation should be reported to *two* senior personnel within a few minutes. If possible, the Safety Officer should be informed — even if the event seems trivial and if it means breaking into a meeting, lecture or meal. In educational establishments, students should be told to report incidents to *any* member of staff (e.g. secretaries, porters) if academics or technicians are not immediately available. This is because regular employees are usually better placed to inform the appropriate people and raise the alarm.

In a true emergency where human life is threatened, then it may be necessary to adopt waste disposal procedures which would normally be completely unacceptable. For example, if a person is splashed with a chemical, it should be flushed off immediately without regard to whether it is permitted in the drains. Similarly, burning or detonation of unstable chemicals on site may be justified as a means of disposing of old stock which would be dangerous to transport, but cannot be condoned as a routine measure. Permitting volatile material to evaporate (preferably via a fume cupboard) is sometimes the only safe option after a spill, though it exceeds pollution limits, and may even result in prosecution.

15.2 SPILLAGES

15.2.1 Forward Planning

Anything used in a laboratory may be knocked down or spilled. Many incidents are minor but are made a great inconvenience because of the lack of simple things such as a broom. When truly hazardous materials are spilled, then it is not time to go searching for items or trying to invent a procedure.

It is recommended that there should be available in every laboratory (or possibly shared by adjoining ones) a few items of cleaning equipment (see Table 15.1) especially for spillages. These should be entirely separate from the materials used for routine cleaning, and this fact should be very obvious. For example, the items can be painted red and placed in a special corner. Brooms, brushes or mops kept for this purpose should have their heads in a plastic bag which can be quickly removed (e.g. held on by a rubber band). After use, they should be washed, cleaned (or if necessary discarded) and a new bag put on. This is to ensure the residue of one spill is not retained to react with another or provide a reservoir of infection. In the most careful laboratories, the person cleaning up will sign a label on the bag before the brush etc. is put away.

Some other useful items will be normally available in many laboratories, e.g. plastic bags and clean screw-capped jars. The former are required to contain solids, the latter for liquids, solids and small, sharp items. It is a not infrequent occurrence for a bottle or jar of chemical to become cracked but still stay together (especially if it has a special plastic coating). With care, the contents may often be transferred to a clean container, avoiding a spillage altogether. Any laboratory using glass thermometers will eventually break one: the parts should not go into an ordinary glass bin, but should go as a temporary measure into a screw-topped jar. The same principle can be applied to other glass items which require special treatment owing to hazardous contamination.

Table 15.1. Basic items for dealing with laboratory spillages

---

It is suggested that the following are kept separate from routine cleaning equipment, and reserved solely for spillages and breakages within the laboratory. Additional items may be required for special hazards (e.g. disinfectants, neutralizers, protective clothing).

> Broom — soft, polypropylene
> Hand brush — soft, polypropylene
> Dustpan — polypropylene
> Rubber scraper — on broom handle
> Hand rubber scraper, e.g. as for car windows
> Mop — cotton, on broom handle
> Mop bucket — polypropylene
> Bucket — rubber or polypropylene
> Jar — 2 litre, wide neck with screw cap
> Inert absorbent — minimum 5 kg
> Plastic bags — minimum 15 cm square
> Ties and self-adhesive labels for bags
> Disposable gloves
> 'Warning' signs — see below

Note: 'Warning' signs. A suitable arrangement is a few A4-sized cards with two holes punched and a length of string so that they can be hung from door handles etc. A few can be prepared for probable uses, e.g. 'DANGER — CHEMICAL SPILL'; 'CAUTION — BROKEN GLASS'; 'DO NOT ENTER'; 'CAUTION — WET FLOOR'. Some cards should be kept blank for other messages to be written as circumstances require.

Other items will depend on the materials in use, and the type of work. One aspect often overlooked is the matter of containment. Water-using items such as stills can be sited so that their contents will tend to flow into a sink or floor drain should they break. A small, raised strip (e.g. plastic beading) on a bench may be all that is necessary to direct flow into a sink. Apparatus containing liquids or powders should be mounted above or stand in trays or troughs which will hold the entire volume. Plastic trays sold for laboratory trolleys are a cheap and effective means for many purposes. The greater the volume, or the greater the trouble of collection (e.g. mercury is very difficult to pick up) the more valuable is the tray. For very hazardous materials, downward containment should be as routine as a safety shield.

Small drips and spills can be contained by the use of absorbent paper (preferably plastic-backed to prevent penetration). Minor modifications to handling technique are practised in radiochemical laboratories, which could be adopted elsewhere. These include wrapping a filter paper or tissue around the front of containers before pouring, placing a glass rod across a beaker to direct the jet of liquid when pouring, and general careful attention to prevent splashing by violent movements. Absorbent paper does not reduce vapour hazards, but is effective in temporarily preventing airborne release of infective agents. Invisible spills can often be seen on paper by inspection under an ultra-violet lamp (alternatively a developing agent such as ninhydrin can be used). This is an excellent test of the general working procedure, which is recommended for particularly hazardous substances.

An absorbent polypropylene sheet is now available, which may be used in place of absorbent paper, and may be better for some circumstances.

A laboratory should in principle have the means to deal with the spillage of its two largest containers together. This is the best argument for keeping stocks of materials in laboratories down, which includes waste. Where solvents are used in 100 ml amounts it is ludicrous to keep a 10 gallon drum of waste in the same room (it is often also illegal). Spill control agents (absorbers, neutralizers: see section 15.2.2) must be available to completely deal with a complete bottle (e.g. a winchester). A small packet is a token amount which will only waste time. Many people do not realize the amount of treatment agent required — usually at least several times the volume of spillage.

Infective materials may require a large volume of disinfectant to be available, or a fumigation procedure to be ready. If particularly hazardous materials are in use, then treatment should be available actually by the experimenter, or in the controlled enclosure. For example, a plastic basin of ferrous sulphate or sodium hypochlorite solution should be made ready before handling cyanides. This can be used to rinse gloves or beakers at the end of a successful experiment, but may also be ready for swabbing down or dumping broken items. Some laboratories keep a container of boiling water simmering for immediate treatment of heat-labile toxins, or aerosol sprays of specific neutralizers/sterilizers.

Many laboratories now give staff fire training in which they actually use an extinguisher to put out a fire. Unfortunately very few give similar practice in dealing with spillages.* A possible routine includes the disposal of 5 litres of water on the floor, a bucket of sand half-on and half-off a bench and a repetition with smaller amounts on the assumption that they are more hazardous materials (using any protective clothing or necessary precautions). If neutralizing agents or special absorbents or kits are kept, then it is most preferable for some staff to try them out under non-emergency conditions (e.g. on a tray in a fume cupboard). In biological laboratories, staff should have the opportunity of cleaning up sterile broth using techniques as if it were a live culture.

If it is intended that staff use breathing protection in any emergency, it is absolutely essential that they have instruction and training in its use, and that there are at least two sets available.

---

* Practical training in dealing with small-scale chemical spillages is given by the J.T. Baker Chemical Co (Phillipsburg) in the USA, by Richardson-Vicks Ltd (Ottawa) in Canada, and by Min-Chem Chemicals Ltd (Dun Laoghaire) in the UK and Eire.

15.2.2 Spill Control Agents (Absorbents)

All laboratories should have available some kind of inert absorbent to control liquid spillages, cover harmful solids or infective material, and possibly to snuff out small chemical fires. This can range from packages specially made for the laboratory and directed to particular substances, to sacks of material sold to deal with oil spills in workshops. Some of these are less universal and less effective than may be thought. (Appendix B-1 lists some proprietary products.)

15.2.2.1 Sawdust, wood flour

This is a traditional agent, but one which should have no place in a modern laboratory. Sawdust is liable to harbour infective organisms unless it is kept dry, when it becomes an important fire hazard. Solvents or even oils may ignite very easily. Oxidizing agents may make an ignitable or even explosive combination which has been responsible for deaths.

15.2.2.2 Inhibited sawdust

This is sterile and non-combustible, non-dusting form of sawdust which is commercially available, and is suitable for oils and aqueous liquids. However, note that when soaked with flammable solvents, the mixture is just as flammable as the original solvent. Oxidizing agents may make an easily ignitable combination, and strong acids may react to give fumes.

15.2.2.3 Paper towels

These are probably the most common treatment for small spills, being readily available in many laboratories. This is not necessarily a bad thing, since prompt action may be better than wasting time searching for the 'correct materials'. However, there is the possibility of spontaneous ignition with strong sulphuric acid or with oxidizing agents: this may be delayed by some hours, and has often occurred in waste bins.

If paper towels are used, they must be treated before disposal. For aqueous chemicals they can be rinsed in the sink (with any neutralization necessary). Volatile liquids may be evaporated off in the fume cupboard. Infective agents may be autoclaved or treated with a chemical disinfectant. If radionuclides are involved they may be disposed as solid waste or extracted with a suitable liquid. Prompt burning is a useful final treatment, but in any case they should not be placed in an ordinary paper bin, even after 'treatment'. They should be sealed in a plastic bag and correctly treated as chemical or biological or radioactive waste. This reduces the chance of hazards to cleaning staff and refuse collectors.

15.2.2.4 Sand

This is a non-combustible and cheap material which rarely does any harm, but is less effective than often supposed. The material should be free-running silver sand, quite dry and kept in closed containers. A plastic screw-capped jar may be kept handy with bench chemicals.

The main defect with sand is its very low absorption capacity, of the order of 10% by weight. Moreover, flammable liquids soaked on sand can be ignited very easily — sometimes more easily than the original liquid. Likewise, sand does not in any way control vapour hazards. Sand should never be used for spills containing hydrofluoric acid since it reacts to give off poisonous fumes, and should never be used for mercury spills, as it merely gets in the way.

Sand can be useful in neutralizing procedures (see section 15.2.3) as a solid diluent to slow down chemical reactions. It can also be used to exclude air as a fire-fighting agent, or to prevent aerosols of harmful organisms being released.

15.2.2.5 Amorphous silicate

This is essentially an expanded form of sand to give a greatly increased surface area, which can absorb several times its own weight of liquid. It has a very low density, so tends to blow away and is not really suitable for use in the free state. It is commonly packaged in the form of 'pillows', i.e. contained in a porous bag (polypropylene or glass-fibre).

As with sand, the absorbed material retains all its hazardous properties, and the pillows must be disposed of as flammable, toxic, infective, etc. The pillows are generally rapid in absorbing common liquids, but take much longer for very viscous ones. They should be put into plastic bags for further treatment. The larger industrial pillows may be in a chemically less resistant cloth which could disintegrate in the presence of some corrosives or solvents.

It is completely unsuitable for hydrofluoric acid or mercury (see 15.2.2.4).

15.2.2.6 Absorbent minerals

Vermiculite, fuller's earth, diatomaceous earth and manufactured expanded minerals are sold under a variety of trade names. They are used for animal litter and for absorbing spills of oil in industry, for which they are highly effective. Powder may be more appropriate on a bench spill, but generally granules are better, particularly on the large scale when dust may be a problem.

The clay minerals can be surprisingly difficult to wet with water, and form a difficult sticky mass unless present in sufficient excess.

As with sand, they do not neutralize corrosives, nor do they inhibit flammable dangers. Although they are harmless-looking dry-seeming solids, they may be capable of giving severe skin burns or of catching fire. This can be an advantage, in that combustible liquids may be burnt in a solids' incinerator when absorbed on the mineral. Infective agents may likewise be bagged and burnt.

The surface is not totally inert: in fact it has considerable catalytic properties. Instances have been reported of spontaneous fires or explosions (usually after a delay of an hour or more) when they have been used to absorb polymerizable, peroxidizable or unstable material (e.g. organic peroxides).

15.2.2.7 Charcoal

This is an important ingredient in some commercial spill kits. What is required is granular (not powder) 'gas' charcoal, which must be kept dry in a suitable container until required.

The major advantage is that most vapours are effectively suppressed. Thus the fire hazard and toxic fumes can be virtually eliminated in many cases. Where flammable solvents are absorbed, the mixture will not catch fire unless the charcoal is hot or saturated.

A few very unstable chemicals may be catalytically decomposed. Oxidizing agents should not be absorbed because a combustible or explosive combination may occur. Charcoal does not neutralize chemicals, nor is it a sterilizing agent, but its special effectiveness in controlling vapour emission is important. Surprisingly, it will absorb a considerable amount of mercury vapour: special treatments enhance this property.

Unfortunately, it is one of the most expensive absorbents.

15.2.2.8 Porous plastic

At least one porous plastic sheet is now sold as a general absorbent for spillages. A principal advantage is that it can be placed in advance on a work surface, and removed if not required. The polypropylene-based sheet has a fair liquid capacity, but is not easily wetted by more viscous aqueous liquids. As with all absorbents apart from charcoal, it does not suppress vapour hazards.

It may be most useful as a not readily combustible alternative to paper towels for bench spillages. It is not totally inert, but is not dangerously affected by hydrofluoric acid, oxidizing agents, and solvents. It is biologically essentially inert and is unlikely to harbour organisms, which makes it attractive for covering small biological spills.

15.2.2.9 Pillows

See section 15.2.2.5.

15.2.2.10 Mercury sponge

This is not actually an absorbent, but a high surface area form of zinc fibres which combine with mercury drops enabling them to be lifted off a surface. It is only effective by direct contact. The zinc is made most active by the presence of acetic acid.

15.2.2.11 Cotton wool swabs, sponges

These items are readily available in many biological laboratories. There is no objection to their use for biological materials, providing they are bagged and disposed of as biological waste. However, they should not be used for chemicals unless the user is certain that they are compatible with the chemical, and the amount is small. No sponge should be re-used.

15.2.3 Neutralizing agents

There is a limit to how much chemistry can be done on the floor. Generally it is better to either collect up material on an absorbent or dilute it to drains, but some limited neutralization can be carried out. It must be understood that the spillage will not be completely neutral, but the problems may be reduced.

15.2.3.1 Sodium carbonate (soda ash)

This is the most popular agent for acids: it can be seen to work by fizzling. It is best used as a mixture with excess sand or vermiculite. Except with dilute acids, stirring and additional water may be required to aid the chemical reaction. For hydrofluoric acid, slaked lime should be included in the mixture.

Soda ash has some fire-inhibiting properties, and is a preferred agent for covering major spills of sodium, potassium, lithium, calcium metals, the alkoxides, and metal alkyls prior to removal for incineration.

15.2.3.2 Sodium bicarbonate (sodium hydrogen carbonate)

This has less neutralizing power than sodium carbonate. It can be used for inorganic acids, but is particularly useful for acid inorganic or organic compounds, e.g. acid halides, to keep fumes down during collection. (Note: it is the basis for some commercial dry powder fire extinguishers. See section 15.3.1.)

15.2.3.3 Calcium hydroxide (slaked lime)

This is too strong a reagent for direct application to strong concentrated acids. It can be used directly on dilute acids, but is most effective as a mixture in excess sand or vermiculite. It gives a solid precipitate with sulphuric and hydrofluoric acids. The resulting sludge will be caustic.

15.2.3.4 Calcium carbonate (chalk)

Its effectiveness is greatly dependent on the particle size. Its advantage over calcium hydroxide is that if excess is added, the resulting sludge is near neutral. Best results are obtained with a fine particle size, diluted with mineral, e.g. sand.

15.2.3.5 Sodium hydrogen sulphate (sodium bisulphate)

This is one of the few solid agents which can generally be applied to neutralize caustics. The reaction is likely to be incomplete, but may reduce the alkalinity sufficiently for flushing into drains to be feasible.

15.2.3.6 Citric acid

This is probably the only mild solid acid cheap enough to keep for spillage treatment; it has been used in some commercial spill control packs for caustic treatment. It may be used with an absorbent such as vermiculite. It has the advantage that it complexes with many metals so that they are less likely to be left on the surface, which can be important where radioactives are involved.

15.2.3.7 Calcium hypochlorite (bleaching powder)

This is effective against cyanides in the presence of water and an alkali, but does not give immediate destruction. It is useful as a prompt treatment for spills of offensive organic sulphur compounds and is a powerful disinfectant, if the affected area is fully treated, but can be dangerous with liquid bleaches and acid.

15.2.3.8 Sulphur

It is widely believed to be a treatment for mercury spillages, but tests have shown that it is of little or no value. More powerful sulphiding agents such as calcium polysulphide may be more effective, but are not an adequate treatment.

CAUTION: sulphur should never be mixed with zinc dust, as this combination is a considerable fire hazard and requires little provocation to explode.

15.2.3.9 Zinc dust

This reacts with elemental mercury to form an amalgam, which may be more easily collected than the mercury. Good contact and clean surfaces are essential for the reaction. though zinc dust in cracks may adsorb some mercury vapour.

CAUTION: zinc dust should never be mixed with sulphur. See section 15.2.3.8.

15.2.3.10 Sodium hydrogen sulphite (sodium metabisulphite)

This is a reducing agent which is mild enough to be used directly on spilled stong oxidizers. The reaction mixture should be acidified with 3 molar H_2SO_4 (*not* other acids) for best results. The resulting slurry will need to be neutralized with sodium carbonate before final disposal.

It also provides a treatment for aldehydes (use solid plus a little water) by formation of an involatile addition compound which may be collected or flushed to sewer. Acid should *not* be used.

15.2.4 Disinfectants

In general, any liquid which is suitable for a discard jar is suitable as an emergency disinfectant. It is the practice in some places to make up additional disinfectant each day when preparing discard baths, the extra being kept available in a wash bottle. It is probably advisable to use a higher strength than the most economic, in order to ensure the fastest possible kill. (See section 10.2.3.)

A 5% bleach solution will keep for several days, as will an activated glutaraldehyde preparation. Normally these only need to be replaced weekly. Alcohol (either ethanol or propan-2-ol: see section 10.2.3.2) is also long lasting and available in the majority of biological laboratories. It is often kept in a special container for swabbing down as a routine precaution, which can be used in the event of a spillage. However, care should be taken not to use it in the vicinity of flames or other sources of ignition.

Although their activity is limited, quaternary ammonium disinfectants can be very effective against spills of susceptible organisms. Their low mammalian toxicity means they can be freely applied, and their cleaning properties aid decontamination.

Bleaching powders (see section 10.2.3.4) have a shelf life of up to a year if correctly stored, and can be applied by a sort of shaker (commercial or home-made) to spills. Note that for dry spills, the biocidal activity only commences on wetting.

15.2.5 Action in the event of a spillage

15.2.5.1 Immediate action

Most spillages and many other accidents are only trivial affairs of no great consequence, if handled responsibly. The spillage of a few cubic centimetres of solvent on a bench is hardly likely to warrant immediate evacuation! However, it could start a fire which becomes serious and causes much damage.

The immediate response should be one of information and containment. This may not require anything more formal than calling out 'Dammit, I've spilt some toluene!' before trying to remedy the situation. A common mistake is not to ask for help immediately — for example if your hands are full, or you are in the middle of some difficult operation.

On hearing a bang and a shout, the natural impulse is to go to look. It is better if other people shut down or make safe operations *elsewhere in the laboratory* which could prove hazardous if unattended, before forming an on-the-spot committee. Generally it is a good idea to isolate gas, water and electric supplies to the accident area, but to leave on any ventilation.

CAUTION: switching on a fan or fume cupboard can cause an electrical spark to ignite flammable vapours.

The spillage of more than 100 cm^3 of flammable solvent may be counted an immediate fire risk, requiring immediate removal of people from the vicinity. The release of significant amounts of volatile toxic agents or pathogens or known carcinogens likewise requires evacuation of at least part of the laboratory (depending on the size).

With all spillages, people in the vicinity must be adequately informed, and the extent must be contained as far as possible. Even with relatively innocuous substances such as paint, it is best not to have the mess made worse by people treading through it. A physical barrier such as a piece of wood, string, chairs, etc. is more effective than spoken or written warnings. If humanly possible, there should be no traffic through the area at all until the spill has been dealt with.

15.2.5.2 Prolonged situations

If a contaminated area has to be left (e.g. overnight) then signs should be put up and clear instructions given to the supervisors responsible for overnight security and cleaning staff, maintenance workers, and others. It is not sufficient to ask for the area of spill to be avoided, the room (or corridor) should be out of bounds for all activities. If it is a storage or delivery area, arrangements should be made forthwith for deliveries to be directed to another suitable point.

If there is any significant hazard involved, it is the clear moral duty of senior staff members to remain in attendance, even at the personal inconvenience of staying late. Even if the room is not their normal responsibility, specially qualified people may be expected to remain, and should not leave until they have checked either that the situation has been controlled or that someone else is remaining. Technicians or hourly-paid staff who remain to help should be entitled to overtime payments, even if this is not the normal practice.

15.2.5.3 Containment

If the experimental work has been planned with sufficient foresight, then many spills will be contained in a tray or an absorbent paper. If not, then the first action must be not to make things worse (e.g. by vigorous brushing) and limit the spread. Solids may be pushed to a heap with a piece of card or a rubber squeegee. Liquids may be surrounded by a dyke of sand or inert absorber.

A small spill may be entirely absorbed onto a paper towel, filter paper or special absorber (see section 15.2.2). In general, harmful micro-organisms should be covered with a paper towel or suitable absorbent before they come into the breathing zone.

Where a spill covers a large area, a trickle of sand can be used to draw a line around the probable splash zone, to mark the area to be contaminated.

With aqueous liquids on benches, it is usually very reasonable to sweep them into a sink rather than allowing them on the floor, but they should generally receive some kind of treatment before being flushed away. Liquids which are totally unsuitable for the drains (e.g. solvents) should be dyked (contained by a wall of absorbent) and taken up on a suitable absorber and/or neutralized in place.

15.2.5.4 Treatment

Where an effective absorber (e.g. a pillow or a proprietary material for a specific chemical) is used then it may well be possible to remove the material for treatment and/or disposal elsewhere.

In the case of biologically infective material it is almost always advisable to use a suitable disinfectant at the site of the spillage, preferably by spraying over the absorbent (pillow, paper towel, absorbent mineral, charcoal, etc.) which had been used to contain the spill. The volume of disinfectant should be rather more than that which would be applied if the material has been not spilt but ready for disinfection. A period of 30 minutes should be allowed for the disinfectant to act, unless there are more pressing reasons to remove the spillage. In the latter case, particular efforts must be made to protect people against infective hazards.

For treatment of chemical spillages, a period of 30 minutes is also advised as a rule of thumb. Many neutralizing reactions are slower than may be thought, e.g. many acid/base reactions. If there is no danger in doing so, then periodic mixing of the spill plus neutralizer often aids the reaction. A skilled chemist will often be able to add some indicator for the reaction which will show completion by a colour change.

It should always be assumed that on-the-spot treatment has only been partially successful. The collected waste should be considered hazardous until shown otherwise.

15.2.5.5 Decontamination

Removal of the visible spillage may not actually free the room of harmful material. Where any kind of impact is involved there is a tendency for tiny particles to be thrown out.

These may make invisible but breathable aerosols. Where infective material is involved, it should be assumed that these airborne particles have been released. Normally they will settle in about half an hour to 2 hours, so people should be kept out of the area for this time.

Droplets of volatile material will of course evaporate: time should be allowed for the laboratory ventilation to clear the vapours. Solids and less volatile liquids will remain longer. This can be a particular problem if they are substances giving skin burns (e.g. phenol, acrylamide, phosphoric acid, fluoracetic acid, sodium hydroxide). As a general rule, a much larger area than one would expect must be carefully washed down by an appropriate technique.

For radioactive substances or carcinogens, then complete removal will probably require harsher chemical or physical action. Of course, radioactive substances have the advantage that they can be monitored by a suitable detector. Many carcinogens can in fact be observed under ultra-violet illumination. Colour reagents can sometimes be used to detect chemical or biological residues.

Mercury droplets are especially difficult to remove from anything other than the most impervious crevice-free surface. Surprisingly, they adhere to vertical surfaces and even the undersides of items. These are often overlooked in cleaning. Mercury can be very harmful to electrical and electronic devices. It is even possible for a droplet to evaporate from one place and condense in another, causing short-circuits. Clearly, modern micro-electronic devices may be especially vulnerable, and must be closely inspected and removed from the area.

Bulk mercury can be removed by suction or sweeping. Small droplets are best picked up on zinc dust or zinc sponge (see section 15.2.3.9).

Traces of mercury and other particles can often be removed by binding in some sticky material such as a mud of fuller's earth, a grease, a polymer or an adhesive. The material is rubbed or scrubbed well onto the surface, left for some time, then peeled, scraped or scratched off (e.g. using wire wool) possibly with the aid of a detergent. This removes a lot of dirt, some of the surface material, and usually most of the contaminant. It may well leave behind traces of the binding agent.

Infective agents and some chemicals which remain after a spill may sometimes be neutralized without removal. The principal difficulty in doing this is getting the neutralizing agent to penetrate and contact the residue.

For example, if an aqueous reagent is used on a hydrophobic surface such as a waxed bench, it will not completely wet it. Contact can be improved by the use of a surfactant and by using an absorbent solid (paper, polypropylene, mineral) completely soaked in the reagent rather than a pool. This is particularly effective for disinfectants with some (even a little) vapour phase action. Alternatively, an effective agent may be available which is oil- or spirit-soluble.

Oxidizing agents can leave dangerous residues, particularly if spills are repeated in the vicinity of combustible material such as wood. Even dilute solutions can eventually accumulate to a point where a portion of (say) bench becomes literally an explosive liable to ignition. Any oxidizing agent provides this hazard, but there is no doubt that perchloric acid is the most dangerous owing to the extreme sensitivity of some perchlorate compounds. For any area which is believed to have (possibly over many years) significant perchlorate contamination, then specialist advice should be sought for a detailed method of decontamination, probably by an outside contractor.

Note that drainage systems may become contaminated after a spillage, perhaps providing a reservoir of infection or an accumulation of explosive material. (See Chapter 5.)

Utensils which have been used for clearing up a spillage must also be decontaminated or disposed of. For this reason, polypropylene items are recommended for biological laboratories, since this plastic can be autoclaved. (See Table 15.1 and section 10.2.2.3.)

15.3 FIRES

15.3.1 Extinguishing

It is vital that laboratory staff receive proper training in the use of extinguishers, and know which type to select for a particular fire. However, a small bench fire can be contained much better in many cases by the use of a solid spill control product, if this is ready to hand. For example, a burning filter paper may be snuffed out by pouring on sand or soda ash, whereas an extinguisher is likely to blow items off the bench.

Sand is cheap and widely effective. Damp sand should be available if there is substantial work with phosphorus; otherwise it should be dry.

CAUTION: sand should not be used on fires involving sodium, potassium or lithium metal, as a chemical reaction occurs.

Soda ash is often kept for acid spills, but has good fire-inhibiting properties. It is the agent commonly used to extinguish sodium or potassium metal fires.

CAUTION: soda ash should not be used on lithium fires, as a chemical reaction occurs.

Activated charcoal is very effective on solvent fires if applied promptly, but may be ineffective once the fire has caught hold. A good excess may be used to snuff out fires involving paper etc.

CAUTION: charcoal should not be applied to fires or spills involving oxidizing agents, as the charcoal can then itself become highly combustible.

Inhibited sawdust is a commercial product which is not readily combustible, and may be used to snuff out ordinary paper or cloth fires, providing it is applied promptly and in large amounts. It will sometimes extinguish a solvent or oil fire, but the resulting mixture can be easily ignited. There is at present no data on its behaviour in chemical fires, so it should *not* be used for this purpose.

Mineral absorbents powders are more effective than granules in excluding air, but a large amount of either will often snuff out a small fire.

CAUTION: the absorbent acts like a wick, drawing up liquid, so that it is easy for solvent or oil fires to re-ignite.

Graphite is not an absorbent and must not be confused with other forms of carbon. It is used to extinguish fires involving reactive metals.

CAUTION: it must be completely dry in use, or a secondary fire is likely to occur owing to a chemical reaction.

Sodium chloride is an essentially inert material available in large quantities in many laboratories, which can be used to snuff out a wide range of fires, including those involving reactive metals. It should be dry and free flowing.

CAUTION: sodium chloride should not be used on fires involving lithium metal, as a chemical reaction occurs. Lithium chloride should be used instead.

15.3.2 Small Fires

With many laboratory fires, there is a real danger that the materials may re-ignite some time after being extinguished. It is therefore important that the waste is not simply added to normal refuse, or left overnight in a waste bin.

For mixed paper waste, or solvents which have been absorbed onto a solid, the best option is to burn the waste under safe conditions, if this is possible (see Chapter 7). If not, the waste should be rendered less reactive. Solvents may be allowed to evaporate off under safe conditions (see section 9.3.4). Paper and cloth may be soaked in water, making due allowance for any chemicals present. Plastics and mixed waste can be put in a metal container with a substantial quantity of soda ash to reduce the chance of ignition. Where chemicals have been extinguished, then the mixture of chemical and extinguishing agent must be collected and chemically neutralized. (See Chapter 9 and take expert advice.)

15.3.3 Large Fires

A large fire in a laboratory is likely to cause problems of contamination (biological, chemical and radioactive) in addition to structural damage and destruction of valuable goods. It therefore requires more than ordinary care in making safe and preparing for restoration. It is strongly recommended that experienced contractors be brought in. These are usually employees of hazardous waste disposal companies or of certain consulting agencies.

If the work is carried out in-house, then it should be supervised by scientific staff who have the necessary training and facilities to monitor contamination. Typically, large items can be collected in a skip; smaller items for disposal in metal drums. Most or all of it may have to be disposed of as hazardous waste to a licensed site.

All workers should wear adequate protective equipment, and be properly instructed in its use. A disposable coverall with hood and a respirator or breathing apparatus will usually be needed, in addition to safety helmet, safety shoes and gloves.

CAUTION: once a fire is extinguished, it is the normal practice of the fire service to ventilate the area (by opening or breaking windows and doors) to allow smoke to escape. This aids inspection (searching for people or residual smouldering). The officer in charge should be informed if there is any reason why this should not be done, e.g. the presence of radioactive, infective or highly toxic substances.

15.3.4 Biological Materials

CAUTION: a typical fire does not destroy micro-organisms. It merely spreads them around.

If it is believed that there has been a loss of containment of harmful organisms, then the area should be sealed off and an expert consulted as to an effective method of fumigation. Fumigation is also required where insects or nuisance organisms (e.g. dry rot) have been released.

Fumigation will not penetrate closed vessels. These should be collected, placed in a sealed bag and sterilized. Incineration is preferred; autoclaving will do. Chemical methods should be used only where the others are not practical.

Microbiological tests should be carried out to ensure that the room has been made safe to work in. These should include swabs taken from underneath items, not just from visible surfaces.

15.3.5 Chemicals

Where containers of chemicals have been exposed to heat, smoke or water, then they may be partially converted to other substances. This renders them unfit for further use, and possibly means that they have deteriorated to a dangerous condition. The safest procedure is for them to be discarded to a licensed hazardous waste site, with as little handling as possible. (See Chapter 9 and Appendix C.)

Exceptionally expensive materials may be put on one side for possible recovery (see Chapter 14), but only if the economic benefits are truly worth while.

Where chemicals have been spilt, washed around or vaporized, then there will be considerable contamination. As a rule of thumb, the area affected is twice what it appears to be. The more hazardous the substance, the greater the caution. Removal and destruction of affected items are always more effective than trying to wash for re-use. Particular care should be taken to prevent contamination spreading on shoes or gloves to supposedly clean areas.

See, also, section 12.4.

15.3.6 Gas Cylinders

It is the practice of firemen to remove gas cylinders from fire areas as soon as possible, but some cylinders may remain in the fire zone.

CAUTION: gas cylinders which have been exposed to heat should never be put back into service.

If there is any confusion (which there often is) it is best to regard all the gas cylinders as potentially unsafe.

The gas supply company should be contacted to arrange collection. Each cylinder should have a large label firmly taped to it with the following words (or similar information)

'CAUTION: Suspect Cylinder FULL of ---------. Not to be refilled until tested and re-certified. Exposed to fire in XYZ Laboratories (telephone ----------).'

The gas cylinder pressure regulators should also be regarded as suspect. They should be examined by a competent person, and if necessary refurbished. The gas supply company may offer this service.

CAUTION: the details of construction of pressure regulators are rather unusual. It should not be assumed that a skilled person (scientist, technician, welder, pipe-fitter, machinist) is fully conversant unless he or she has special training or experience.

15.3.7 Mercury

It is possible that mercury from instruments could be released during a fire. If this has occurred, then great care must be taken to assess the extent of the contamination. Most particularly, hot mercury can vaporize and condense elsewhere, even inside apparatus such as electronic instruments. Any electrical apparatus in the very close vicinity should be assumed to be contaminated. Electrical apparatus some distance from a mercury release should be opened and carefully inspected before further use. As a precaution, even if there is no mercury visible, electrical devices should be operated for some hours in a well-ventilated place (preferably a fume cupboard) to allow any mercury traces to escape. Contaminated apparatus should *not* be operated beause of the likelihood of a short-circuit occurring.

A portable continuous-reading mercury detector should be obtained. This can be used to mark out the area of contamination, by sweeping the sensor head close to the surfaces. The most practical way of decontaminating a surface is simply to remove the surface, or the whole object. Despite what is occasionally advocated in the literature, there is no way of 'washing down' which will successfully remove or neutralize mercury. The degree of trouble required will depend on the future use of the laboratory, and whether a small background level of mercury can be tolerated. See, also, sections 14.4.5 and 15.2.

15.3.8 Animals

Animals which have been involved in a fire will certainly have been distressed, and may have been harmed by smoke and fumes. From both a scientific and a humane point of view it is probably better to destroy the animals promptly, as lung injuries may not be apparent for some time. However, where exposure was minimal (e.g. if they were evacuated from rooms near the fire) then it is economically sensible to maintain the animals, but under careful observation for ill-effects.

Food and bedding which have been (or may have been) exposed to smoke should be discarded, even if they appear unaffected to human senses. Likewise, a considerable effort must be made to clean off traces of smoke from rooms and furniture where animals are to be kept. Water supply points should be very well flushed, since the fire-fighting operation can disturb water mains, allowing dirt to get in.

CAUTION: some species of animal are very susceptible to toxic effects from fire extinguishing agents.

Any room where a halon (e.g. BCF) fire extinguisher has been used should be ventilated for at least two days before animals are brought in.

15.3.9 Radioactives

Certain types of large source may need to be inspected by the supplier to check that the container is still functional.

With smaller sources, it is first of all necessary to check by a suitable monitoring procedure whether or not there has been any release. If the sources are secure, they should be removed as soon as possible, to permit salvage work to continue.

The enforcement agency will direct that any waste containing radioactivity should be taken to a particular site. This may not be the usual one, but will take into account the local situation after the fire.

The laboratory radiological protection officer should ensure that everyone involved in clearing up is monitored (by film badge, dosemeter, etc.), which may include people who are not employees of the organization (e.g. contractors). For any serious contamination, it is strongly recommended that an experienced and specialist radioactive decontamination team be bought in.

If unsealed sources are involved in a fire, the senior responsible person present should as soon as possible estimate the likely amount and type of airborne release. It may be necessary to consider the balance of risks in either sealing off the area, or ventilating it. See, also, section 15.5.3.

APPENDIX A

Addresses

There follows a select list of government agencies, trade associations and other organizations which may be able to provide information relevant to the disposal of waste from laboratories. See Chapter 4 for other sources of information.

    A-1 Addresses in the United Kingdom
    A-2 Addresses in the Republic of Eire
    A–3 Addresses in the United States of America
    A-4 Addresses in Canada

APPENDIX A-1 Addresses in the United Kingdom

Asbestos Information Centre
40 Piccadilly
LONDON W1V 9PA
01–439.9231

Asbestosis Research Council
P.O. Box 18
Cleckheaton
W. YORKS BD19 8UJ
0274–875711

Asbestos Removal Contractors Association
45 Sheen Lane
LONDON SW14
01–876.4415

Association for Science Education
College Lane
Hatfield
HERTS AL10 9AA

Chemical Cancer Hazard Information Service
Cancer Research Campaign Laboratories
Dept of Cancer Studies
Medical School
University of Birmingham
BIRINGHAM B15 2TJ
021–472.1301

Chemical Industries Association
93 Albert Embankment
LONDON SE1 7TU
01–735.3001

Chemical Recovery Association
697 Warwick Road
Solihull
WEST MIDLANDS B
021–705.0111

Department of Education & Science
39 York Road
LONDON SE1 7PH
01–928.9222

Department of Health & Social Security
Scientific & Technical Branch
14 Russell Square
LONDON WC1B 5EP
01–636.6811

Department of Health & Social Security (Northern Ireland)
27 Adelaide Street
BELFAST BT2 8FH
Belfast 224431

Harwell Laboratories: Hazardous Waste Disposal
Building 151
A.E.R.E.
Harwell
OXON OX11 0RA
0235–24141

Harwell Laboratories: Radioactive Waste Disposal
Building 329
A.E.R.E.
Harwell
OXON OX11 0RA
0235–24141

Health & Safety Executive
1–13 Chepstow Place
LONDON W2 4TF
01–229.3456

Her Majesty's Stationery Office (enquiries)
St Crispin
Duke Street
NORWICH NR.3 1PD
01–928.6977

Home Office
50 Queen Anne's Gate
LONDON SW1H 9AT
01–213.3000

MRC Laboratory Animal Centre
Woodmansterne Road
Carshalton
SURREY SM5 4EF
01–643.8000

National Radiological Protection Board
Chilton
Didcot
OXON OX11 0RQ
0235–831600

Research Defence Society
Grosvenor Gardens House, Grosvenor Gardens,
LONDON SW1W 0BS
01–828.8745

Scottish Health Service
Scientific & Technical Branch
Trinity Park House
South Trinity Road
EDINBURGH EH5 3SH
031–552.6255

TUC Centenary Institute of Occupational Health
London School of Hygiene & Tropical Medicine
Gower Street
LONDON WC1E 7HT
01–636.8636

UK Waste Materials Exchange
Institute of Purchasing & Supply
Easton House
Easton on the Hill
LINCS PE9 2NZ
0780–56777

Universities Federation for Animal Welfare
230 High Street
Potters Bar
HERTS
0707–58202

APPENDIX A-2 Addresses in the Republic of Eire

Association of Consulting Engineers of Ireland
4 Northbrook Road
Leeson Park
DUBLIN 6
01–975716

Department of Energy
Clare Street
DUBLIN 2
01–715233

Department of the Environment
Custom House
DUBLIN 1
01–742961

Department of Health
Custom House
DUBLIN 1
01–742961

APPENDIX A-3 Addresses in the United States of America

American Chemical Society
1155 Sixteenth Street N. W.
Washington DC 20036

Animal Welfare Institute
P.O. Box 3650
Washington DC 20007

Chemical Manufacturers' Association
2501 M Street
Washington DC 20037

Environmental Protection Agency
Office of Solid Waste (WH 565)
401 M Street S. W.
Washington DC 20460
800–424.9346 (toll-free, out-of-town callers)
202–382.3000 (toll-free, Washington callers)

National Association of Solvent Recyclers
P.O. Box 1288
Dayton
OHIO 45407

National Cancer Institute
Office of Biohazard and Environmental Control
National Institute of Health
Bethesda MD 2025

National Institute for Occupational Safety and Health
Parklawn Buildings
5600 Fishers Lane
Rockville MD 20857

National Technical Information Service
5285 Port Royal Road
Springfield VA 22151

Nuclear Regulatory Commission
1717 H Street N. W.
Washington DC 20555

Occupational Safety and Health Administration
200 Constitution Avenue N. W.
Washington DC 20210

US Government Printing Office
Washington DC 20402

APPENDIX A-4 Addresses in Canada

Canadian Chemical Producers' Association
350 Sparks Street, Suite 505
Ottawa
ONTARIO K1P 7S8

Canadian Federation of Biological Sciences
Room 124
118 Veterinary Road
Saskatoon
SASKETCHEWAN S7N 2R4

Canadian Government Publishing Centre
Supply and Services Canada
Ottawa
ONTARIO K1A 0S9

Canadian Hospital Association
17 York Street 100
Ottawa
ONTARIO K1N 9J6

Canadian Society of Animal Science
151 Slater Street, Suite 1105
Ottawa
ONTARIO K1P 5H3

Canadian Society of Laboratory Technologists
Box 830
Hamilton
ONTARIO L8N 3N8

Canadian Waste Materials Exchange
Ontario Research
Sheridan Park Research Community
Mississanga
ONTARIO L5K 1B3

Enquirey Centre
Environmental Canada
Ottowa
ONTARIO K1A 0H3

Environmental Protection Service
Department of the Environment
Ottawa
ONTARIO K1A 1C8

Health and Welfare Canada
Health Protection Branch
Health Protection Building
Holland Avenue
Tunney's Pasture
Ottawa
ONTARIO K1A 0L2

APPENDIX B

Commercial Products

The following lists give a selection of products which may be useful in the disposal of laboratory waste. All information is given in good faith, but the presence of a company or product does not constitute an endorsement. Likewise, the absence of any company or product does not imply any criticism. Technical comments are for general guidance only: the suppliers should be approached for current specifications and performance of their products.

B-1   Proprietary Spill Control Products
B-2   Sharps' Collection Containers
B-3   Liquid Waste Containers
B-4   Laboratory Cleaning Agents
B-5   Disinfectants
B-5.1 Hypochlorite Liquid Disinfectants
B-5.2 Solid Disinfectants (Bleaching Powders)
B-5.3 Glutaraldehyde Liquid Disinfectants
B-5.4 Clear Phenolic Disinfectants
B-6   Labels etc.
B-7   Silver Recovery Companies

APPENDIX B-1 Proprietary Spill Control Products

Manufacturer or principal supplier	Product names	Comments
3M USA Aldrich UK	3M Sorbent All-Purpose	Absorbent plastic blanket
Absorbent Products UK	Pillo-Dri	Amorphous mineral in pillow
		Solvents and acids may affect cover
Aldrich UK	Spillow	Amorphous silicate in pillow
J.T. Baker USA American Hospital Supply UK Diamed Diagnostics UK Linton Products UK	Cinnasorb, CN-Plex, Fluoril, Liquisorb, Mercury Sponge, Neutracit, Neutrasol, Neutrasorb, Resisorb, Solusorb, Spill Control Center	Kits for absorption/neutralization Specifically for acids (except HF), alkalies, cyanides, hydrofluoric acid, mercury, solvents
Lab Safety Supply USA Bennett & Co UK	Spill Control Pillow	Amorphous silicate in pillow
Mallinckrodt USA Camlab UK	Mercury Tamer, Spill Tamer	Standard kit consists of absorbent, acid neutralizer, caustic neutralizer, mercury treatment and bags, gloves, etc.
Philip Harris UK	Harris Spill Pack	Contains absorbent for 6 $\times$ 500 cm^3 liquid, plus soda ash plus bags
Roth USA Aldrich UK	Mercurisorb	Mercury collection kit, absorber
Sellstrom USA	Spill Control Station	Contains absorbent pads, absorbent silicate pillows, plus special pillows for HF. Protective clothing and bags

APPENDIX B-2 Sharps' Collection Containers

The following companies can supply containers which comply with the requirements of the UK Department of Health and Social Security, Specification TSS/S/330.015. Some products may also be available through general laboratory and hospital supply houses.

DRG Hospital Supplies
1 Dixon Road
Brislington
BRISTOL BS4 5QY

Labco Ltd
54 Marlow Bottom Road
Marlow
BUCKS SL7 3NF

Frontier Medical Products
North Blackvein Industrial Estate
Crosskeys
GWENT

Lawtons of Liverpool
60 Vauxhall Road
LIVERPOOL L69 3AU

IPS Hospital Services
Victoria Mill
Lower Vickers Street
MANCHESTER M10 7LY

Metal Box Composite Containers
Stokes Street
Clayton
MANCHESTER

APPENDIX B-3 Liquid Waste Containers

The following companies supply high-quality containers, specially made for the safe collection and transport of flammable, corrosive or toxic liquid chemical waste. (These and comparable products may be available through other general supply houses.)

B/R Instrument Corporation
P.O. Box 7
Pasadena
Maryland 21122
USA

Justrite Manufacturing Co.
2454 Dempster Street
Des Plaines
ILLINOIS 60016
USA

Flametamers International Ltd
Clock House
Laindon Centre
Basildon
ESSEX
ENGLAND

Walter Page Safeways Ltd
46 Lower Shelton Road
Marston
Moretaine
BEDS MK43 0LW
ENGLAND

Safety Unlimited
107 Eleanor Cross Road
Waltham Cross
HERTS EN8 7NT
ENGLAND

UKAM
White Lodge
31 Ledborough Lane
Beaconsfield
BUCKS HP9 2DB
ENGLAND

APPENDIX B-4 Laboratory Cleaning Agents

There follow some alkaline agents for soak cleaning of laboratory apparatus. As a 1–5% solution in water, most are effective in removing chemical, biological and radioactive dirt, though some may be more effective than others in particular applications. The lysing effect will kill most micro-organisms, especially if used hot, but for critical work (e.g. with human pathogens) the use of a separate sterilizing procedure is recommended.

Trademark	Manufacturer or principal supplier	Notes
Alconox	Alconox (USA) PC International (UK)	(1)
Contrad	Decon (USA)	
Decon 75, Decon 90	Decon (UK)	(2)
Dri-Decon	Decon (UK)	
Extran 100, Extran 300	BDH (UK)	
Haemasol	A.R. Howell (UK)	(3)
Lipsol	L.I.P. (UK)	
Liquinox	Alconox (USA) PC International (UK)	
Pyroneg	Diversey (UK)	
Quadralene	Fisons Scientific (UK)	
RBS 35, RBS 50	Chemical Concentrates (UK) Pierce (USA)	(4)
Terg-a-Zyme	Alconox (USA) PC International	(3)

Notes:
(1) Powder.
(2) Also BDH, Fisons.
(3) Contains protease.
(4) May be used in alcohol.

APPENDIX B-5 Disinfectants

Appendix B-5.1 Hypochlorite Liquid Disinfectants

There follows a selection of liquid products containing hypochlorite or chlorite, suitable for use in discard jars, and available in the UK.

Name	Manufacturer	Available chlorine %	Notes
Alcide-Exspor	Alcide Corp. (USA) Life Science Laboratories (UK)		(1)
Chloros	ICI Mond Div.	10	
Deosan Green Label Sterilizer	Diversey Ltd	10	
Domestos	Lever Bros.	10	
Eusol	Thornton & Ross Ltd	0.25	(2), (3)
Milton	Richardson-Vicks Ltd	1	
Milton-2	Richardson-Vicks Ltd	2	
Parozone	Jeyes Ltd	5 or 10	
Parozone Plus	Jeyes Ltd	10	(4)
Sodium hypochlorite	Various	10 to 14	(2)
Teepol Bleach	Shell	4	(5)

Notes:
(1) A two-part formulation releasing chlorine dioxide instead of chlorine. See section 10.2.3.4.
(2) Non-proprietary name.
(3) British Pharmacopeia specification is 1.25% calcium hypochlorite and 1.25% boric acid in water.
(4) Plus detergent.
(5) Distributed by BDH Ltd.

Appendix B-5.2 Solid Disinfectants (Bleaching Powders)

There follows a selection of solid materials which release halogen agents on contact with water. They may be used directly on spills, to make up solutions for discard jars, or for scrubbing surfaces to eliminate microbial contamination. Unfortunately, the terms 'bleaching powder', 'calcium hypochlorite', and 'chlorinated lime' are used very loosely, even by chemical suppliers. In the UK, 'Calcium hypochlorite' usually refers to $CaOCl_2$ whigh gives about 35% available chlorine, whereas true calcium hypochlorite is CaO_2Cl_2, the pure form being called 'high test hypochlorite' which gives 99% available chlorine. (The term 'available chlorine' is a somewhat arbitrary measure of bleaching power, but is a useful guide to relative strengths. Sodium chlorite has a power of 125% available chlorine.)

In recent years, other halogen release agents have been introduced. The most common are sodium or calcium dichloroisocyanurate, typically giving 27% available chlorine.

Name	Manufacturer	Notes
Bleaching powder	Various	(1)
Calcium hypochlorite	Various	(2)
Chlorinated lime	Various	(3)
High test hypochlorite	Various	(4)
Sodium chlorite	Various	(5)
Sodium dichloroisocyanurate	Various	(6)
ACL	Monsanto UK	(7)
Diversol BX	Diversey Ltd	(8)
Fi-Clor Clearon	Fisons plc	(7) (9)
Kirbychlor	Kirby-Warrick Ltd	(7)
Milton Solid	Richardson-Vicks	(8)
Multichlor	Spearhead Ltd	(7)
Reddichlor	Reddish Detergents	(7)
Simpla	Ashe Ltd	(7)
Tricidechlor	Wingfield Ltd	(7)

Notes:
(1) Composition varies: 10 to 35% available chlorine.
(2) See above: 35% available chlorine.
(3) 35% available chlorine.
(4) 99% available chlorine.
(5) 125% available chlorine.
(6) 27% available chlorine.
(7) Dichloroisocyanurate products.
(8) Mixture: 3% available chlorine.
(9) Supplied by Griffin & George Ltd, not Fisons Scientific.

Appendix B-5.3 Glutaraldehyde Liquid Disinfectants

There follows a selection of liquid products containing glutaraldehyde, suitable for use in discard jars, and available in the UK. Stability is that claimed by the manufacturer for an unused solution of 2% final concentration of glutaraldehyde. See section 10.2.3.6.

Name	Manufacturer	Stability (days)	Notes
ASEP	Galen Ltd	14	(1)
Cidex	Surgikos Ltd	14	(2)
Cidex Long-life	Surgikos Ltd	28	(2)
Glutaraldehyde	Various	0.5	(3)
Pantasept	Samoore Ltd		(4)
Totacide 28	Tennoco Organics Ltd	28	(1)

Notes:
(1) Liquid activator.
(2) Solid activator.
(3) Normally sold as 25% or 50% in water. Dilute to 2 or 3% and add 0.3 NaHCO$_3$.
(4) A 30% solution in water.

Appendix B–5.4 Clear Phenolic Disinfectants

A selection of phenol, cresol or xylenol mixtures with surface agents, suitable for use in discard jars, and available in the UK.

Name	Manufacturer	Notes
Clearsol	Tenneco Organics Ltd	
Hycolin	William Pearson Ltd	
Lyseptol	Philip Harris Ltd	
Lysol	Various	(1)
Medol	William Pearson Ltd	(2)
Printol	Tenneco Organics Ltd	(3)
Stericol	Sterling Winthrop Group	(4)
Sudol	Tenneco Organics Ltd	

Notes:
(1) Lysol

This is made by various manufacturers and the composition and properties vary. It is typically a mixture of cresols and soap, described as a 'clear phenolic'. In the UK it is non-proprietary. Manufacturers are Robert Young & Co Ltd, Tenneco Organics Ltd, Thornton & Ross Ltd, William Pearson Ltd, Walker Lunt & Co. Ltd. A product may be labelled 'B.P. 1968' to show that it conforms with the British Pharmacopeia specification.

In the USA, 'Lysol' is the trade mark of Lehn & Fink. In West Germany it is the trade mark of Schulke u. Mayr.

(2) Similar to Hycolin.
(4) Contains terpenes.
(4) Alcoholic.

APPENDIX B-6 Labels etc.

Many laboratory suppliers stock some labels, labelled bags and other ancillary items. However, the following companies have a particularly good range of labels and miscellaneous items which are useful for waste disposal.

Jencons (Scientific) Ltd
Cherrycourt Way Industrial Estate
Stanbridge Road
Leighton Buzzard
BEDS LU7 8UA
ENGLAND

Lab Safety Supply Company
Janesville
WI 53545
USA

APPENDIX B-7 Silver Recovery Companies

The following companies buy precious metals and/or operate a silver recovery service for laboratories and photographic or X-ray departments in the UK.

DC Refining Ltd
Unit 22
Enterprise Trading Estate
Brierly Hill
Dudley
WEST MIDLANDS DY5 1TX

John Betts Refiners Ltd
61 Charlotte Street
BIRMINGHAM B3 1PY

Johnson Matthey Chemicals Ltd
Orchard Road
Royston
HERTS SG8 5HE

Photographic Silver Recovery Ltd
Saxon Way
Melbourn
Royston
HERTS SG8 6DN

Lokas Ltd
Winstanley Industrial Estate
Scot Lane
Blackrod
Bolton
LANCS BL6 5SL

Myson Group Ltd
Light Industrial Estate
Ongar
ESSEX

Parkerdell Refining Ltd
Rushenden Road
Queenborough
KENT ME11 5HY

APPENDIX C

Chemical Tables

C-1   UK Special Waste
C-2   US Hazardous Waste
C-2.1 Acute Hazardous Waste
C-2.2 Hazardous Waste
C-3   Known Carcinogens
C-3.1 UK Controlled Substances
C-3.2 US Controlled Substances
C-3.3 Human Carcinogens (ACGIH)
C-3.4 Known Carcinogens (HSE)
C-3.5 Some Experimental Carcinogens
C-3.6 Other Carcinogens
C-4   Explosive Chemicals
C-5   Named Reagents
C-6   Deteriorated Chemicals
C-7   Low-hazard Solids for Land Disposal
C-8   Properties of Gases for Venting
C-9   List of Chemical Tables Elsewhere in the Book

APPENDIX C-1 UK Special Waste

Waste to be transported from non-domestic premises is considered 'special' if it falls into a definition given by the Control of Pollution (Special Waste) Regulations 1980 (Statutory Instrument no. 1709, made under the provisions of Section 17 of the Control of Pollution Act 1974).

Pharmaceutical substances which are only available on a doctor's prescription are 'special waste'. Other materials are 'special waste' if they are 'listed substances' *and also* flammable, corrosive or toxic. Note that 'laboratory chemicals' is included in the list, as well as more specific chemical terms.

A 'listed substance' is flammable (and thus 'special') it if has a flash point of 21 °C or less.

A 'listed substance' is considered dangerous to life if it is thought likely (on available toxicological data) that a child would be seriously harmed by eating 5 cubic centimetres of the material (i.e. the material is toxic and thus Special). A 'listed substance' is also dangerous to life and hence 'special' if a 15 minute exposure is likely to cause serious tissue damage if the material is breathed in or contacts the skin or eyes.

Listed substances

Acids and alkalis
Antimony and antimony compounds
Arsenic compounds
Asbestos (all chemical forms)
Barium compounds
Beryllium and beryllium compounds
Biocides and phytopharmaceutical substances
Boron compounds
Cadmium and cadmium compounds
Copper compounds
Heterocyclic organic compounds containing oxygen, nitrogen or sulphur
Hexavalent chromium compounds
Hydrocarbons and their oxygen, nitrogen and sulphur compounds
Inorganic cyanides
Inorganic halogen-containing compounds
Inorganic sulphur-containing compounds
Laboratory chemicals
Lead compounds
Mercury compounds
Nickel and nickel compounds
Organic halogen compounds, excluding inert polymeric materials
Peroxides, chlorates, perchlorates and azides
Pharmaceutical and veterinary compounds
Phosphorus and its compounds
Selenium and selenium compounds
Silver compounds
Tarry materials from refining and tar residues from distilling
Tellurium and tellurium compounds
Thallium and thallium compounds
Vanadium compounds
Zinc compounds

APPENDIX C-2 US Hazardous Waste, as defined by the Environmental Protection Agency

The following list is in two parts. The first part is those substances designated 'acute hazardous waste' by the Environmental Protection Agency, and given an EPA Hazardous Waste Number beginning with P. The second part is those substances designated toxic, reactive, corrosive or ignitable, and given an EPA Hazardous Waste Number beginning with U. Most substances are listed because of their toxicity. For other hazards or multiple hazards, the following letter codes are used: T = toxic, R = reactive, C = corrosive, I = ignitable. The same chemical substance may be listed under different names, but always has the same Hazardous Waste Number.

Mixtures of listed substances are usually 'hazardous waste' (unless, for example, a chemical reaction has neutralized the hazardous properties).

Non-listed substances are also counted 'hazardous waste' if they meet EPA criteria for ignitability (which includes liquids with a flash point below 60 °C, but also solids, gases and oxidizing agents), corrosivity (which includes aqueous liquids of pH 2 or less, or pH 12.5 or greater), reactivity (which includes explosives, substances reacting violently or giving hazardous fumes with water, and normally unstable compounds), or toxicity. For details of these criteria and test methods, refer to the Code of Federal Regulations 40 CFR 261, or to the Environmental Protection Agency.

Non-listed substances are given EPA Hazardous Waste Numbers beginning with D, as follows:

Ignitable = D001
Corrosive = D002
Reactive = D003
Toxic   = D004–D017, depending on the most significant toxic component.

C-2.1 Acute Hazardous Waste

Hazardous waste No.	Substance	Hazardous waste No.	Substance
P023	Acetaldehyde, chloro-	P070	Aldicarb
P002	Acetamide, N-(aminothioxomethyl)-	P004	Aldrin
P057	Acetamide, 2-fluoro-	P005	Allyl alcohol
P058	Acetic acid, fluoro-, sodium salt	P006	Aluminium phosphide
P066	Acetimidic acid, N-[(methylcarbamoyl)oxy]thio-, methyl ester	P007	5-(Aminomethyl)-3-isoxazolol
		P008	4-Aminopyridine
		P009	Ammonium picrate (R)
P001	3-(alpha-acetonylbenzyl)-4-hydroxycoumarin and salts	P119	Ammonium vanadate
		P010	Arsenic acid
P002	1-Acetyl-2-thiourea	P012	Arsenic(III) oxide
P003	Acrolein	P011	Arsenic(V) oxide

C-2.1 Acute Hazardous Waste *cont.*

Hazardous waste No.	Substance
P011	Arsenic pentoxide
P012	Arsenic trioxide
P038	Arsine, diethyl-
P054	Aziridine
P013	Barium cyanide
P024	Benzenamine, 4-chloro-
P077	Benzenamine, 4-nitro-
P028	Benzene, (chloromethyl)-
P042	1,2-Benzenediol, 4-[1-hydroxy-2-(methylamino)ethyl]-
P014	Benzenethiol
P028	Benzyl chloride
P015	Beryllium dust
P016	Bis(chloromethyl)ether
P017	Bromoacetone
P018	Brucine
P021	Calcium cyanide
P123	Camphene, octachloro-
P103	Carbamimidoselenoic acid
P022	Carbon bisulfide
P022	Carbon disulfide
P095	Carbon chloride
P033	Chlorine cyanide
P023	Chloroacetaldehyde
P024	p-Chloroaniline
P026	1-(o-Chlorophenyl)thiourea
P027	3-Chloropropionitrile
P029	Copper cyanides
P030	Cyanides (soluble cyanide salts), not elsewhere specified
P031	Cyanogen
P033	Cyanogen chloride
P036	Dichlorophenylarsine
P037	Dieldrin
P038	Diethylarsine
P039	O,O-Diethyl S-[2-(ethylthio)ethyl] phosphorodithioate
P041	Diethyl-p-nitrophenyl phosphate
P040	O,O-Diethyl O-pyrazinyl phosphorothioate
P043	Diisoproyl fluorophosphate
P044	Dimethoate
P045	3,3-Dimethyl-1-(methylthio)-2-butanone, O-[(methylamino)carbonyl] oxime
P071	O,O-Dimethyl O-p-nitrophenyl phosphorothioate
P082	Dimethylnitrosamine
P046	alpha, alpha-Dimethylphenethylamine
P047	4,6-Dinitro-o-cresol and salts
P034	4,6-Dinitro-o-cyclohexylphenol
P048	2,4-Dinitrophenol
P020	Dinoseb
P085	Diphosphoramide, octamethyl-
P039	Disulfoton

Hazardous waste No.	Substance
P049	2,4-Dithiobiuret
P109	Dithiopyrophosphoric acid, tetraethyl ester
P050	Endosulfan
P088	Endothail
P051	Endrin
P042	Epinephrine
P046	Ethanamine, 1,1-dimethyl-2-phenyl-
P084	Ethenamine, N-methyl-N-nitroso-
P101	Ethyl cyanide
P054	Ethylenimine
P097	Famphur
P056	Fluorine
P057	Fluoroacetamide
P058	Fluoroacetic acid, sodium salt
P065	Fulminic acid, mercury(II) salt (R,T)
P059	Heptachlor
P051	1,2,3,4,10,10-Hexachloro-6,7-epoxy-1,4,4a,5,6,7,8,8a-octahydro-endo,endo-1,4:5,8-dimethanonaphthalene
P037	1,2,3,4,10,10-Hexachloro-6,7-epoxy-1,4,4a,5,6,7,8,8a-octahydro-endo,exo-1,4:5,8-demethanonaphthalene
P060	1,2,3,4,10,10-Hexachloro-1,4,4a,5,8,8a-hexahydro-1,4:5,8-endo, endo-dimeth- anonaphthalene
P004	1,2,3,4,10,10-Hexachloro-1,4,4a,5,8,8a-hexahydro-1,4,5,8-endo, exodimethanonaphthalene
P060	Hexachlorohexahydro-exo, exo-dimethanonaphthalene
P062	Hexaethyl tetraphosphate
P116	Hydrazinecarbothioamide
P068	Hydrazine, methyl-
P063	Hydrocyanic acid
P063	Hydrogen cyanide
P096	Hydrogen phosphide
P064	Isocyanic acid, methyl ester
P007	3(2H)-isoxazolone, 5-(aminomethyl)-
P092	Mercury, (acetato-O)phenyl-
P065	Mercury fulminate (R,T)
P016	Methane, oxybis(chloro-
P112	Methane, tetranitro-(R)
P118	Methanethiol, trichloro-
P059	4,7-Methano-1H-indene, 1,4,5,6,7,8,8-heptachloro-3a,4,7,7a-tetrahydro-
P066	Methomyl
P067	2-Methylaziridine
P068	Methyl hydrazine
P064	Methyl isocyanate
P069	2-Methyllactonitrile
P071	Methyl parathion

C-2.1 Acute Hazardous Waste *cont.*

Hazardous waste No.	Substance
P072	alpha-Naphthylthiourea
P073	Nickel carbonyl
P074	Nickel cyanide
P074	Nickel(II) cyanide
P073	Nickel tetracarbonyl
P075	Nicotine and salts
P076	Nitric oxide
P077	p-Nitroaniline
P078	Nitrogen dioxide
P076	Nitrogen(II) oxide
P078	Nitrogen(IV) oxide
P081	Nitroglycerine (R)
P082	N-Nitrosodimethylamine
P084	N-Nitrosomethylvinylamine
P050	5-Norbornene-2,3-dimethanol, 1,4,5,6,7,7-hexachloro, cyclic sulfite
P085	Octamethylpyrophosphoramide
P087	Osmium oxide
P087	Osmium tetroxide
P088	7-Oxabicyclo[2.2.1]heptane-2,3-dicarboxylic acid
P089	Parathion
P034	Phenol, 2-cyclohexyl-4,6-dinitro-
P048	Phenol, 2,4-dinitro-
P047	Phenol, 2,4-dinitro-6-methyl-
P020	Phenol, 2,4-dinitro-6-(1-methylpropyl)-
P009	Phenol, 2,4,6-trinitro-, ammonium salt (R)
P036	Phenyl dichloroarsine
P092	Phenylmercuric acetate
P093	N-Phenylthiourea
P094	Phorate
P095	Phosgene
P096	Phosphine
P041	Phosphoric acid, diethyl p-nitrophenyl ester
P044	Phosphorodithioic acid, O,O-dimethyl S-[2-(methylamino)-2-oxoethyl]ester
P043	Phosphorofluoric acid, bis(1-methylethyl)-ester
P094	Phosphorothioic acid, O,O-diethyl S-(ethylthio)methyl ester
P089	Phosphorothioic acid, O,O-diethyl O-(p-nitrophenyl)ester
P040	Phosphorothioic acid, O,O-diethyl O-pyrazinyl ester
P097	Phosphorothioic acid, O,O-dimethyl O-[p-((dimethylamino)-sulfonyl)phenyl]ester

Hazardous waste No.	Substance
P110	Plumbane, tetraethyl-
P098	Potassium cyanide
P099	Potassium silver cyanide
P070	Propanal, 2-methyl-2-(methylthio)-, O-[(methylamino)carbonyl]oxime
P101	Propanenitrile
P027	Propanenitrile, 3-chloro-
P069	Propanenitrile, 2-hydroxy-2-methyl-
P081	1,2,3-Propanetriol, trinitrate-(R)
P017	2-Propanone, 1-bromo-
P102	Propargyl alcohol
P003	2-Propenal
P005	2-Propen-1-ol
P067	1,2-Propylenimine
P102	2-Propyn-1-ol
P008	4-Pyridinamine
P075	Pyridine, (S)-3-(1-methyl-2-pyrrolidinyl)-, and salts
P111	Pyrophosphoric acid, tetraethyl ester
P103	Selenourea
P104	Silver cyanide
P105	Sodium azide
P106	Sodium cyanide
P107	Strontium sulfide
P108	Strychnidin-10-one, and salts
P018	Strychnidin-10-one, 2,3-dimethoxy-
P108	Strychnine and salts
P115	Sulfuric acid, thallium(I) salt
P109	Tetraethyldithiopyrophosphate
P110	Tetraethyl lead
P111	Tetraethylpyrophosphate
P112	Tetranitromethane (R)
P062	Tetraphosphoric acid, hexaethyl ester
P113	Thallic oxide
P113	Thallium(III) oxide
P114	Thallium(I) selenite
P115	Thallium(I) sulfate
P045	Thiofanox
P049	Thiomidodicarbonic diamide
P014	Thiophenol
P116	Thiosemicarbazide
P026	Thiourea, (2-chlorophenyl)-
P072	Thiourea, 1-naphthalenyl-
P093	Thiourea, phenyl-
P123	Toxaphene
P118	Trichloromethanethiol
P119	Vanadic acid, ammonium salt
P120	Vanadium pentoxide
P120	Vanadium(V) oxide
P001	Warfarin
P121	Zinc cyanide
P112	Zinc phosphide (R,T)

C-2.2 Hazardous Waste

Hazardous Waste No	Substance
U001	Acetaldehyde (I)
U034	Acetyaldehyde, trichloro-
U187	Acetamide, N-(4-ethoxyphenyl)-
U005	Acetamide, N-9H-fluoroen-2-yl-
U112	Acetic acid, ethyl ester (I)
U144	Acetic acid, lead salt
U214	Acetic acid, thallium(I) salt
U002	Acetone (I)
U003	Acetonitrile (I,T)
U004	Acetophenone
U005	2-Acetylaminofluorene
U006	Acetyl chloride (C,R,T)
U007	Acrylamide
U008	Acrylic Acid (I)
U009	Acrylonitrile
U150	Alanine, 3-[p-bis(2-chloroethyl)amino] phenyl-, L-
U011	Amitrole
U012	Aniline (I,T)
U014	Auramine
U015	Azaserine
U010	Azirino(2′,3′:3,4)pyrrolo(1,2-a)indole-4,7-dione, 6-amino-8-[((aminocarbonyl) oxy)methyl]-1,1a,2,8,8a,8b-hexahydro-8a-methoxy-5-methyl-,
U157	Benz[j]aceanthrylene, 1,2-dihydro-3-methyl-
U016	Benz[c]acridine
U016	3,4-Benzacridine
U017	Benzal chloride
U018	Benz[a]anthracene
U018	1,2-Benzanthracene
U094	1,2-Benzanthracene, 7,12-dimethyl-
U012	Benzenamine (I,T)
U014	Benzenamine, 4,4′-carbonimidoylbis(N,N-dimethyl-
U049	Benzenamine, 4-chloro-2-methyl-
U093	Benzenamine, N,N′-dimethyl-4-phenylazo-
U158	Benzenamine, 4,4′-methylenebis(2-chloro-
U222	Benzenamine, 2-methyl-, hydrochloride
U181	Benzenamine, 2-methyl-5-nitro
U019	Benzene (I,T)
U038	Benzeneacetic acid, 4-chloro-alpha-(4-chlorophenyl)-alpha-hydroxy, ethyl ester
U030	Benzene, 1-bromo-4-phenoxy-
U037	Benzene, chloro-
U190	1,2-Benzenedicarboxylic acid anhydride
U028	1,2-Benzenedicarboxylic acid, [bis(2-ethylhexyl)] ester

Hazardous waste No.	Substance
U069	1,2-Benzenedicarboxylic acid, dibutyl ester
U088	1,2-Benzenedicarboxylic acid, diethyl ester
U102	1,2-Benzenedicarboxylic acid, dimethyl ester
U107	1,2-Benzenedicarboxylic acid, di-n-octyl ester
U070	Benzene, 1,2-dichloro-
U071	Benzene, 1,3-dichloro-
U072	Benzene, 1,4-dichloro-
U017	Benzene, (dichloromethyl)-
U223	Benzene, 1,3-diisocyanatomethyl-(R,T)
U239	Benzene, dimethyl-(I,T)
U201	1,3-Benzenediol
U127	Benzene, hexachloro-
U056	Benzene, hexahydro- (I)
U188	Benzene, hydroxy-
U220	Benzene, methyl-
U105	Benzene, 1-methyl-1-2,4-dinitro-
U106	Benzene, 1-methyl-2,6-dinitro-
U203	Benzene, 1,2-methylenedioxy-4-allyl-Benzene, 1,2-methylenedioxy-4-
U141	propenyl-
U090	Benzene, 1,2-methylenedioxy-4-propyl-
U055	Benzene, (1-methylethyl)- (I)
U169	Benzene, nitro- (I,T)
U183	Benzene, pentachloro-
U185	Benzene, pentachloro-nitro-
U020	Benzenesulfonic acid chloride (C,R)
U020	Benzenesulfonyl chloride (C,R)
U207	Benzene, 1,2,4,5-tetrachloro-
U023	Benzene, (trichloromethyl)-(C,R,T)
U234	Benzene, 1,3,5-trinitro- (R,T)
U021	Benzidine
U202	1,2-Benzisothiazolin-3-one, 1,1-dioxide
U120	Benzo[j,k]fluorene
U022	Benzo[a]pyrene
U022	3,4-Benzopyrene
U197	p-Benzoquinone
U023	Benzotrichloride (C,R,T)
U050	1,2-Benzphenanthrene
U085	2,2′-Bioxirane (I,T)
U021	(1,1′-Biphenyl)-4,4′-diamine
U073	(1,1′-Biphenyl)-4,4′-diamine, 3,3′-dichloro-
U091	(1,1′-Biphenyl)-4,4′-diamine, 3,3′-dimethoxy-
U095	(1,1′-Biphenyl)-4,4′-diamine, 3,3′-dimethyl-
U024	Bis(2-chloroethoxyl) methane

C-2.2   Hazardous Waste *cont.*

Hazardous waste No.	Substance
U027	Bis(2-chloroisopropyl) ether
U244	Bis(dimethylthiocarbamoyl) disulfide
U028	Bis(2-ethylhexyl) phthalate
U246	Bromine cyanide
U225	Bromoform
U030	4-Bromophenyl phenyl ether
U128	1,3-Butadiene, 1,1,2,3,4,4-hexachloro-
U172	1-Butanamine, N-butyl-N-nitroso-
U035	Butanoic acid, 4-[Bis(2-chloroethyl)amino] benzene-
U031	1-Butanol (I)
U159	2-Butanone (I,T)
U160	2-Butanone peroxide (R,T)
U053	2-Butenal
U074	2-Butene, 1,4-dichloro- (I,T)
U031	n-Butyl alcohol (I)
U136	Cacodylic acid
U032	Calcium chromate
U238	Carbamic acid ethyl ester
U178	Carbamic acid, methylnitroso-, ethyl ester
U176	Carbamide, N-ethyl-N-nitroso-
U177	Carbamide, N-methyl-N-nitroso-
U219	Carbamide, thio-
U097	Carbamoyl chloride, dimethyl-
U215	Carbonic acid, dithallium (I) salt
U156	Carbonochloridic acid, methyl ester (I,T)
U033	Carbon oxyfluoride (R,T)
U211	Carbon tetrachloride
U033	Carbonyl fluoride (R,T)
U034	Chloral
U035	Chlorambucil
U036	Chlordane, technical
U028	Chlornaphazine
U037	Chlorobenzene
U039	4-Chloro-m-cresol
U041	1-Chloro-2,3-epoxypropane
U042	2-Chlorethyl vinyl ether
U044	Chloroform
U046	Chloromethyl methyl ether
U047	beta-Chloronaphthalene
U048	o-Chlorophenol
U049	4-Chloro-o-toluidine, hydrochloride
U032	Chromic acid, calcium salt
U050	Chrysene
U051	Creosote
U052	Cresols
U052	Cresylic acid
U053	Crotonaldehyde
U055	Cumene (I)
U246	Cyanogen bromide
U197	1,4-Cyclohexadienedione
U056	Cyclohexane (I)

Hazardous waste No.	Substance
U057	Cyclohexanone (I)
U130	1,3-Cyclopentadiene, 1,2,3,4,5,5-hexa-chloro-
U058	Cyclophosphamide
U240	2,44-D, salts and esters
U059	Daunomycin
U060	DDD
U061	DDT
U142	Decachlorooctahydro-1,3,4-metheno-2H-cyclobuta[c,d]-pentalen-2-one
U062	Diallate
U133	Diamine (R,T)
U221	Diaminotoluene
U063	Dibenz[a,h]anthracene
U063	1,2:5,6-Dibenzanthracene
U064	1,2:7,8-Dibenzopyrene
U064	Dibenz[a,i]pyrene
U066	1,2-Dibromo-3-chloropropane
U069	Dibutyl phthalate
U062	S-(2,3-Dichloroallyl) diisopropyl-thiocarbamate
U070	o-Dichlorobenzene
U071	m-Dichlorobenzene
U072	p-Dichlorobenzene
U073	3,3'-Dichlorobenzidine
U074	1,4-Dichloro-2-butene (I,T)
U075	Dichlorodifluoromethane
U192	3,5-Dichloro-N-(1,1-dimethyl-2-propynyl) benzamide
U060	Dichloro diphenyl dichloroethane
U061	Dichloro diphenyl trichloroethane
U078	1,1-Dichloroethylene
U079	1,2-Dichloroethylene
U025	Dichloroethyl ether
U081	2,4-Dichlorophenol
U082	2,6-Dichlorophenol
U240	2,4-Dichlorophenoxyacetic acid, salts and esters
U083	N,N-Diethylhydrazine
U084	1,2-Dichloropropane
U085	1,3-Dichloropropene
U108	1,2:3,4-Diepoxybutane (I,T)
U086	1,4-Diethylhydrazine
U087	O,O-Diethyl-S-methyl-dithiophosphate
U088	Diethyl phthalate
U089	Diethylstilbestrol
U148	1,2-Dihydro-3,6-pyradizinedione
U090	Dihydrosafrole
U091	3,3'-Dimethoxybenzidine
U092	Dimethylamine (I)
U093	Dimethylaminoazobenzene
U094	7,12-Dimethylbenz[a]anthracene
U095	3,3'-Dimethylbenzidine
U096	alpha,alpha-Dimethylbenzylhydroperoxide (R)

C-2.2 Hazardous Waste *cont.*

Hazardous waste No.	Substance
U097	Dimethylcarbamoyl chloride
U098	1,1-Dimethylhydrazine
U099	1,2-Dimethylhydrazine
U101	2,4-Dimethylphenol
U102	Dimethyl phthalate
U103	Dimethyl sulfate
U105	2,4-Dinitrotoluene
U106	2,6-Dinitrotoluene
U107	Di-n-octyl phthalate
U108	1,4-Dioxane
U109	1,2-Diphenylhydrazine
U110	Dipropylamine (I)
U111	Di-N-propylnitrosamine
U001	Ethanal (I)
U174	Ethanamine, N-ethyl-N-nitroso-
U067	Ethane, 1,2-dibromo-
U076	Ethane, 1,1-dichloro-
U077	Ethane, 1,2-dichloro-
U114	1,2-Ethanediylbiscarbamodithoic acid
U131	Ethane, 1,1,1,2,2,2-hexachloro-
U024	Ethane, 1,1'-[methylenebis(oxy)]bis(2-chloro-
U003	Ethanenitrile (I,T)
U117	Ethane, 1,1'-oxybis- (I)
U025	Ethane, 1,1'-oxybis(2-chloro-
U184	Ethane, pentachloro-
U208	Ethane, 1,1,2-tetrachloro-
U209	Ethane, 1,1,2,2-tetrachloro-
U218	Ethanethioamide
U247	Ethane, 1,1,1-trichloro-2,2-bis(p-methoxyphenyl)
U227	Ethane, 1,1,2-trichloro-
U043	Ethene, chloro-
U042	Ethene, 2-chloroethoxy-
U078	Ethene, 1,1-dichloro-
U079	Ethene, trans-1,2-dichloro-
U210	Ethene, 1,1,2,2-tetrachloro-
U173	Ethanol, 2,2'-(nitrosoimino)bis-
U004	Ethanone, 1-phenyl-
U006	Ethanoyl chloride (C,R,T)
U112	Ethyl acetate (I)
U113	Ethyl acrylate (I)
U238	Ethyl carbamate (urethan)
U038	Ethyl 4,4'-dichlorobenzilate
U114	Ethylenebis(dithiocarbamic acid)
U067	Ethylene dibromide
U077	Ethylene dichloride
U115	Ethylene oxide (I,T)
U116	Ethylene thiourea
U117	Ethyl ether (I)
U076	Ethylidene dichloride
U118	Ethylmethacrylate
U119	Ethyl methanesulfonate
U139	Ferric dextran
U120	Fluoranthene
U122	Formaldehyde

Hazardous waste No.	Substance
U123	Formic acid (C,T)
U124	Furan (I)
U125	2-Furancarboxaldehyde (I)
U147	2,5-Furandione
U213	Furan, tetrahydro- (I)
U125	Furfural (I)
U124	Furfuran (I)
U206	D-Glucopyranose, 2-deoxy-2(3-methyl-3-nitrosoureido)-
U126	Glycidylaldehyde
U163	Guanidine, N-nitroso-N-methyl-N'nitro-
U127	Hexachlorobenzene
U128	Hexachlorobutadiene
U129	Hexachlorocyclohexane (gamma isomer)
U130	Hexachlorocyclopentadiene
U131	Hexachloroethane
U132	Hexachlorophene
U243	Hexachloropropene
U133	Hydrazine (R,T)
U086	Hydrazine, 1,2-diethyl-
U098	Hydrazine, 1,1-dimethyl
U099	Hydrazine, 1,2-dimethyl
U109	Hydrazine, 1,2-diphenyl-
U134	Hydrofluoric acid (C,T)
U134	Hydrogen fluoride (C,T)
U135	Hydrogen sulfide
U096	Hydroperoxide, 1-methyl-1-phenylethyl- (R)
U136	Hydroxydimethylarsine oxide
U116	2-Imidazolidinethione
U137	Indeno[1,2,3-cd]pyrene
U139	Iron dextran
U140	Isobutyl alcohol (I,T)
U141	Isosafrole
U142	Kepone
U143	Lasiocarpine
U144	Lead acetate
U145	Lead phosphate
U146	Lead subacetate
U129	Lindane
U147	Maleic anhydride
U148	Maleic hydrazide
U149	Malononitrile
U150	Melphalan
U151	Mercury
U152	Methacrylonitrile (I,T)
U092	Methanamine, N-methyl- (I)
U029	Methane, bromo-
U045	Methane, chloro- (I,T)
U046	Methane, chloromethoxy-
U068	Methane, dibromo-
U080	Methane, dichloro-
U075	Methane, dichlorodifluoro-
U138	Methane, iodo-

C-2.2   Hazardous Waste *cont.*

Hazardous waste No.	Substance
U170	p-Nitrophenol
U171	2-Nitropropane (I)
U172	N-Nitrosodi-n-butylamine
U173	N-Nitrosodiethanolamine
U174	N-Nitrosodiethylamine
U111	N-Nitroso-N-propylamine
U176	N-Nitroso-N-ethylurea
U177	N-Nitroso-N-methylurea
U178	N-Nitroso-N-methylurethane
U179	N-Nitrosopiperidine
U180	N-Nitrosopyrrolidine
U181	5-Nitro-o-toluidine
U193	1,2-Oxathiolane, 2,2-dioxide
U058	2H-1,3,2-Oxazaphosphorine, 2 [(bis(2-chloroethyl)amino]tetra-hydro-, oxide 2-
U115	Oxirane (I,T)
U041	Oxirane, 2-(Chloromethyl)-
U182	Paraldehyde
U183	Pentachlorobenzene
U184	Pentachloroethane
U185	Pentachloronitrobenzene
U242	Pentachlorophenol
U186	1,3-Pentadiene (I)
U187	Phenacetin
U188	Phenol
U048	Phenol, 2-chloro-
U039	Phenol, 4-chloro-3-methyl-
U081	Phenol, 2,4-dichloro-
U082	Phenol, 2,6-dichloro-
U101	Phenol, 2-4-dimethyl-
U170	Phenol, 4-nitro-
U242	Phenol, pentachloro-
U212	Phenol, 2,3,4,6-tetrachloro-
U230	Phenol, 2,4,5-trichloro-
U231	Phenol, 2,4,6-trichloro-
U137	1,10-(1,2-phenylene)pyrene
U145	Phosphoric acid, Lead salt
U087	Phosphorodithioic acid 0,0-diethyl-, S-methylester
U189	Phosphorus sulfide (R)
U190	Phthalic anhydride
U191	2-Picoline
U192	Pronamide
U194	1-Propanamine (I,T)
U110	1-Propanamine, N-propyl- (I)
U066	Propane, 1,2-dibromo-3-chloro-
U149	Propanedinitrile
U171	Propane, 2-nitro- (I)
U027	Propane, 2,2'-oxybis[2-chloro-
U193	1,3-Propane sultone
U235	1-Propanol, 2,3-dibromo-, phosphate (3:1)
U126	1-Propanol, 2,3-epoxy-
U140	1-Propanol, 2-methyl- (I,T)
U002	2-Propanone (I)

Hazardous waste No.	Substance
U119	Methanesulfonic acid, ethyl ester
U211	Methane, tetrachloro-
U121	Methane, trichlorofluoro-
U153	Methanethiol (I,T)
U225	Methane, tribromo-
U044	Methane, trichloro-
U121	Methane, trichlorofluoro-
U123	Methanoic acid (C,T)
U036	4,7-Methanoindan, 1,2,4,5,6,7,8,8-octachloro-3a,4,7,7a-tetrahydro-
U154	Methanol (I)
U155	Methapyrilene
U247	Methoxychlor.
U154	Methyl alcohol (I)
U029	Methyl bromide
U186	1-Methylbutadiene (I)
U045	Methyl chloride (I,T)
U156	Methyl chlorocarbonate (I,T)
U226	Methychloroform
U157	3-Methylcholanthrene
U158	4,4'-Methylenebis(2-chloroaniline)
U132	2,2'-Methylenebis(3,4,6-trichlorophenol)
U068	Methylene bromide
U080	Methylene chloride
U122	Methylene oxide
U159	Methyl ethyl ketone (I,T)
U160	Methyl ethyl ketone peroxide (R,T)
U138	Methyl iodide
U161	Methyl isobutyl ketone (I)
U162	Methyl methacrylate (I,T)
U163	N-Methyl-N'-nitro-N-nitrosoguanidine
U161	4-Methyl-2-pentanone (I)
U164	Methylthiouracil
U010	Mitomycin C
U059	5,12-Naphthacenedione, (8S-cis)-8-acetyl-10-[(3-amino-2,3,6-trideoxy-alpha-L-lyxo-hexopyranosyl)oxyl]7,8,9,10-tetrahydro-6,8,11-trihydroxy-1-methoxy-
U165	Naphthalene
U047	Naphthalene, 2-chloro-
U166	1,4-Naphthalenedione
U236	2,7-Naphthalenedisulfonic acid, 3,3'-[(3,3'-dimethyl-(1,1'-biphenyl)-4,4'diyl)]-bis (azo)bis(5-amino-4-hydroxy)-, tetrasodium salt
U166	1,4-Naphthaquinone
U167	1-Naphthylamine
U168	2-Naphthylamine
U167	alpha-Naphthylamine
U168	beta-Naphythylamine
U026	2-Naphthylamine, N,N'-bis(2-chloromethyl)-
U169	Nitrobenzene (I,T)

C-2.2   Hazardous Waste *cont.*

Hazardous waste No.	Substance	Hazardous waste No.	Substance
U007	2-Propenamide	U232	2,4,5-T
U084	Propene, 1,3-dichloro-	U207	1,2,4,5-Tetrachlorobenzene
U243	1-Propene, 1,1,2,3,3,3-hexachloro-	U208	1,1,1,2-Tetrachloroethane
U009	2-Propenenitrile	U209	1,1,2,2-Tetrachloroethane
U152	2-Propenenitrile, 2-methyl- (I,T)	U210	Tetrachloroethylene
U008	2-Propenoic acid (I)	U212	2,3,4,6-Tetrachlorophenol
U113	2-Propenoic acid, ethyl ester (I)	U213	Tetrahydofuran (I)
U118	2-Propenoic acid, 2-methyl-, ethyl ester	U214	Thallium (I) acetate
U162	2-Propenoic acid, 2-methyl-, methyl ester (I,T)	U215	Thallium (I) carbonate
		U216	Thallium (I) chloride
U233	Propionic acid, 2-(2,4,5-trichlorophenoxy)-	U217	Thallium (I) nitrate
		U218	Thioacetamide
U194	n-Propylamine (I,T)	U153	Thiomethanol (I,T)
U083	Propylene dichloride	U219	Thiourea
U196	Pyridine	U244	Thiram
U155	Pyridine, 2[(2-(dimethylamino)-2-thenylamino]-	U220	Toluene
		U221	Toluenediamine
U179	Pyridine, hexahydro-N-nitroso-	U223	Toluene diisocyanate (R,T)
U191	Pyridine, 2-methyl-	U222	O-Toluidine hydrochloride
U164	4(1H)-Pyrimidinone, 2,3-dihydro-6-methyl-2-thioxo-	U011	1H-1,2,4-Triazol-3-amine
		U226	1,1,1-Trichloroethane
U180	Pyrrole, tetrahydro-N-nitroso-	U227	1,1,2-Trichloroethane
U200	Reserpine	U228	Trichloroethene
U201	Resorcinol	U228	Trichloroethylene
U202	Saccharin and salts	U121	Trichloromonofluoromethane
U203	Safrole	U230	2,4,5-Trichlorophenol
U204	Selenious acid	U231	2,4,6-Trichlorophenol
U204	Selenium dioxide	U232	2,4,5-Trichlorophenoxyacetic acid
U205	Selenium disulfide (R,T)	U234	sym-Trinitrobenzene (R,T)
U015	L-Senine, diazoacetate (ester)	U182	1,3,5-Trioxane, 2,4,5-trimethyl-
U233	Silvex	U235	Tris(2,3-dibromopropyl) phosphate
U089	4,4'-Stilbenediol, alpha,alpha'-diethyl-	U236	Trypan blue
		U237	Uracil, 5[bis(2-chloromethyl)amino]-
U206	Streptozotocin	U237	Uracil mustard
U135	Sulfur hydride	U043	Vinyl chloride
U103	Sulfuric acid, dimethyl ester	U239	Xylene (I)
U189	Sulfur phosphide (R)	U220	Yohimban-16-carboxylic acid, 11,17-dimethoxy-18-[(3,4,5-trimethoxy-benzoyl)oxy]-, methyl ester.
U205	Sulfur selenide (R,T)		

APPENDIX C-3 Known Carcinogens

Note: This is a selection of legally-controlled or highly potent or common materials which might give a special risk in the waste disposal operation. Fuller lists of substances with a possible cancer hazard in regular use are available from the occupational health agencies, and the cancer research institutes. Some materials not yet tested may in future prove to be carcinogens.

1. Substances controlled under the UK 'Carcinogenic Substances Act 1967'

   4-Aminobiphenyl
   Benzidine
   Dianisidine
   3,3'-Dichlorobenzidine
   1-Naphthylamine
   2-Naphthylamine
   4-Nitrobiphenyl
   o-Toluidine

2. Substances controlled under US Occupational Safety and Health Administration Regulations, 1974 onwards.

   2-Acetylaminofluorene
   Acrylonitrile
   4-Aminobiphenyl
   Asbestos
   Benzidine
   bis-Chloromethyl ether
   3,3'-Dichlorobenzidine (and its salts)
   4-Dimethylaminoazobenzene
   Ethylenimine
   Inorganic arsenic
   Methyl chloromethyl ether
   4,4'-Methylene-bis-2-chloroaniline
   α-Naphthylamine
   β-Naphthylamine
   4-Nitrobiphenyl
   N-Nitrosodimethylamine
   β-Propiolactone
   Vinyl chloride

3. Substances listed by the American Conference of Governmental Industrial Hygienists as human carcinogens (in Appendix A to the list of Threshold Limit Values)

Acrylonitrile
4-Aminodiphenyl   (p-Xenylamine)
Asbestos (all forms)
Benzidine
bis(Chloromethyl) ether
Chloromethyl methyl ether
Coal tar pitch volatiles
1,2-Dibromoethane   (Ethylene dibromide)
β-Naphthylamine
4-Nitrodiphenyl
Vinyl Chloride

4. Known human carcinogens, for which the U.K. Health and Safety Executive has assigned a Control Limit or Occupational Exposure Limit

Acrylonitrile
Asbestos (all forms)
Coal tar pitch volatiles
Vinyl chloride

5. Some experimental carcinogens. Chemicals not given in lists 1 to 4, which are used to induce tumours in experimental animals, and which should therefore be considered potent carcinogens.

Benz(a)pyrene
Diazomethane
1,4-Dichlorobutene
7,12-Dimethylbenzanthracene
Ethyl carbamate
Ethyl diazoacetate
Ethylenethiourea
Ethyl N-nitrosocarbamate
Hexamethylphosphoramide
Iodomethane
3-Methylcholanthrene
Methyl ethanesulphonate
N-Nitrosomethylurea
Propyleneimine

6. Any chemical labelled by the manufacturer as a carcinogen for experimental purposes.

APPENDIX C-4 Explosive Chemicals

The following may be supplied as laboratory reagents, pharmaceuticals, or polymer components. However, they are in fact explosives. Appropriate care should be taken in storage and disposal, especially if they have deteriorated in any way.

acetylene	picramide
acetyl peroxide	picric acid
ammonium nitrate	picryl chloride
ammonium picrate	picryl sulphonic acid
benzoyl peroxide	propargyl bromide
cumene peroxide	succinic peroxide
dinitrophenylhydrazine	trinitroanisole
dipicrylamine	trinitrobenzene
dipicryl sulphide	trinitrobenzene sulphonic acid
ethylene oxide	trinitrobenzoic acid
lauric peroxide	trinitrocresol
methyl ethyl ketone peroxide	trinitronaphthalene
nitrogen trifluoride	trinitrophenol
nitroglycerin	trinitroresorcinol
nitroguanidine	trinitrotoluene
nitromethane	urea nitrate

APPENDIX C-5 Named Reagents

A select list of compounds and mixtures used in biological and chemical work, which may be stored and labelled with a non-chemical name, e.g. Bouin's Soution, Tollen's Reagent. The following have special hazards which must be controlled in the waste disposal operation. The hazards are given as a letter code after the name. The magnitude of the hazard may vary with the amount, condition, and exact formulation of each sample.

Adams (F)	Flemmings (O)	Millon (M)
	FWA Fluid (O)	
Bensley (M)		Nessler (M)
Bouin (P)	Hager (P)	
Bray (E)	Hanus (c)	Raney Nickel (F)
Brucke (M)	Hubl (M)	Rossman (P)
Caro (E, V)	Jeffery (V)	Schauding (M)
Champy (O)		Schulze (V)
Cour = La Cour (O)	Karl Fischer (C)	Susa (M)
	Kharasch (E)	
Denige (M)		Tauber (B)
	La Cour (O)	Tollen (E)
Esbach (P)		Wij (C)
	Maclean (M)	
Fischer (C)	Mayer (M)	Zenker (M)

Key
(B) Contains benzidine, a severe carcinogen.
(C) Corrosive by skin contact; very harmful fumes: halogens in organic solvents.
(E) Explosive, possibly shock-sensitive if stored for any length of time.
(F) May be spontaneously flammable in air — possibly after an induction period.
(M) Contain highly toxic mercury compounds.
(O) Contain osmic acid, which is a severe poison; dangerous by skin and eye contact.
(P) Contain picric acid. Dried-up reagents may be explosive. Disposal to drains may build up explosive deposits in the drain.
(V) Strong oxidants. Highly corrosive and liable to react violently with organic material.

APPENDIX C-6 Deteriorated Chemicals

The following is a selection of chemical substances which can deteriorate to a dangerous condition with age, under common storage conditions. The degree of the hazard will vary considerably with age and the exact situation, but it is advisable to take precautions when discarding, recycling or otherwise handling old samples.

Acetal (3)
Acetaldehyde diethyl acetal (3)
2-Acetyl furan (3)
Acetyl peroxide (1)
Aluminium chloride (5)
Aluminium lithium hydride (5)
Ammonia solution (5)
Ammonium dichromate (4)
Ammonium hyroxide (5)
Ammonium persulphate (5)
Anethole (3)
Anisaldehyde (3)
Anisole (3)
Anisyl chloride (5)
Aqua regia (5)

Benzenesulphonyl chloride (5)
Benzoyl peroxide (1)
Bleach (5)
Bleaching powder (5)
2-(2-Butoxyethoxy)ethyl acetate (3)

2-Butoxyethyl acetate (3)
t-Butyl hydroperoxide (4)
iso-Butyl ether (2)
n-Butyl ether (3)
n-Butyl glycidyl ether (3)

Calcium carbide (5)
Calcium hydride (5)
Calcium hypochlorite (5)
Cellosolve (3)
Chloroform (5)
Chromic acid (5)
Chromium trioxide (4)
Cleaning mixtures (5)
Cumene (3)
Cumene hydroperoxide (5)
Cyclohexene (3)
Cyclopentadiene (3)
Cyclopentene (3)
Decahydronaphthalene (3)
Decalin (3)

Di-allyl ether (3)
Di-iso-amyl ether (3)
Dibenzyl ether (3)
Di-iso-butyl ether (2)
Di-n-butyl ether (3)
Dicyclopentadiene (3)
1,1-Diethoxyethane (3)
Diethylacetal (3)
Diethyl azidoformate (4)
Diethyl azodicarboxylate (1)
Diethylene glycol dimethyl ether (3)
Diethyl ether (3)
Diglyme (3)
Dihydropyran (3)
1,2-Dimethoxyethane (3)
Dimethoxymethane (3)
Dimethylamine (5)
2,4-Dinitrophenol (1)
2,4-Dinitrophenylhydrazine (1)
1,4-Dioxan (3)
Diphenyl ether (3)
Di-iso-propyl ether (2)
Di-n-propyl ether (3)

Ether (3)
Ethyl cellosolve (3)
Ethylene glycol dimethyl ether (3)
Ethylene glycol ethyl ether acetate (3)
Ethylene glycol monobutyl ether (3)
Ethylene glycol monoethyl ether (3)
Ethylene glycol monomethyl ether (3)
Ethyl ether (3)
2-Ethoxyethanol (3)
2-Ethoxyethyl acetate (3)
Ethyl vinyl ether (2)

Furan (3)

Glycidyl n-butyl ether (3)
Glyme (3)

Hydrogen peroxide (5)

Iodine pentoxide (4)
Isoamyl ether (3)

Isobutyl ether (2)
Isopentyl ether (3)
Isopropyl alcohol (3)
Isopropyl ether (2)
Isopropyl benzene (3)

Lauroyl peroxide (5)
Lithium aluminium hydride (5)
Lithium hydride (5)

Magnesium perchlorate (4)
Mercury fulminate (1)
2-Methoxyethanol (3)
Methylal (3)
Methyl cellosolve (3)
Methyl iso-butyl ketone (3)
Methyl ethyl ketone peroxide (1)
Methyl vinyl ketone (3)

Nitric acid (5)
Nitromethane (1)
Nitrosoguanidine (5)

Peracetic acid (1,4,5)
Perchloric acid (4)
Phosphorus trichloride (5)
Picric acid (1)
Picryl chloride (1)
Picryl sulphonic acid (1)
Potassium (metal) (1)
Potassium amide (1)
Potassium chlorate (4)
Potassium perchlorate (4)
Potassium persulphate (5)
Propan-2-ol (3)
Propargyl bromide (1)
Propargyl chloride (1)

Silicon tetrachloride (5)
Silvering solution (1)
Sodamide (1)
Sodium amide (1)
Sodium borohydride (5)
Sodium chlorate (4)
Sodium chlorite (4)

Sodium dithionite (5)
Sodium hydride (5)
Sodium hydrosulphite (5)
Sodium hypochlorite (5)
Sodium metal dispersions (1)
Sodium perchlorate (4)
Sodium peroxide (5)
Sodium persulphate (5)
Styrene (3)

Tetrahydrofuran (3)
Tetralin (3)

Thionyl chloride (5)
Trinitrobenzene (1)
Trinitrobenzene sulphonic acid (1)

Urea nitrate (4)
Urea peroxide (5)

Vinyl acetate (3)
Vinylidene chloride (1)
Vinyl pyridine (3)

Zinc (5)

See, also, Appendices C-4 and C-5

Key
(1) Can deteriorate to a shock-sensitive explosive. Take exceptional care if there is evidence of drying out, crystallization or contamination. It may be very dangerous to attempt to open the container.
(2) Forms peroxides, especially on exposure to air and light, making the material liable to explode. This class is so dangerous that it should not normally be distilled unless it has been very well controlled. Material more than one year old should be discarded, even if unopened. Containers should not be opened if there is any solid visible around the closure or any evidence of crystals inside.
(3) Also forms peroxides. If very old or obviously in poor condition treat as (2). Otherwise take care to test for peroxides before use or recovery procedures.
(4) High energy materials which are sensitive to the presence of dust. Clean the outside of containers before opening. If in doubt, do not open. Mixtures of the material with dust, paper or organics may ignite or detonate when exposed to friction, e.g. on the threads of a screw-capped container.
(5) Containers may have a high internal gas pressure, owing to decomposition. Open carefully behind a safety shield in a fume cupboard.
     (N.B. A high internal pressure can also result from biological decay, radioactive decay or corrosion of metal containers.)

APPENDIX C-7 Low-hazard Solids for Land Disposal

The following substances are supplied as laboratory chemicals, but are of such low-hazard potential that current legislation is unlikely to restrict their disposal. They are of such low solubility that they are unsuitable for disposal to drains, but should be acceptable for land disposal along with ordinary refuse. This may *not* apply to substances used as absorbents or otherwise contaminated with other materials. See section 9.3.2.

Agar
Agarose
Alumina

Aluminium*
Aluminium hydroxide
Aluminium oxide

Aluminium silicate
Aluminium stearate
Anhydrite
Anti-bumping granules

Bauxite
Beeswax
Bentonite
Borax

Calcite
Calcium carbonate
Calcium fluoride
Calcium phosphate
Calcium stearate
Calcium sulphate
Canada balsam
Carbon
Carborundum
Carnauba
Casein
Cellulose
Cement
Charcoal
Chromatographic supports
Clay
Cotton wool

Diatomaceous earth
Dutch metal

Emery
Eschka's Mixture

Ferric oxide
Ferric phosphate
Ferrosilicon
Ferrous oxide
Filter aid
French chalk
Fuller's earth

Gelatin
Glass (beads, wool)
Graphite
Gypsum

Iceland spar
Ion-exchange resin
Iron*
Iron(II) oxide
Iron(III) oxide
Iron ore

Kaolin
Kieselguhr

Lampblack
Lanolin
Lecithin
Lithium carbonate
Lloyd's reagent
Lycopodium

Magnesia
Magnesium carbonate
Magnesium fluoride
Magnesium oxide
Marble
Methyl cellulose
Mica
Molecular sieve

Nylon

Paraffin wax
Pectin
Polycarbonate
Polyester
Polyethylene
Poly(methyl methacrylate)
Polypropylene
Polystyrene
Polytetrafluorethylene
Polyvinyl acetate
Polyvinyl alcohol
Polyvinyl chloride
Pumice

Quartz*

Rouge
Rubber

Sand
Shellac
Silica*
Silica gel
Silicon
Silicon carbide
Silicon dioxide*
Silicon monoxide
Soda glass (beads, wool)
Sodium metaphosphate (insoluble)
Sodium metasilicate (solid)
Sodium silicate
Soya bean meal
Starch

Talc
Titanium dioxide

Wax: beeswax
      carnauba
      paraffin
      polyester
      polyethylene

Yeast

Zinc carbonate
Zinc oxide
Zinc phosphate

---

*Not acceptable in the form of a fine powder

APPENDIX C-8 Properties of Gases for Venting

The following are commercial gases, with the hazards involved in venting and reference to the disposal section 6.5.3.

Gas	Hazards	Disposal
acetylene	F (C)	4
allene	F (C)	4
ammonia	C	6
argon	—	3
arsine	T F	5
boron trichloride	C	6
boron trifluoride	C	6
bromomethane	T (C)	6
bromotrifluoroethylene	F T (C)	5
bromotrifluoromethane	(T)	3
buta-1,3-diene	F (T)	4
butane	F	4
but-1-ene	F	4
cis-but-2-ene	F	4
trans-but-2-ene	F	4
but-1-yne	F (C)	4
carbon dioxide	(T)	3
carbon monoxide	T F	5
carbon tetrafluoride	(T)	3

Gas	Hazards	Disposal
carbonyl fluoride	C	6
carbonyl sulphide	F T	5
chlorine	C	6
chlorine trifluoride	C	6
1-chloro-1,1-difluoroethane	F (T)	4
chlorodifluoromethane	(T)	3
chloroethane	F (T)	4
chloromethane	F C	6
chloropentafluoroethane	(T)	3
2-chloropropene	F (T)	4
chlorotrifluoroethylene	F T	5
chlorotrifluoromethane	(T)	3
cyanogen	T F	5
cyclopropane	F T	5
deuterium	F	4
diborane	F T	5
dibromodifluoromethane	(T)	3
dichlorodifluoromethane	(T)	3
dichlorofluoromethane	(T)	3
1,2-dichlorotetrafluoroethane	(T)	3
difluorodiazine	T	5
1,1-difluoroethane	F (T)	4
1,1-difluoroethylene	F (T)	4
dimethylamine	F C	6
dimethyl ether	F (T)	4
2,2-dimethylpropane	F	4
ethane	F	4
ethylacethylene	F (C)	4
ethylamine	F C	6
ethyl chloride	F (T)	4
ethylene	F	4
ethylene oxide	F T	5
ethyl fluoride	F (T)	4
fluorine	C	6
fluoromethane	F (T)	4
fluoroethane	F T	5
freon	(T)	3
germane	F T	5
halocarbon (halon)		
— included in Table 6.3	F (T)	4
— not included in Table 6.3	(T)	3

Gas	Hazards	Disposal
helium	—	3
hexafluoroacetone	T	5
hexafluoroethane	(T)	3
hexafluoropropene	T	5
hydrogen	F	4
hydrogen bromide	C	6
hydrogen chloride	C	6
hydrogen cyanide	T F	5
hydrogen fluoride	C	6
hydrogen iodide	C	6
hydrogen selenide	T F	5
hydrogen sulphide	T F (C)	5
hydrogen telluride	T F	5
iodine pentafluoride	C	6
isobutane	F	4
isobutylene	F (T)	4
krypton	—	3
MAPP gas	F	4
methane	F	4
methanethiol	T F	5
methyl acetylene	F (C)	4
methylamine	F C	6
methyl bromide	T (C)	6
2-methylbut-1-ene	F (T)	4
2-methylbut-2-ene	F (T)	4
3-methylbut-1-ene	F (T)	4
methyl chloride	F C	6
methyl fluoride	F T	5
methyl mercaptan	T F	5
2-methylpropane	F	4
2-methylpropene	F	4
methyl vinyl ether	F (T)	4
natural gas	F	4
neon	—	3
nitric oxide	C	6
nitrogen	—	3
nitrogen dioxide	C	6
nitrogen trifluoride	C F	6
nitrogen trioxide	C	6
nitrosyl chloride	C	6
nitrous oxide	(T)	3

Gas	Hazards	Disposal
octafluorobut-2-ene	(T)	3
octafluorocyclobutane	(T)	3
octafluoropropane	(T)	3
oxygen	—	3
ozone	T C	6
perfluorobut-2-ene	T	5
perfluoropropane	(T)	3
phosgene	T C	6
phosphine	T F	5
phosphorus pentafluoride	C	6
propane	F	4
propene	F	4
propylene	F	4
propylene oxide	T F	5
propyne	F (C)	4
silane	F T	5
silicon tetrachloride	C	6
silicon tetrafluoride	C	6
stibine	T F	5
sulphur dioxide	C	6
sulphur hexafluoride	(T)	3
sulphur tetrafluoride	C	6
sulphuryl fluoride	T (C)	5
tetrafluoroethylene	F (T)	4
tetrafluorohydrazine	F T	5
town gas	F T	5
trichlorofluoromethane	(T)	3
trifluoroiodomethane	T	5
trifluoromethyliodide	T	5
trifluoromethane	(T)	3
trimethylamine	F (T) (C)	6
tungsten hexafluoride	T	5
vinyl bromide	F (T) (C)	5
vinyl chloride	F T (C)	5
vinyl fluoride	F (T) (C)	5
vinyl methyl ether	F (T)	4
xenon	—	3

Key	*Hazards*		*Disposal*	
	C	corrosive	3	see section 6.5.3.3
	(C)	corrosive to a few materials	4	see section 6.5.3.4
	F	flammable	5	see section 6.5.3.5
	T	toxic	6	see section 6.5.3.6
	(T)	toxic in larger amounts		

APPENDIX C-9 List of Chemical Tables Elsewhere in the Book

Table	Title	Page
6.3	Flammable Halocarbon Gases of Low to Moderate Toxicity	101
9.1	Possible Hazardous Reactions in Drains	140
9.2	Quality of Water Discharged to Sewer	142
9.3	Concentrations of Chemicals to be Poured Down a Laboratory Drain	143
9.4	Oxidizing Conditions for Selected Groups of Substances	158
14.1	Hazards in Solvent Recovery	272
14.4	Solvents Which Should Not be Recovered by Distillation	285

APPENDIX D

Glossary of Abbreviations

There follows a selection of abbreviations in use in the UK and the USA, in publications relating to waste disposal.

ACGIH	American Conference of Governmental Industrial Hygienists
ACS	American Chemical Society
AEC	Atomic Energy Commission (US)
AERE	Atomic Energy Research Establishment (UK)
AIChE	American Institute of Chemical Engineers
ANSI	American National Standards Institute
ASE	Association for Science Education (UK)
ASTM	American Society for Testing and Materials
BOD	Biochemical Oxygen Demand
BSI	British Standards Institute
CAG	Carcinogen Assessment Group, EPA (USA)
CAS	Chemical Abstracts Service
CCBW	Chemically Contaminated Biological Waste (USA)
CDC	Center for Disease Control (USA)
CFR	Code of Federal Regulations (USA)
CIA	Chemical Industries Association (UK)
CMA	Chemical Manufacturers Association (USA)

COD	Chemical Oxygen Demand
CoP	Code of Practice
CoPA	Control of Pollution Act 1974 (UK)
CSA	Cancer Suspect Agent
DE	Department of Energy (USA)
DES	Department of Education and Science (UK)
DHEW	Department of Health, Education, and Welfare (USA)
DHSS	Department of Health and Social Security (UK)
DoE	Department of the Environment (UK)
DoT	Department of Transportation (USA)
EEC	European Economic Community
EPA	Environmental Protection Agency (USA)
FDA	Food and Drug Administration (USA)
FP	Flash Point
FR	Federal Regulations (USA)
HAZCHEM	Label system for Hazardous Chemicals (UK)
HEPA	High Efficiency Particulate Air Filter
HMSO	Her Majesty's Stationary Office (UK)
HMTA	Hazardous Materials Transportation Act 1975 (USA)
HSC	Health and Safety Commission (UK)
HSE	Health and Safety Executive (UK)
HSWA	Health and Safety at Work Act 1974 (UK)
HWMF	Hazardous Waste Management Facility (USA)
HWN	Hazardous Waste Number (USA)
IAEA	International Atomic Energy Agency
IARC	International Agency for Research on Cancer
IChemE	Institution of Chemical Engineers (UK)
ICRP	International Commission on Radiological Protection
ILO	International Labour Office
IMS	Industrial Methylated Spirits (ethanol)
IPA	Iso-Propyl Alcohol (propan-2-ol)
LC_{50}	Lethal Concentration for 50% of exposed subjects
LD_{50}	Lethal Dose for 50% of exposed subjects
lfm	Linear feet per minute (velocity)
LLW	Low Level Waste (radio-active)
MAC	Maximum Allowable Concentration (USA)*
MCA	Manufacturing Chemists' Association (now CMA) (USA)
MRC	Medical Research Council (UK)
MUF	Material Unaccounted For

NA	North American number of chemical substance
NBS	National Bureau of Standards (USA)
NCRP	National Committee on Radiation Protection and Measurement (USA)
NFPA	National Fire Protection Association (USA)
NHS	National Health Service (UK)
NIH	National Institutes of Health (USA)
NIOSH	National Institute for Occupational Safety and Health (USA)
n.o.s.	not otherwise specified
NRC	(1) National Research Council (USA)
	(2) National Response Center (USA)
	(3) Non-Reusable Container (USA)
	(4) Nuclear Regulatory Commission (USA)
NRPB	National Radiological Protection Board (UK)
NSF	National Sanitation Foundation (USA)
NTIS	National Technical Information Service (USA)
OEL	Occupational Exposure Limit (UK)*
ORM	Other Regulated Material (USA)
OSH Act	Occupational Safety and Health Act 1970 (USA)
OSHA	Occupational Safety and Health Administration (USA)
PCB	(1) Poly-Chlorinated Biphenyl
	(2) Printed Circuit Board
PEL	Permitted Exposure Limit (USA)*
PHLS	Public Health Laboratory Service (UK)
PHS	Public Health Service (USA)
POHC	Principal Organic Hazardous Constituent
PV	Permanganate Value
RCRA	Resource Conservation and Recovery Act 1976 (USA)
RQ	(minimum) Reportable Quantity (USA)
RSC	Royal Society of Chemistry (UK)
s.s.	suspended solids
STC	Single Trip Container
STEL	Short Term Exposure Limit*
TDI	Tolyl Di-Isocyanate
t.d.s.	total dissolved solids
TDSF	Treatment, Storage or Disposal Facility (USA)
TLV	Threshold Limit Value (USA)*
t.o.c.	total organic carbon
TOSCA TSCA	Toxic Substances Control Act 1976 (USA)

t.s.s.	total suspended solids.
TWA	Time Weighted Average*
UFAW	Universities Federation for Animal Welfare (UK)
UN	United Nations number for chemical substance
USPHS	U.S. Public Health Service
WA	Water Authority (UK)
WDA	Waste Disposal Authority (UK)
WHO	World Health Organization

*For discussion of these terms, see section 6.1.2.

Index

Key page references for particular entries are distinguished by bold type. (t) is put after a page number when the reference is to a table.

abbreviations, 344–347
absorbents, 229, 293–296, 303, 315
accidents, 255–256
   see also emergencies, spillages
acetone, 33, 271, 273, 282, 285
acetonitrile, 168
acids, 141, **160–162**, 296–298
acrylic
   chemical hazard, 272
   plastic, 67, 88
acrylonitrile, 76, 156, 168, 332, 333
activity, specific, 208
addresses, 308–313
adsorbents, 149, 216, 274
aerodynamic fume cupboards, 82
aerosol cans, 41–44
agar (autoclave hazard), 180
agencies, Government, 49–50, 133, 206
AIDS (Auto-Immune Deficiency Syndrome), 191, 205
air
   changes, 79
   conditioning, 80
   velocity, *see* face velocity
alcohol, 12, 182–183, 282–283
aldehydes, 141, 272, 298
   see also formaldehyde, glutaraldehyde

algae, 176, 195
alkali, 141, **160–162**, 223, 296–298
alkaline cleansers, 182, 317(t)
allergens, 133, **189–190**, 199
alpha radiation, 207, 208, 214, 220
alumina, activated, 149, 223, 274, 285–
   286
aluminium, 33, 41, 164, 337
   hazards, 33, 41, 141(t)
amides, 141(t), 156, 165–166
amines, 140–141(t), 273(t)
ammonia, 103, 140(t), 142(t), 145, 161,
   166
ammonium salts, 140–141(t)
amorphous silicate, 294, 315(t)
ampholytic detergents, 182, 195
ampoules, 239
anaesthetics, 174, 195, 232
aniline, 287
   see also amines
animals, 194–202
   bedding, **195–200**, 226, 233, 306
   carcasses, 201–202, 226–227
   house, 195, 198–199
   in fires, 306
   killing, 201, 306
   laws, 194

live, 200–201
antimony, 141(t), 166
anti-neoplastic drugs, 233–234
anti-pollution laws, 11–15
aqueous liquids:
    biological, 188–189
    chemical, 137–145
    radioactive, 217–221
arsenic, 141(t), 166
arson, 26, 43
asbestos, 87, 234–237
ash, incinerator, 110, 112, 114, 115, 117,
    215–216
ashtrays, 25
atomic absorption spectrophotometer
    as incinerator, 120–121
    fumes from, 81
    liquid from, 143–145
authorities
    control, **12–15**, 133, 194, 206, 237, 307
    source of information, 49–50, 131
autoanalysers, 143–145
autoclaves, steam, 27, **177–180**, 212, 253
azeotropes, 270, 271, **281–283**
azides, 64, 138, 140–141(t), 149

bacteria, 176, 177, 183–186
bagging, double, 27, 117, 199–200
bags, 21–22, **26–32**, 202, 290–291
    autoclave, 20–21, 27, 29(t), 200
    colour code, 29(t)
baling, 39, 193
basins, hand, 22–23, **54–56**, 67, 69
    radioactive, 217–219
batch distillation, 275–281, 283–284
batteries, scrap, 260
becquerel (unit), 208
bedding, animal, **195–200**, 226, 233, 306
bedpan macerators, **59–62**, 193, 200, 201
benchwork, 34–35
beta radiation, 207, 208, 214
bibliography, *see end of each chapter*
bin, waste, **26–32**, 34, 41, 208, 212–214
    liners, 26–32
    truck, 39
biochemicals, 133
biological materials, 174–205
    in fires, 304
    radioactive, 226–227
biological oxygen demand (BOD), 72,
    138, 142
biological safety cabinets, **82–84**, 86,
    90–94, 98
    fumigation, 93–94
birds, 194, 195, 198, 202

bisulphate, sodium, 161, 297
bisulphite, sodium, 157–160, 186, 267,
    298
bleach
    disinfectant, 181, **184–185**, 219, 298,
        318–319(t)
    hazards, 140–141(t), 219
    oxidant, 156–157, 158–159(t)
bleach-fix, 264
bleaching powder, 141(t), **297**, 298, 318–
    319(t)
blockages, drains, 57, 62, 69, 71
boilers, for waste disposal, 119–121
boiling water, 177, 190, 292
    in drains, 54, 65–67
bombs, oxygen, 121
bonfires, 106–108
borohydride, sodium
    destruction, 166
    for metals recovery, 261
bottle crusher, 39, 224
bottles, 32–34, 37, 39, 222
bottle traps, 56–58, 70
boxes
    return to supplier, 45
    waste collection, **32**, 36–37, 117
    see also packaging
brass
    drain fittings, 64
    scrap, 260
breathing apparatus, 77, 171, 198, 227,
    241
broken glass, 29, 35, 39, **41–44**, 192,
    202–203
bromine, 103, 186
bubble flow-meter, 146
burial
    chemicals, 133–137, 162
    radioactives, 210, 211, **214–215**, 217
burning and incineration, 106–122
    open, 106–108, 165, 233
    chemicals, 106–108, 109, **147–148**, 161–
        162, 165, 166, 167, 169, 233, 234,
        237
    radioactives, 210, **215–216**, 221–223
    see also fire, incinerators

cabinets, safety, 82–84, 86, 90–94, 98
    laminar flow, 84–85
cadmium, 135, 140(t), 261
calcium, 161–164
    carbonate, 161, 164, 297
    hydroxide, 161, 296
    hypochlorite, 297, 298, 318–319
cancer, 231–237, 3   , 331–333(t)

cans
  aerosol, 36, 41–44
  solvent, 32–34, 42, 43
captor fume systems, 80–81
carbon
  carbon-14, 214, 215, 219, 223, 258
  charcoal, 117, 227, 267, 295, 302
  graphite, 262, 303
carcasses, animal, 201–202
carcinogens, 77, **231–237**, 299, 301, 331–
    333(t)
cartridges
  filter, 96, 99, 100, 235, 267
  respirators, 77, 171, 198, 227, 235,
    241
  silver recovery units, 264
cast iron pipes, 63
catalysts, 152, **164–165**, 260, 268
catchpots, 56–58, 60
catering waste, 29(t), 109, 191–193
caustics, *see* alkali
cerium (III) hydroxide, reductant, 286
chalk (= calcium carbonate), 161, 164,
    297
charcoal, absorbent
  for mercury, 267, 295
  for solvent incineration, 117
  for spill control, 227, 295, 302
chemical oxygen demand (COD), 72,
    138, 142
  silver recovery, 263
chemicals, 128–173, 322–343(t)
  burning, 108, 147–148
    see also incineration
  containers, 45, 108, 133
    opening, 239
    unidentified, 238–240
  deteriorated, 129, 130–131, 333–337(t)
  explosive, 237–238, 238–239, 333–
    337(t)
  fires, 303–305
  laws, 130
  methods of disposal, 133–168
  packaging, 45, 108, 133
  principles of disposal, 129–133, 137–
    138
  reactions, 132, 140–141(t), 149–150,
    151–168
  recipes, 132–133, 161–168
  recovery, 254, 260–288
  separation, 148–151
  spillages, 293–302
  tables, 101, 140–143, 272–273, 285,
    322–343
  unknown, 109, 130, 238–240

water-reactive, 153–155
  *see also individual categories and sub-
    stances*
china sinks, 54
chloramine disinfectants, 186
chlorinated solvents, 33, 35, 273(t), 275
chlorine, 103, 186
chloroform, 33, 195, 272–273(t), 274
chromic acid, 157, 182
chutes, 39, 200
cigarettes, 25
civil liability, 10
citric acid, 134, 297
classes, teaching, 250–256
clay
  absorbents, 229, 294–295
    see also vermiculite
  pipes, 63
clean air stations, 84–85
cleaning:
  agents, 181–182, 224, 317(t)
  materials, 290–291
  mixture (hazard), 157, 182
  staff, 19–26, 32, 299
clear phenolic disinfectants, **183–184**,
    187, 189, 219, 320(t)
clearing out a dead store, 240–244
clinical waste, 29(t), 190–191, 192(t)
clothing
  contaminated, 199, 212–213, 256
  protective, 77, 93, 97, 116, 146–147,
    198, 227, 235, 241
cocktails, scintillation, **223–225**, 271
Code of Practice
  asbestos, 247
  autoclaves, 178–179
  carcinogens, 247
  chemicals, 132–133
  clinical waste, 47, 190–191, 204
  disinfectants, 181
  fume cupboards and safety cabinets,
    92(t)
  fumigation, 93
  radioactives, 92(t), 229
  solid waste, 47
coffee-break, 19, 24–25, 70, 191
collection of waste, 19, 26–35, 41–44, 269
colleges, 249–257
colour code, bags, 28, 29(t)
column, distillation, 276–284
combustion
  definition, 106
  principles, 110–111
  standards, 113–114(t)
  starved air, 112

various techniques, 119–121
wet, 121
see also burning, fire, incineration
commercial disposal companies, *see* con-
    tractors
compaction (= baling), 39, 193
concentration of waste
    adsorbents, 149
    evaporation, 145
    extraction, 150–151
    precipitation, 149–150
    radioactive, 211, 216, 221, 225
    separation procedures, 148–151
    solvent stripping, 148–149
consent, drainage, 14, 18
consignment notes, 14
constant-boiling mixtures, *see* azeotropes
containers, chemical, *see* chemical con-
    tainers
containment of spillages, 291–292, 299,
    300
contamination
    animals, 194–200, 306
    biological, 190–193, 301
    chemical, 132, 133, **268–269**, 301–302,
        304–305
    clothing, 212–213, 256
    drains, 69–70
    fire, 303–305, 305–307
    mercury, 256, **265–267**, 301, 305–306
    radioactive, 212–214, 227–228, 306–307
contractors, 13, 34, 47, 123–127, 304
contracts, 18, 125–126, 127
Control of Pollution Act, 1974 (UK), 13
copper
    metal, 260, 287
    pipe, 64, 218
corrosion, corrosives
    drains, 62–67, 71, 137–138
    gases, 86–88
    liquids, 103–104
    solvent cans, 33
    see also chemicals
cost, *see* economics
cotton wool, 216
crates, 45
cresol, disinfectants, **183–184**, 189, 195
criminal liability, 10
Crown Property, 14
crusher, 39, 224
cryogenic liquids, 100
curie (unit), 207
customs duty, 12, 19
cyanides, 156, 166–168
cylinders, gas, 168–171

economics, 44
in fires, 305
cytotoxic drugs, 231–234

dampers, in extract ducts, 89–90
dangerous substances, security, 26, 168
date stamping, 129
dead store, clearing out, 240–244
decontamination
    biological, 300–302
    chemical, 244, 300, 302
    clothing, 212–213, 256
    radioactive, **227–229**, 300, 302, 307
    see also cleaning agents, fumigation
delay tanks, 58–59, 212
Department of Transport (USA),
        approved containers, 13, 34, 136
destroying animals, 201, 306
destruction, *see* carcinogens, chemicals,
        incineration
detergent, *see* cleaning agents
Devarda's Alloy, 141(t)
dichromate, 160, 264, 286
dilution
    procedures, 139–143, 153–155
    tanks, 58–59
dioxin, 107, 168
discard jars, 180–186, 187–188
disinfection, 175, **180–189**, 195, 298, 300
disposables:
    clothing, 212, 235
    labware, 178, 192(t), 202–203, 223,
        226, 253
    economics, 46, 253
disposal, *see individual category*
disposal unit, sink waste, 39, 57, **59–62**,
        193, 200, 201, 253
distillation, 148, 270–275, 275–287
    azeotropes, 270, 271, 281–283
    examples, 222, 271, 274
    hazards, 272–273(t), 284, 285–287
    operation, 283–284
    peroxide removal, 160, 285–287
    pre-treatment, 274–275
    radioactives, 222, 224, 271
    theory and practice, 275–281
dithionite, 141(t)
DOT containers (USA), 13, 34, 136
double bagging, 27, 117, 199–200
drainage, drains, 41, 51–73
    arrangement, 67–70
    blockages, 57, 62, 69, 71
    chemicals disposal, 137–145
    components, 53–62
    corrosion, 62–67, 71, 137–138

flooding, 51, 53, 57, 61, 68–70, 190
hazards, 59, 64, 68–70, 71, 137–138,
  140–141(t), 219
layout and management, 67–71
leaks, 51, 53, 56, 59, 63, 69–71
limits on effluent quality, 71–72, 142(t)
maintenance, 70–71
pipes, 53–54, 62–70
radioactive, 51, 53, 58–59, 65, 69–71,
  216–221
repairs, 64, 65, 66, 69–70
drip cups, 54–56
drugs, 12, 26, 48, 191, 192(t), 20, **232–234**
drums, 32, 34, 35, 38, 39, 135–136
dust
  asbestos, 235–236
  biohazard, 79, 99, 190, 194–195
  filters, 96, 236
  metal, 141(t), 164
  radioactive, 97, 100, 211
dustbins, 34, 44, *see also* bin, waste

eating, 24–25, 191
economics, 45–47
  contractors, 123, 125, 126, 127
  disposables, 46, 253
  gas cylinders, 44
  incineration, 110, 113–116
  liquid scintillation, 224
  recycling, 46, 258–260, 274, 288
  teaching laboratories, 253–254
educational institutions, 249–257
effluent, 52–53, 137–138, 142(t)
electronic apparatus, 258, 260, 265, 305–
  306
emergencies
  legal aspects, 14
  procedures, 289–307
  services, 127
  spillages, 290–302
    radioactive, 227–229
    teaching laboratories, 255–256
endotoxins, 189–190 ‾
environmental legislation, 11–12
ethanol
  customs duty, 12, 19
  disinfectant, 182–183
  distillation, 282–283
ethers, 129, 285(t)
  disposal, 145, 222
  hazards:
    explosion, 160, 222, 224, 272–273(t),
      285–287
    rat in refrigerator, 174
    toxic, 232

peroxide removal, 160, 285–287
  recycling, 222, 224, 285(t)
ethyl alcohol, *see* ethanol
ethyl ether, *see* ethers
ethylene oxide, 102, 175, 177, 198, 203
evaporation and venting, 97–104, 145–
  147
examinations (teaching laboratories),
  254–255
excreta
  animal, 194, 198–200, 226
  human, 23–24, 184, 189, 211, 216, 219
excursions, 14
exemptions and exceptions (law), 14–15
explosives, **237–238**, 285, 333–337(t)
  see also ethers
exposure limits, 74–76
extraction, fume, 74–105
extraction, liquid, 150–151

faeces
  animal, 194, 198–200
  human, 23–24, 184, 189, 211, 216, 219
face velocity, 82, 90–91, 92(t)
  substandard, 95
  testing, 90–91, 93
fans, 61, 81, 85–86, 91, 93–95, 98
film, silver from, 260, 321
filters
  asbestos hazards, 235, 236
  cartridges, 96, 99, 100, 235, 267
  fume cupboard, 83–84, 93
  in-line, 99, 100, 267
  mercury, 266, 267
  safety cabinet, 93, 96–97, 99, 236
fire
  animals in, 306
  brigade, 98, 127, 304, 305
  clearing up after, 303–307
  combustion and incineration, 106–122
  extinguishing, 302–303
  pits, 107–108
  service, 98, 127, 304, 305
fixer baths, 263–265, 321
flammables
  gases, 101–103, 339–343(t)
  liquids:
    collection, 26, 33, 269–270
    evaporation, 145–148
    fires, 302–303
    incineration, 117–118, 119, 222
    landfill, 134–136
    recycling, 222–225, 269–288
    spillages, 293–295, 298–300
  see also ethers

flash arrestor for solvent cans, 33
flies, 28, 194, 195
flooding, 51, 53, 57, 61, 68–70, 190
floor drains, 53, 68–69
fluorescent lights, 38
food
    eating, 24–25, 191
    laboratories, 191–193
formaldehyde, 93–94, 185
formalin, 93–94, 185
foul sewer, 52–53, 54, 189, 209, 211,
    219
freezers, 202, 211, 226
freons, 99, 100–101
fume cupboards, 77–80, 82–84
    portable, 95–96
    radioactive, 92, 98, 100, 219, 225
    waste venting, **97–104**, 290, 339–342(t)
fume ducts, 85–90
    checking, 98–99, 103
    dampers, 89–90
    filters, 83–84, 93, 96–97, 99, 236
    flexible, 89
    local extract, 80–81
    maintenance, 90–97, 103
    materials, 87–89
    radioactive, 95, 98, 100, 225
    servicing, 90–97
    substandard, 95–96
    testing, 90–93, 98–99
fume extraction, 74–105
fume hazards, 74–78, 101–103, 339–
    342(t)
fume hoods, *see* fume cupboards
fume respirators, 77, 171, 198, 227, 235,
    241, 304
fume ventilation, 79–80, 304
fumigation
    cabinets, 93–94, 97, 185–186
    rooms, 185–186, 195, 304
fungi, 176, 177, 190
furnaces
    incinerators, 108–119
    laboratory, 120

galvanized iron/steel, 63, 88
gamma radiation, 187, 214
garbage, *see* refuse
gas
    cylinders, 168–171
        economics, 44
        in fires, 305
    flammable, 101–103, 339–343(t)
    toxic, 102–103, 339–343(t)
    venting, 100–103, 169–171, 339–342(t)

    radioactive, 225
    see also chemicals, fume extraction,
        fumigation
glass
    broken, 29, 35, **41–44**, 192, 202–203
    crusher, 39, 224
    drains, 65, 218
    radioactive, 213–214, 218
    repairs, 259
    vials, 223–224
glass-reinforced plastic (GRP), 87
glove
    bag, box, 79, 83, 168, 236, 253
    radioactive, 217
    see also clothing, protective
glutaraldehyde, 93, **186**, 319–320(t)
gold, 45, 267–268
Government agencies, 49–50, 133, 206
graphite
    electrodes, 262
    fire extinguishing, 303
grinder, sink, 39, **59–62**
    installation, 57, 67–68, 71
    uses, 193, 200, 201, 216, 253

half-life, radioactive, 207
    long, 210, 217
    short, 209, **210–212**, 213, 217, 220,
        221, 226
halogenated solvents, 33, 35, 273(t), 275
halons, 100–101, 306
handling equipment, 39, 242
hand washing, 22–24, 54–56, 69
    radioactive, 217–219
Hazardous Waste (USA law), 13, 237,
    324–331(t)
hazards — indicated by **bold** cautionary
    notices in the text. *See also* the
    bibliography to each chapter, especi-
    ally Chapters 9, 10 and 11, plus the
    tables in Appendix C. *See also* health
    hazards
health and safety laws, 11
health hazards
    asbestos, 235–236
    biological, 174, 189–192, 202
    chemical, 74–78, 231–233
    radioactive, 207
HEPA filters, **82–83**, 86, 93, 96–97
hepatitis, 191, 205
high efficiency particulate air filters, *see*
    HEPA filters
histology, 274
    see also biological materials, chemicals
    hoods, fume, *see* fume cupboards

holidays, 47, 94, 220
hospitals, 190–191, 192(t), 232–234
    colour code, 28, 29(t)
    drains, 51, 69
    see also drugs, excreta, hypodermics,
        macerators
Howie Report, 92, 93, 179, 181, **190–191**
hydrides, 141(t), 165–166
hydrolysis, 155–156
hygiene
    biological, 175, 190–195, 199
    personal, 22–25, 27, 190, 199, 228
    radioactive, 212–213, 228
hypochlorite
    disinfectant, 181, **184–185**, 189, 219,
        298, 318–319(t)
    oxidant, 156–157, 158–159(t)
hypodermic syringes, 26, 192(t), **202–203**,
    233–234

ignitables (USA law), 135, 324
incineration, incinerators, 106–122
    biological waste, 189, 200, 201
    economics, 113–116
    laboratory and institutional units, 110–
        119
    operation and management, 116–118
    public and commercial units, 108–110
    radioactive waste, 215, 222, 226
    sharp objects, 117, 203
    standards, 113–114(t)
infective waste, 27, 29(t), 116, **190–191**,
    192(t), 198–200, 202–203
information, sources of, 18, 48–50
inflammables, *see* flammables
instructions, 19, 116, 126, 139, 178
insurance, 11, 14
interceptors, 58–59
iron piping, 63
iron (II) sulphate (reductant), 157, 160,
    286
iso-propyl alcohol (IPA), 183, 272–
    273(t), 285(t)

jars, discard, 180–186, 187–189
joints
    drains, 63–70
    glassware, 259
junk, hoarding, 46, 258–259

ketones, 273(t)
killing animals, 194, 201
kitchen
    food laboratories, 177, 191–193
    staff rooms, 25
waste disposal unit, *see* grinder, sink

'lab packs' (USA chemical disposal),
    135–136
labels, 31, 37, 38, 45, 49
laboratories
    biological, 58, 82–83, 93, 174–205,
        226
    chemical, 128–173, 231–248, 258–288,
        288–306
    clinical, *see* medical
    educational, 240–257
    engineering, 57, 58, 71
    food, 65, 177, 191–193
    medical, 190–191, 202–203, 231–234
    mobile, 245–256
    pilot, 53, 57, 58, 59, 65, 138
    radioactive, 65, 206–230
    small, 244–245
    teaching, *see* educational
laminar flow cabinets, 84–85
land disposal, 12–13, 44
    biological waste, 192(t), 193, 200
    chemicals, 133–137, 162
    radioactives, 210, 211, **214–215**, 217
lanthanides, 268
laundry, 212–213
lavatory, *see* toilet
law, **10–13**, 135, 136, 194, 233, 236–237
lead
    metal, 260
    pipe, 64, 218
leaks
    chemical, 190, 285, 289–302
    drains, 63, 69–71, 140(t)
    fume duct, 85, 96, 99, 190
    gas cylinder, 170–171
    radioactive, 227–229, 301
    water, 53, 61, 68–70, 190
legislation, *see* law
licences, 12, 13, 17–18
    animal, 194
    radioactive, 207, 210, 215, 217, 221–
        222
light bulbs, 38
lime, *see* calcium hydroxide
lithium, 163, 269, 302, 303
litter, animal, *see* animal bedding
low-waste procedures, 46, 252–254
Lysol, 183, 320

macerators, 59–62, 193, 200, 201, 216,
    253
magnesium, 161–164
management, 7, 17–22, 45–47
    chemicals, 128–130
    drains, 67–71

incinerator, 116–119
  mercury, 245, 265–266
  solvents, 32–34, 35–38, 269–271
manufacturers, 44–45, 49
materials of construction
  drains, 62–67, 218–219
  fume extracts, 87–89
materials recovery, 26 39 46 131
mechanical aids, 39
medical laboratories, 190–191, 192(t)
  see also hospitals
mercaptan, 156, 158(t)
mercury, 265–267
  cleaning, 266–267
  hazards, 64, 140, 265, 305–306
  in drains, 64, 140
  in fires, 305–306
  management, 245, 265–267
  recovery, 261–267
  spillage, 291, 295–296, 297–298, 301,
    315(t)
metal waste, 41–44, 164–165, 260–268
methylated spirit, *see* alcohol
micro scale experiments, 254
microbiology
  laboratories, 174–193, 202–203
  safety cabinets, 82–84, 86, 90–94, 98
microwave ovens, 176–177
Ministry of Defence (UK), 14
mobile laboratories, 245–246
monitoring radioactivity, 212–213, **214**,
  216, 217, 226, 307
monkeys, 200
moulds, 176, 186, 198
mouse, 195, 201

National Health Service (UK), 14
National Research Council (USA), 10,
  132–133
needles, hypodermic, 26, 192(t), 202–
  203, 233–234
neutralization, 152, 160–161, 296–298
  tank, 59, 139
nickel, 142(t), 164
nitriles, 156, 158(t), 166–168
nitro-compounds, 272–273(t), 285(t)
notification, 13, 14, 135, 237

Occupational Safety and Health Act 1970
  (USA), 11
offensive substances, 27, 28, 79, 147
  see also excreta, mercaptans, nitriles
oil, 288
organization, 17–47
oxidation, 156–157, 158–159(t)

packages
  contaminated, 45, 108, 133
  returns to supplier, 44–45
  waste transport, 13, 28, 29(t), 34,
    133–137, 199–200, 243
palladium, 260–263, 267–268
pallets, 45
paper
  absorbent, 227, 291–292
  bags, 27, 117
  tissues, 24, 227
  towels, **23**, 24, 25, 108, 212, 213, 217,
    227, **293**, 300
  waste, 27, 29(t), 38, 212, 227
paraformaldehyde (fumigation), 93–94,
  185
Pasteur pipettes, *see* disposables
pathogens, 179, **190–191**, 192, 200, 299,
  304
PCBs, 234
pedal bins, 30, 32
people, organization, 19–26, 47
perchloric acid, 86, 153, 302
peroxides, 140(t), 149, 153, 163, 237, 285
  in solvents, 129, 145, 160, 272–273(t),
    285–287
Petri dishes, 178, 185, 226
  see also disposables
pH, effluent limits, 143(t)
pharmaceuticals, 26, 231–234
phenols (disinfectants), 183–184, 219,
  320(t)
phosphorus
  element, 107, 165, 302
  oxide, 161
  phosphorus-32, 213, 218
picric acid, picrates, 64, 140(t), 237
pillows, absorbent, 294, 300
pilot laboratories, 53, 57, 58, 59, 65, 138
pipes, *see* drains
planning, 46
  drains, 51–52, 67–71
  fume extracts, 85–86
  permission, 12, 18
  spill control, 290–292
plastic
  absorbent sheet, 295
  bags, 20–21, **26–32**, 178
  drains, 54, 61, **65–67**, 219
  fume ducts, 87–88
  incineration, 107, 115, 117, 118
plates
  distillation, 276–280
  Petri dishes, 178, 185, 226
    see also disposables

platinum, 260–263, 267–268
plutonium, 168, 208
poisons
  legally controlled, 12, 18, 26
  severe, 26, 168, 231–234
pollution
  educational topic, 249–250
  laws, **9–11**, 113(t), 116, 118–119
polychlorinated biphenyls (PCBs), 234
polymers in solvents, 285–287
polypropylene
  absorbent sheet, 295
  bags, 21, 27
  bottles, 33, 266
  drains, 54, 60–61, **66–67**
polythene
  bags, 27
  bottles, 33, 266
  drains, 66
polyvinyl chloride (PVC)
  drains, 54, 56, **65–66**, 219
  fume ducts, 87, 88
  incineration, 115, 118
potassium, **161–164**, 269, 302
precipitation, **149–150**, 261, 262, 267, 268
precious metals, 260–268
pre-filters, 96–97
preservation of carcasses, temporary, 202, 226–227
primates, 200
projects, 46, 76–77, 256
propan-2-ol
  disinfectant, 183
  distillation hazard, 272(t), 285(t)
protection factor, 91
protective clothing, 77, 93, 97, 116, 146–147, 198, 227, 235, 241
protective enclosures, 77, 78, 82–84
protozoa, 176
PVC, *see* polyvinyl chloride
pyrogens, 190
pyrolysis, 110–113

quaternary ammonium disinfectants, 181, 298
quinol (inhibitor), 287
quotations for services, 125, 127

radioactive substances, 206–230
  biological waste, 226–227
  clothing, 212–213
  collection, 213–214
  decontamination, **227–229**, 300–302, 306–307

drains, 51, 53, 58–59, 65, 69–71, **216–221**
  fires, 306–307
  fume extraction, 92, 95, 98, **100**, 219, **225**
  incineration, 215–216, 221–223
  land disposal, 13, 210, 211, **214–215**, 217, 221
  liquids, 217–225
  monitoring, 212–213, **214**, 216, 217, 226, 307
  records, 208–209
  scintillation vials, 223–225
  sealed sources, 209–210
  sinks and basins, 217–219
  solids, 213–217
  spillages, 227–229
  storage, 211, 213, 217, 220, **221**, 225, 226
  units (measurement), 207–208
rats, 174, 195
receptacles, 25, 26–44, 252
  see also bags, bin, bottles
receptor fume extract, 81
recipes, 8, 132–133, 161, 233, 288
reclamation, *see* recovery
record-keeping, 10, **18–19**, 44, 240
  radioactives, 19, 208–209
recovery of materials, 26, 39, 46, 131, 253, 254, **258–288**
recycling, *see* recovery
reduction
  chemical, **157–160**, 260–261, 286, 287, 298
  volume, 46, 145, 148, 193, 211, 252–254
reflux ratio, 276–280, 283–284
refreshment facilities, 24–25
refrigerator, 25, 147, 174, 202, 211, 226
refuse, 24, 28, 29(t), **34**, 35, 38, 44, 211, 235
  biological waste, 193, 200, 203
  chemicals to, 133–134, 337–339(t)
  chute, 39, 200
  collectors, 11, 18, 26, 34, 38, 200
  compactor, 39, 193
  incinerator, 44, **108–109**, 200, 203
  radioactives to, 211, 214–215
relative volatility, 279–281
repairs, 124, 209, 259
resins, ion exchange and macroporous, 149, 262
Resource Conservation and Recovery Act 1976 (USA), 13
respirators, 77, 171, 198, 227, 235, 241

responsibilities
   legal, 11, 25
   personal, 19–22, 26, 198
returns to manufacturers and suppliers,
   44–45
rhodium, 260, 261
ricin, 168
rotary evaporators, 148–149
running traps (= S-bends), 56–58, 62, 219
ruthenium, 260, 261

sacks, 27–32, 41, 117, 199
   see also bags
   holders, 20–21, 28
safety
   biological, 174–175, 190–191, 202–203,
      231–233
   cabinets, 77–79, 82–83
   cans, 33, 42
   chemical, 129–130, 168, 231–233, 237–
      238, 331–337(t)
   distillation, 272–273(t), 285–287
   laws, 9, 11, 126
   officer, 22, 50, 289
   security, **25–26**, 32, 34, 195, 203,
      233–234
   shower, 53, 241
   see also accidents, asbestos, emergen-
      cies, hazards, hygiene
salvage, 46, 253, 254, 258–288
salmonella, 192
sand, 293–294, 302
sanitary piping, 52
sanitary procedures, *see* hygiene
sanitization, *see* disinfection
sawdust
   animal bedding, 198
   for incineration of liquids, 117, 221
   for spillages, 292, 303
S-bends, 56–58, 62, 219
scalpel blades, *see* sharp objects
schools, 200, 249–257
scintillation vials, 223–225, 271
scrap, 45, 260
scrubber, fume duct, 86–87
sealed sources, radioactive, 209–210
Section 17 Regulations (UK), 13, 322
security, 19, **25–26**, 32, 34, 195, 203,
   233–234, 299
segregation, 19, **35–38**, 116–117
semi-micro experiments, 254
separation, chemical, 148–151
sewage, sewers, *see* drains, excreta
sharp objects, 27, 36, **41–44**, 117, **202–203**
shredder, *see* macerator

shower, safety, 53, 241
silicate, amorphous, 294, 315(t)
silver, 140(t), 263–265
sinks, 22–23, 54–56, 67–70
   biological, 54
   chemical, 137–139, 142–144
   grinders, 39, 57, **59–62**, 67–68, 71, 193,
      200, 201, 216, 253
   radioactive, 217–219
   traps, 56–58
skips, 34, 43
slime moulds, 176
slop-hoppers, 54
sluices, 54–56, 69, 189, 219
small laboratories, 244–245
small generator exemption (USA), 15
smoke
   incinerator limits, 113(t)
   test for fume ducts, 98–99
smoking, 25
soda ash (= sodium carbonate), 161, 296,
   302, 303
sodium (metal)
   disposal, 161–163
   fires, 302
   recovery, 269
   solvents, 160, 273
sodium (compounds)
   borohydride, 166, 261
   carbonate
      drain cleaning, 71
      fire suppressant, 296, 302, 303
      neutralizer, 161, 296
   hydroxide, 160, 161
   hydrogen carbonate, 296
   hydrogen sulphate, 161, 297
   hydrogen sulphite, 156, 160, 186, 298
soil pipes, 52
solid waste
   collection, 26, 27, 193, 213–214, 226
   radioactive, 213–217, 223, 226
   USA legal term, 135
solvents
   cans, 32–34
   chlorinated, 33, 35, 275
   collection, 32–38, 269–270
   distillation, 275–288
   evaporation, 145–147
   halogenated, 33, 35, 275
   hazards, 272–273(t), 285–287
   incineration, 112, **117–118**, 120, 222
   peroxides, 129, 145, 160, 272–273(t),
      285–287
   removal, 160, 285–287
   purification, 287–288

recovery, 269–288
stripping, 148–149
sources
information, 48–50
radioactive, 207, 209–210
Special Waste (UK legal term), 13,
322–323(t)
specific activity, radioactive, 208
spillages, 290–302
absorbents, 293–296, 315(t)
control kits, 39–41, 245, 292, 315(t)
educational laboratories, 255–256
mercury, 265, 291, **295–298**, 301, 305–
306
precious metal, 262–263
procedures, 47, 290–302
radioactive, 227–229
sponge, 296
mercury, 296–297, 301
spores, 99, **175–177**, 183–186, 190
squeegee, 227, 291, 300
stainless steel
drains, 64–65, 218
fume ducts, 88
sinks, 54, 64–65, 218
solvent cans, 33
starch-iodide test, 160, 286
steam autoclaves, 27, **177–180**, 212, 253
sterilization, 175–190, 198
see also autoclaves, disinfection
stills
mercury, 265
solvent, 39, 148–149, 222, 224–225,
275–285
water, 52, 53, 60, 67–69
stock control, 12, 18, **128–129**, 265–266
storage
biological, 117, 191, **200**, 202, 211,
226–227
chemical, 12, 25–26, 77, **128–129**, 131,
146, 157, 245, 292
clearing out, 240–244
disinfectants, 184, 186, **299**, 320(t)
food, 25, 193, 200
radioactives, 209–213, 220–227
security, 25–26, 117, 292
segregation, 35–38, 117, 193, 200
unsafe, 25–26, 32, 117, 200, 292
stripping, solvent, 148–149
strontium-90, 207, 214
students, *see* educational institutions
sulphate in effluent, 63
sulphur, 297
swabs
collection, 35, 174, 183, 187

disinfection, 183, 185, 298
radioactive, 228
spill absorbents, 296
test, 181, 188, 304
syringes, 26, 192(t), **202–203**, 233–234

tables, chemical, 101, 140–143, 272–273,
285, 322–343
tar, 142, 147, 149, 287, 332(t), 333(t)
tea-break, *see* coffee-break!
teaching
laboratories, 249–257
waste disposal, 249–250
see also training
technicians, 41, 251, 252, 256, 266, 299
tenders for service, 125, 127
teratogens, 231
test burn, EPA, 116
theoretical plates, distillation, 276–280
thermometers, 38, 245, 290
thioacetamide, 164–165
thiols (= mercaptans), 156, 158(t)
thorium, 210, 214
tin
cans, *see* cans
(II) chloride (reductant), 286
scrap, 260
tips, pipette, 35–37
tips, refuse, *see* land disposal
tissues, paper, 24, 227
toilets, 22, **23–24**, 217
towels, 23
paper, 23, 24, 57
collection, 24, 28, 214
radioactive, 213, 214, 217, 227, 228
spill absorbent, 147, 227, 228, **293**,
300, 303
Tox Box, 136
toxic substances, 12, 46, 168
toxins, 189–190
trade associations, 50, 124
trade waste, 18
training, 19, 27, 77, 139
evaporation, 146
incineration, 116
spillages, 292
see also education institutions
transport, 39, 125, 193, 199–200, 242–243
law, **13–14**, 125, 136
traps, drain, 56–59 140
trash, *see* refuse
trashcans, *see* dustbins
trash compacters, 39, 193
trays, 266, 291
tritium, 214, 215, 219, 223, 224

trolleys, 242, 243, 291
troughs, 54–56
Tyndallization, 177

ultra-violet light, 187
unidentified materials, 109, 130, **238–240**
United Kingdom
  landfill practice, 134–135
  laws, 9–16, 194, 206
United States of America
  contractors, 127
  landfill practice, 135–136
  laws, 9–16, 127, 194, 206
uranium, 210
urine, 189, 198, 200, 219
  containers, 189, 192(t)

vacuum autoclave, 177–178
vacuum cleaners, 39, 195, 236
vacuum pumps, 55, 99–100, 267
vanes, fume duct, 89–90
vapours, see evaporation, fume, gas
vehicle
  engineering laboratory, 58
  waste transport, 14
velocity, air, see face velocity
ventilated furniture, 77, 78, **82–85**, 90–95,
  97–98
ventilation, 74, **79–80**, 194, 254
  after fire, 304, 306, 307
venting wastes, 97–104
  autoclaves, 179

chemicals, 100–104, 145–147
  gases, **100–103**, 169–171, 339–343(t)
  radioactives, 100, 225
vermiculite, 117, 135, 136, **294–295**, 296
viruses, 175–176, 185
vinyl chloride, 101, 331(t), 332(t), 333(t)

washbasins, see basins, hand
waste disposal units, see sink grinders
water
  dilution, 153–155
  drainage, 52–53, 137–138, 142
  drinking, 195, 303
  flood, 53, 68–69
  hot, 23
  hydrolysis, 153–155, 190
  leaks, 53, 61, 63, 68–70
  quality, 71–72, 138, 142
WC, see toilet
wood, 153, 244, 302, 304
  flour, 293

X-ray plates, 260
xylene, 274

yeast, 186

Zero Effluent Laboratory, 254
zinc, 140–141(t), 142(t), 263, 264, 287,
  295–298
  dithiol, 261